HOW TO DESIGN, BUILD & PROGRAM YOUR OWN ADVANCED WORKING COMPUTER SYSTEM

Other TAB books by the author:

No. 792 *Build-It Book of Miniature Test & Measurement Instruments*
No. 905 *Build-It Book of Digital Electronic Timepieces*
No. 975 *The "Compulator" Book—Building Super Calculators / Minicomputer Hardware with Calculator Chips*
No. 1111 *How To Design, Build & Program Your Own Working Computer System*

No. 1332
$16.95

HOW TO DESIGN, BUILD & PROGRAM YOUR OWN ADVANCED WORKING COMPUTER SYSTEM

BY ROBERT P. HAVILAND

TAB BOOKS Inc.
BLUE RIDGE SUMMIT, PA. 17214

FIRST EDITION

SECOND PRINTING

Printed in the United States of America

Library of Congress Cataloging in Publication Data

Haviland, Robert P.
How to design, build & program your own advanced working computer system.

Bibliography: p.
Includes Index.
1. Electronic digital computers—Design and construction—Amateurs' manuals. I. Title. II. Title: How to design, build, and program your own advanced working computer system.
TK9969.H37 621.3819'582 81-9178
ISBN 0-8306-0022-1 AACR2
ISBN 0-8306-1332-3 (pbk.)

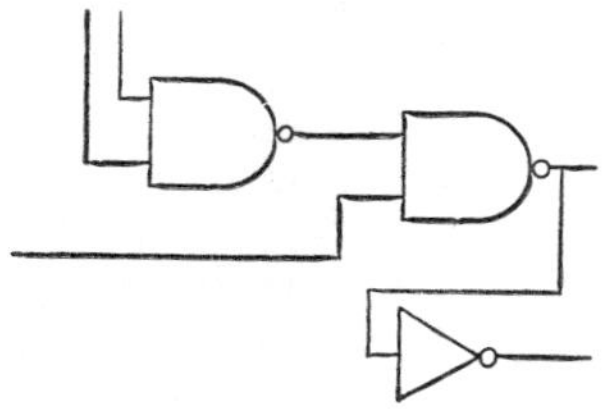

Contents

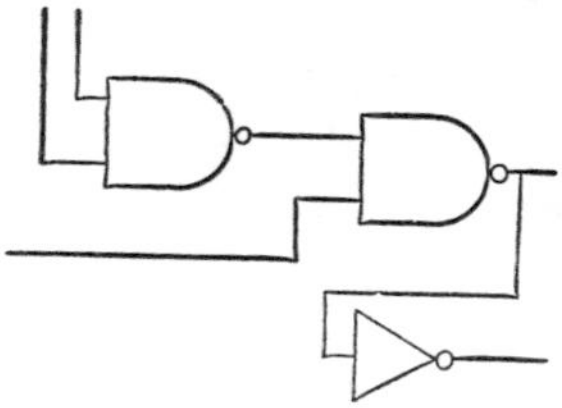

Introduction

This book continues the instruction in design, construction and operation of home computer systems that began with *How to Design, Build and Program Your Own Working Computer System*. As in the first volume, emphasis is on home construction, with explanation of the reasons for design choices in the prototype, and includes alternate choices that may be better for particular problems. Approaches to very advanced systems are included.

Actually, for reasons covered in the first chapter, home construction of a basic computer is less attractive than it was several years ago. But this statement does not apply to advanced systems covered in this volume. For one thing, very substantial savings in cost are possible, and can be increased by substituting time for cost. The approach outlined here can create an advanced system at one-fourth to one-tenth the cost of purchased units—very real savings. In addition, if you are really pioneering—you must design and build from scratch.

Because quality is always a problem in home construction, board construction techniques have been developed further. Using the techniques described and reasonable care, home-constructed boards that are equal in quality and appearance to manufactured boards, and nearly equal in packing density, can be produced singly or in limited lots. Usability and operating lives of these boards seem to be equal to commercial boards.

In the first volume work concentrates on the SC/MP microprocessor chip, and on the NIBL version of Tiny BASIC. Two lines of extension are covered. One is based entirely on NIBL as a basis for the operating system.

The second volume covers an extension of NIBL, the NSC BASIC intended for the INS-807X series of chips successors to the SC/MP. One of these chips is of special interest because it includes NSC in one chip ROM. In addition, a separate extension of NIBL, the Z80-based ZIBL, is reviewed.

With processor chips becoming so common and so inexpensive, multiprocessor operation is an inexpensive way of extending capability. Basic principles of multiprocessor operation are covered for designs based on the SC/MP, and for the SC/MP in combination with other chips, including the Z80. Some data on other basic computer bus systems is included. Basic data on advanced operating systems for multitasking and multiprogramming is also covered.

For enhanced computer power, a substitute for floppy disk data recording is described. While not as fast, this cartridge tape-based system has nearly the convenience, and at least the same storage capacity, as common disk designs. Further, a special form of multiprocessing, using either calculator chips or the number-cruncher type of auxiliary processors is included. For those doing extensive and complex numerical work, these special forms give greatly increased computational power.

As in the first volume, emphasis is placed on hard copy for output. Use of low cost Baudot teleprinter equipment continues, with some thought given to changing to the more modern ASCII code devices as opportunity arises. There is essentially no coverage of the TV-typewriter approach common in commercial computers—there is ample data on this elsewhere. However, one memory board includes design features to permit easy use of the memory-tap version of TV display.

A final note: The transcriber, editor, publisher, and I have tried to keep all errors out of this material. But despite this work, errors can occur. It is suggested that you cross check each item against other data in the volume and against the manufacturers' literature. It is a good way to check your own work as well. I would appreciate receiving a note regarding any errors found in the text.

Robert P. Haviland

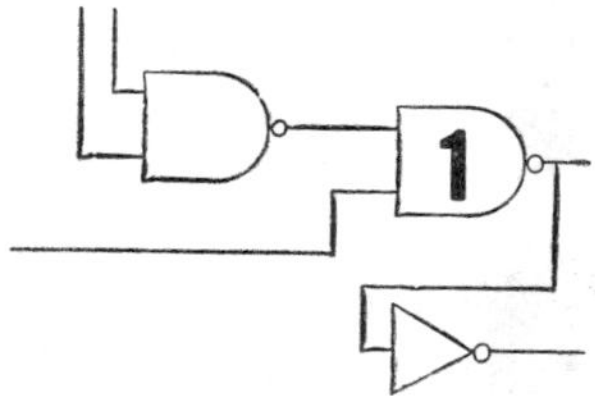

Review Of The Microprocessor Field: Advanced System Concepts

In the few years which have elapsed since the first book of this build it-yourself computer series was started, the commercial microprocessor field has changed considerably. Some of the developments have made home building less attractive, and some others have actually made it more difficult. But equally, others have emphasized the desirability and need for home construction. Home construction still has its place.

MICROCOMPUTER TRENDS

The dominant factor in microcomputer development has been the emergence of the *single-board computer*. While the name isn't strictly true, a typical single-board device, as shown in Fig. 1-1, will include on the board, the central processor, serial input-output, the video display generator, memory (typically one to 4K ROM and four to 8K RAM), and a key board.

Some designs increase the memory, and others are based on an external keyboard. A few provide an output display, and one provides a simple printer. There are many variations in detail.

A minimum working computer is obtained by adding a power supply and, usually, a display. For some, the display is simply a small screen TV set, for others a set modified for video input and better scan characteristics. A cabinet is also a necessity for continued use.

The cost of these single-board units is quite attractive. A minimum design with a hex keyboard input may cost as little as

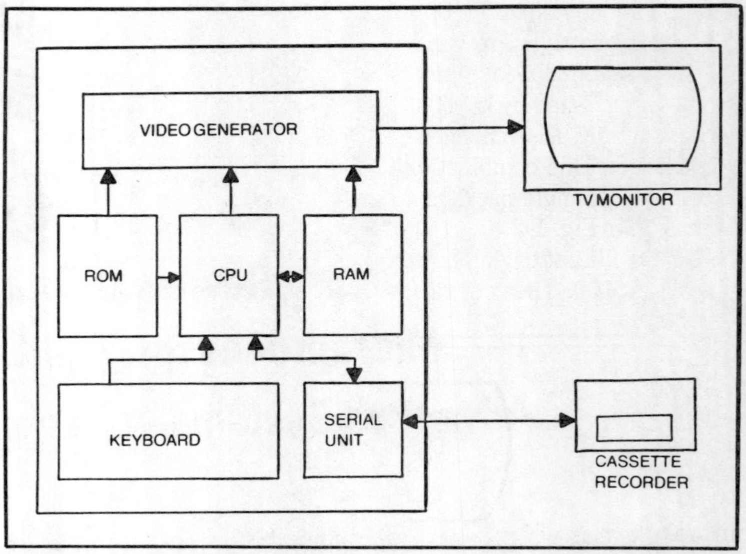

Fig. 1-1. Block diagram of a typical single-board computer. This assumes that RAM is shared between the CPU and the video section. Other designs use separate RAM for video, although both sets of RAM are under CPU control.

$100, and a typical design with considerable ROM (monitor, editor and a BASIC) may be only $300. About $100 more will handle the power supply and cabinet costs, including provisions for expansion. The display may cost $100 to $300 more.

Being limited in memory, this grade of computer can handle only lower level problems. Considerable improvement can be made by adding a cartridge magnetic recorder at a cost of about $50. It is now possible to save programs and data from the board memory, to reuse programs, and to build up a program and data library. The problem solving capability is good, but the computer is slow, and requires many manual operations.

Magnetic Disk

Speed can be greatly increased by taking advantage of another major development, the magnetic disk recorder. The drive, recording and pickoff head mechanisms, and control circuits have reached the point where costs are very low when measured against performance. A basic recording element is now available for around $350, and a complete single-disk package might be only $500.

The combination of a single-board computer and a disk package greatly improves speed and capability. Transfer of a 4K program from disk to memory takes a fraction of a minute, and can be accomplished by keyboard command. Storage capacity per disk is about 50K. Disk changes take no more time than changing a tape cartridge. This combination is so attractive that there are now a number of single-package designs available in kit and complete form, as in Fig. 1-2.

Cost is not trivial, however. Prices range from about $1500 to nearly $5000. The cost varies with the amount of RAM, ROM and

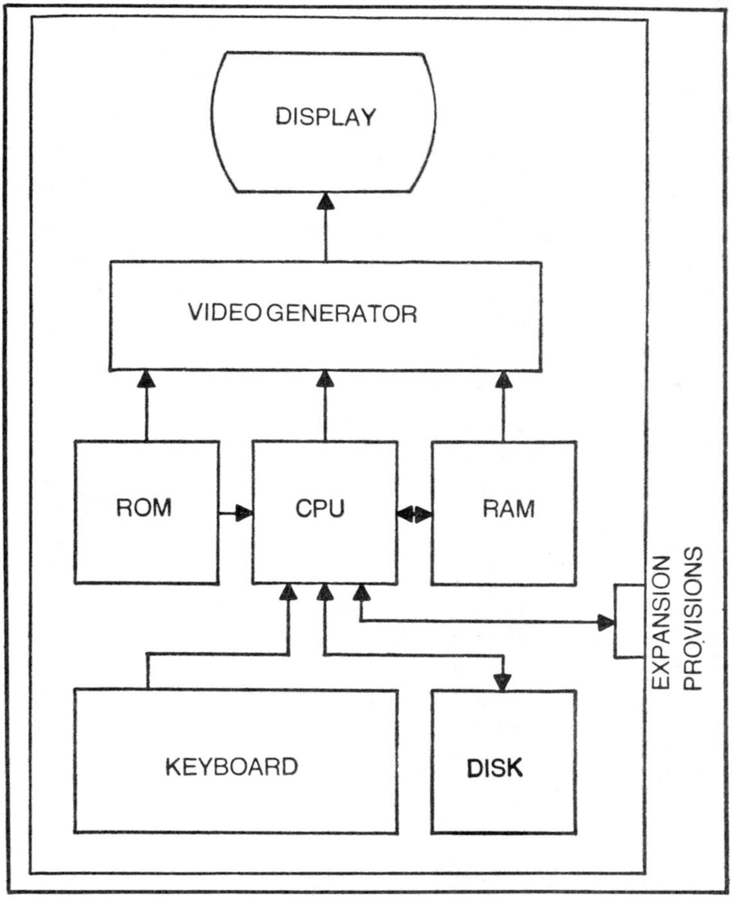

Fig. 1-2. Block diagram of a typical single-package computer, again assuming common RAM. Early designs used a built-in cassette instead of disk for slow memory, and some such designs are still used.

package programming provided in the operating system software. Virtually all of these units are good, and some are very good.

The Importance of Hard Copy

The designs in Figs. 1-1 and 1-2 have a characteristic that some consider to be a serious flaw. There is no inherent provision for hard copy, that is, copy that does not disappear when power is turned off or when another program is run. Having hard copy available might be only a convenience, as when a program is being debugged. Scrolling and a cursor do very well for most program bugs. Hard copy can even be a nuisance, as in a game of tic-tac-toe between a pair of 10-year olds. But it is a necessity for any important computer output, such as business and scientific problems. A truly practical single-package computer should have the format of Fig. 1-3.

The common lack of hard copy has led to poor programming practices by many new computer users. A program of any consequence represents considerable work, and has a definite value. It should be preserved. Of equal importance, the logic of the problem solution should be preserved, so it can be recovered when needed. This is particularly important if a program is ever to be modified or extended. These needs are not satisfied by a magnetic tape or a floppy disk. They are poorly served by the combination of these plus a monitor screen.

The better complete-system makers distribute a range of hard copy devices. They range from modifications of cash-register tape printers to high-speed line printers handling standard computer formats. Prices are not low, however. The very minimum printers are typically around $200, and the least expensive line printers are around $600 to $900. Computer grade printers may run $2000 to $3000—as much as the rest of the system.

Because of the rapid rise in cost linked with increased capability, computer ads and brochures tend to emphasize the low-cost end of the line. They may even be misleading. Neither ads nor sales personnel are really open about the necessity for hard copy in the world of real problems. Even worse, many of the small computer advertisements do not cover the need for an expansion interface if any expansion is made.

Other Major Developments

These are the major developments, but there are several others which affect the home builder. One of these is the great

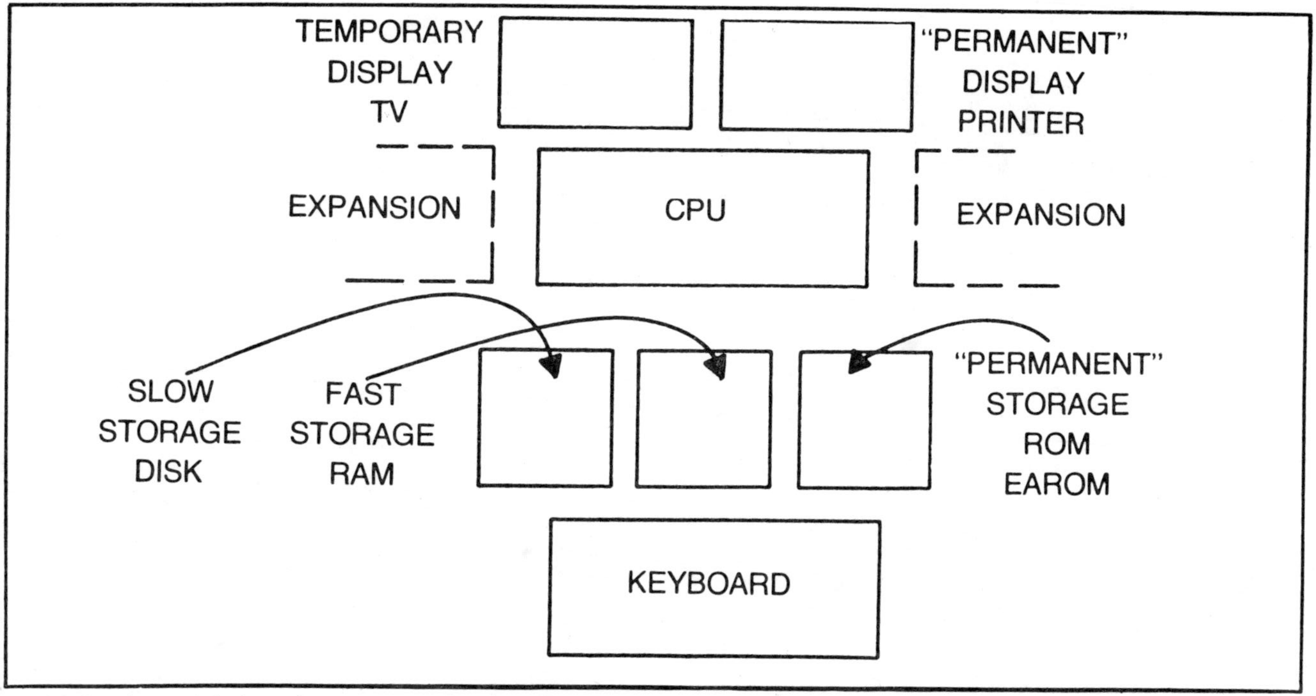

Fig. 1-3. An advanced single-package computer. A major addition to the design of Fig. 1-2 is addition of a hard copy printer. Provisions for EAROM, to allow the operating system to be extended, are also important.

growth in the number of board fabricators, and equal growth in the range of board types. These boards include processor boards, random memory, read-only memory, ROM burners, serial interfaces, a video board, a deck controller, and music boards—plus a limited number of more specialized designs for such purposes as home security.

Many of these boards are for the 100-pin connector, the S-100 bus. But many other types are also available, including 44 pin, 50 pin and 76 pin. Some are designed for use with a specific model computer, others are intended for a range of models.

While there have been, and occasionally still are, problems with signal levels or timing, with poor plating or etching, or with bad components, these boards are generally of good to high quality. Because of competition and pricing policies, it is often possible to buy a board for less than the normal cost of components.

It is difficult for the home constructor to pack components as closely as the commercial photo-etched thru-plated-hole boards. It is also difficult to maintain the same average quality. However, see later comments and construction procedures before deciding that purchase of these boards is the best way to go.

COMPONENTS

In the field of components for the home builder, the situation is not quite as good as earlier. There are still a good number of vendors, and supplies of most standard items are plentiful. Prices, of course, have increased. Coupled with this has been more careful scheduling and purchasing by the large users, so good quality surplus is harder to find.

The field of microprocessor chips has seen several changes. Most manufacturers now have a line of variations on a basic chip, including CPUs which have ROM and RAM in the same chip, in various sizes. These are not standardized, and most are not available in small quantities.

The small microcomputer manufacturers have essentially standardized five chips, the 8080, Z80, 2650, 6502 and 6800. Some others are used, such as the 9900, SC/MP, and so on. (The SC/MP chip used in the basic computer of the first volume of this series is now designated as 8060A. It is found in floppy disk controllers).

All major chip manufacturers are working on 16 and 32-bit processors. Several are on the market. Prices are still high. There is a great amount of work on chip families or support chips, for example, for CRT displays and for interface parts. These will nearly always reduce the number of packages needed to accomplish a given function, making design and construction easier. However, cost may increase over standard package use, and the cost of a chip failure is higher. Further, the specific chip needed may not be available for one reason or another. A common practice with respect to these special chips is not to include them in designs unless there are at least two independent sources (often called *second-sourcing*).

Some of the chips not commonly used with microprocessors are becoming rare. Where most vendors formerly listed calculator and clock chips, only a few do now. However, other special chips are much more available. This includes digital-analog converters, sound and waveform generators, and various TV games.

HARDWARE SUPPORT

The availability of information about hardware has also changed. Most manufacturers continue to supply ample data, but there is likely to a be a document charge for this, sometimes well above reproduction cost. True specification sheets are now more difficult to obtain: Some are no more than advertising or brag folders, and many manufacturers will not supply specification literature to anyone other than an actual purchaser. This can block attempts to determine the functions of a custom chip.

One way of handling the data problem is to obtain short form catalogs which list manufacturers' products and give short descriptions of functions. Specification sheets and application notes can be requested as needed.

For items which are well standardized, many alternate sources are available. Several manufacturers prepare lists of exact and close equivalents. Three have extensive, well-indexed, tables of discrete and integrated circuit interchangability. While these are intended for radio and TV servicemen, they are equally usable in all electronic fields.

SOFTWARE

The software field has also changed. The deluge of new languages continues, but now the emphasis is on greater and greater memory. Where a 2K BASIC was considered acceptable several years ago, and 4K was nearly the standard size, the minimum BASIC is now around 8K. Now, 16K or even 32K interpreters are common. However, these are rarely in the public domain, or available from magazine articles. Instead they must be purchased. Sometimes programs are available only as a cartridge or disk, possibly with only a hex listing. They are virtually impossible to modify for another machine or for adding memory. However, some vendors provide very good documentation, including hints on extending applications.

The number of prewritten object or user programs available has grown enormously. Games remain popular, with some, such as chess, becoming very good. Many vendors offer programs for home finance and for businesses. The number of programming houses writing programs for specialized uses of small computers has increased greatly.

The previous problem of *non-portability* (the fact that a program written for one machine or for a specific interpreter, will not run on another machine) continues. The situation may improve in the future, as the currently available standards take effect, and as new standards are developed.

The IEEE is working on standards for assembly language, for floating point operation, and for various aspects of the high-level languages. However, until much more progress is made, the program user must depend on full documentation. This should include a well annotated program listing.

Because of the high cost of program creation, there is a lot of program pirating. This is partially responsible for the tendency to limit program documentation and listings. If a vendor can't sell a program, he can't recover the cost of writing it. Several vendors are avoiding this problem by selling programs at a price that is more attractive than piracy. Several sources of low-cost programs are listed in the bibliography.

There has also been a change in the type and level of support available from microprocessor manufacturers. A number of these have expanded their software and processor board development into full business lines. As a result, they now charge for software and for board design.

Other manufacturers are changing their market approach, and/or their product mix. National Semiconductor, the producer of the SC/MP (8060) chip, no longer publishes their newsletter *Compute*, which offered an exchange of construction and program ideas. The independent group publishing SCAMPUS continues. See Appendix A.

The larger microcomputer manufacturers do sponsor users' groups, which serve to spread product information, and to exchange ideas and programs.

PUBLISHING

As these changes in the hardware and software field occur, there are also changes in publishing. The number of active houses has increased, and the number of volumes on computers and computing has increased enormously. There is now an overgenerous supply of books on the basics of computer operation, and on the elements of program writing. There are some good hardware construction and maintenance books, plus a few which have useful programs. These are often selections from earlier magazine publications.

The magazines have changed also. Most are entranced by the amount of advertising available. Their emphasis is on new product announcements and reviews of new equipment. And home construction is down-played. Whole issues go by without a construction article; and when one does appear, it is likely to be in general terms. Programs are often reviewed for new features and presented in highlight form. If a listing is provided, it is likely that it is written for a specific configuration of a specific processor.

Probably in an effort to have something new, some of the magazines have been pushing new languages. There is some merit in this, but it is disheartening to go through a series of such articles without finding any information on machine language. It is also disheartening to discover that the minimum memory requirement for a highly touted break-through in programming is 102K bytes, which seems to mean that 99+ percent of the home computer users can't use the material.

Lest these comments seem too harsh, please note that there is much good material being published. It may take a little time to separate it out, but it is there.

THE BUILD-OR-BUY DECISION

The general availability of good quality computers and support and expansion elements, and their relatively low cost, are making the build-or-buy decision more difficult. The following is offered in the hopes that it will simplify reaching a good decision in particular cases.

Some years ago, when asked why anyone would undertake to build a boat, I replied, "It costs less. If there are any flaws present, you know where they are; you put them there. Its good for the waistline." To say it differently, if you are rich, slim and carefree, don't even consider building a boat.

Much the same reasoning applies to building a computer. To save money when building a computer, you must buy carefully over a period of time. The savings are there, but it does take some time. On the other hand, you must be careful when buying complete computers, especially when expansion is likely. The problem here is incompatibility of signals due to timing, noise and other factors.

There have been, and are, physical failure problems. Some are not easy to find. And there is always the problem of failures during the warranty period—most manufacturers take a dim view of opening the case. There is also the factor that few distributors have repair capability; the unit must be returned to the factory with added expense and with delays.

Over all, the significance of flaws seems to depend mostly on individual skill. If you are a good, careful worker, flaws in home-built equipment should be less than in purchased items. If your building skill is not great, a kit is the most you should undertake.

Building a computer, or even computer equipment add ons, probably won't do much for the waistline, but its a great head stretcher. Even kit assembly is a help, in that you have familiarity with all the components, their location, and their relation to other components. But this is nothing to the familiarity that results from designing, building and debugging your own equipment. Not only do you know the location of every item, you know its function, its inputs and outputs, and its relation to the complete assembly. The builder is far ahead of the usual purchaser.

There is, of course, one overriding reason for building. If you can't find the item you need in production, you either build, do without, or face an expensive development contract.

There is no flat guide for this make-or-buy decision. It must be made by each person. To try to simplify the decision, see Table 1-1. If several statements apply, the sum of the various votes indicates the direction to go. Sometimes weighing factors are applied to the statements. For example, if skill level is considered more important than the need to save money, it could be given two votes as compared to one for cost. Each individual needs to work out his own set of questions.

Actually, all computer users should do some building, just as all users should do at least some of their own programming. In the long run, this will save time, money and exasperation, by avoiding down time and hassles with vendors and shippers. It will even give a better and more satisfying system.

Table 1-1. Typical Questions and Responses for the Make-or-Buy Decision.

	Buy	Get Kits	Build	Buy Programs	Write Programs
I am interested only in what the computer can do for me	X		X		
I want to save money		X	X		X
My time is valuable	X			X	
I want to repair my own computer		X	X		
I'm interested in programming				X	X
I want to know how a computer works			X		X
I want to experiment with new applications			X		X
I've never built equipment, but would like to try		X			
I have a particular application problem	X	X	X	X	X
I like to build things		X	X		
I may go into hardware as a business			X		
I may go into software as a business					X
I find logical problems a challenge			X		X
My business needs a computer	X			X	
I'm not a good craftsman	X	X			

Complete building of a system and complete writing of all its programs is another matter. A very real dedication is needed for this. It's very satisfying, and recommended to those who have this dedication. For the average person, in the average situation—build, certainly—and buy, certainly; but use discrimination. Possibly a good rule to follow is that if several vendors offer what you need, buy. Otherwise, expect to build.

Despite this view, the material which follows will be dedicated to the one-hundred percent builder. If you use it and the preceding volume, you will have a working system with fairly advanced capability.

The material doesn't have to be used this way, however. You might wish to use it only for ideas, adding these to your own, or to those from other books, to develop your own designs. Or you might select a particular design element, building this is in the form shown, or building it in another format, as an add-on to your present system.

Whichever course you follow, remember that success follows good workmanship.

BASIC CONCEPTS FOR THE ADVANCED SYSTEM

If you followed the course of the first volume of this series, your starting point for expansion is shown in Fig. 1-4 above the dashed line. This foundation is a complete working computer. It can be widely expanded; the construction directions of the first volume covered only limited abilities. The limits were lack of memory, and the number of manual operations required.

If you were ambitious, you built one or more of the expansion features below the dotted line. Depending on the course followed, the expanded computer might have more memory, perhaps NIBL in firmware, or perhaps greatly expanded calculation or time

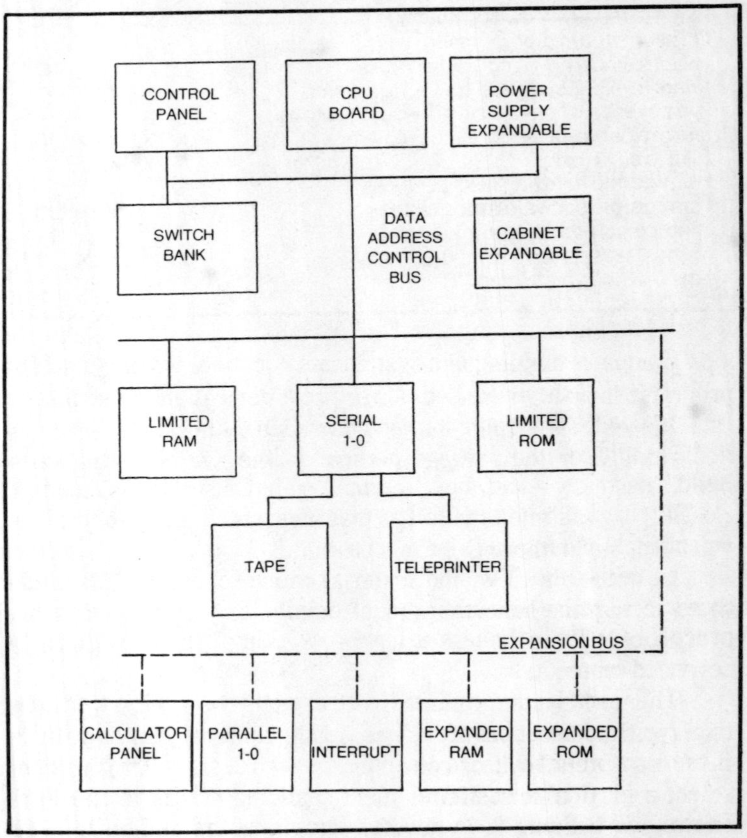

Fig. 1-4. Basic concept of the prototype SCAMP:Y computer of this series. Expansion capability and testability permit self-checking at each stage of construction. The direction of expansion should be determined by individual applications.

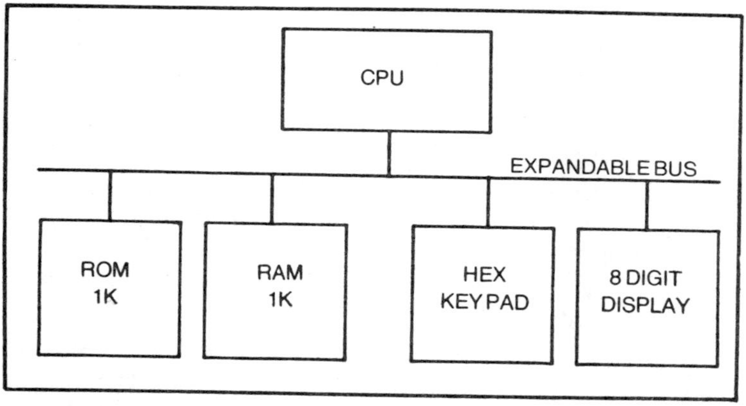

Fig. 1-5. Typical minimum computer system using hex data input and calculator display output. Several good kits are on the market.

capability. Also, if you followed recommendations, you have a program notebook with full listings, well annotated, and a library of programs on paper or magnetic tape. If you were very ambitious, you also have a complete operating system.

Those who are interested in building but are starting from a different foundation may have followed the approach shown in Figs. 1-5 or 1-6, representing the minimum system approach and the most common single-board approach. Figure 1-5 would be representative of the KIM-1, the RCA, ELF, or several others. The keyboard is typically a hex pad, possibly with a few added control keys. Display is likely to be octal. ROM and RAM are typically 2K each.

Figure 1-6 represents a larger selling group, such as the Ohio Scientific Challenger I, or Radio Shack Level I. Keyboard is alphanumeric; the TV display is typically 16 lines of 32 or 64 characters, with some other graphics capability. ROM is 8K or so, with 4K of RAM in the less expensive system, and 16K or more in expanded systems.

Your own system can include features from Figs. 1-4, 1-5 or 1-6. Typical differences are the presence of a hard copy printer, as in the Rockwell AIM; or color display, as in some Cromemco or Compucolor products. Memory capacity is typically similar to the units shown in Fig. 1-6.

This book is primarily written for the home builder, and especially for the start-from-scratch builder of Fig. 1-4. The projects described will be designed to be fully compatible with the system of this figure. But features are included to allow specific board projects to be used with most small computers now known. These features include:

- Wire-wrapping of board interconnections, permitting use of a board with any physical bus structure.
- Provision of both active high and active low signals where at all reasonable.
- Good buffering.
- Attention paid to drive capability, to timing and to loading.
- Single-function board designs.

These won't solve all interconnection problems, of course, but they are expected to keep the number small. Minimum redesign or minimum interface adaptations should take care of particular problems.

Physically, board projects will retain the 44 pin 4½ x 9½ inches format used in the first volume. The reasons for this selection several years ago are equally valid now, and include readily available stock size; its availability in blank, precoated, predrilled and wire-wrap designs; ample number of pins for most projects; and a relatively low cost.

OVERALL APPROACH TO SYSTEM EXPANSION

Before starting actual board design, it is worthwhile taking some time to set up the goals for system expansion. While these can be modified as work goes along, having them available from the start can prevent false starts and scrapping some boards. It's a good idea to know where you are going. It's also a good idea to have some idea of how to get there.

You don't want to restrict yourselves unnecessarily. Your goal should remain a good, general purpose computer, flexible enough to handle calculations and data, and capable of control of external devices. It should be capable of still further expansion.

In addition, the design should allow work towards some of the recent developments. Multiprocessing, using two or more processors, should be kept in mind. Multitask computation, perhaps by time sharing or perhaps by two processors, and multiprogramming are further expansion areas.

One of the most important goals to pursue is increase in computational power. Quite a bit of this will come from memory expansion. But you can go further by adding a more sophisticated microprocessor chip. The SC/MP was originally chosen on the

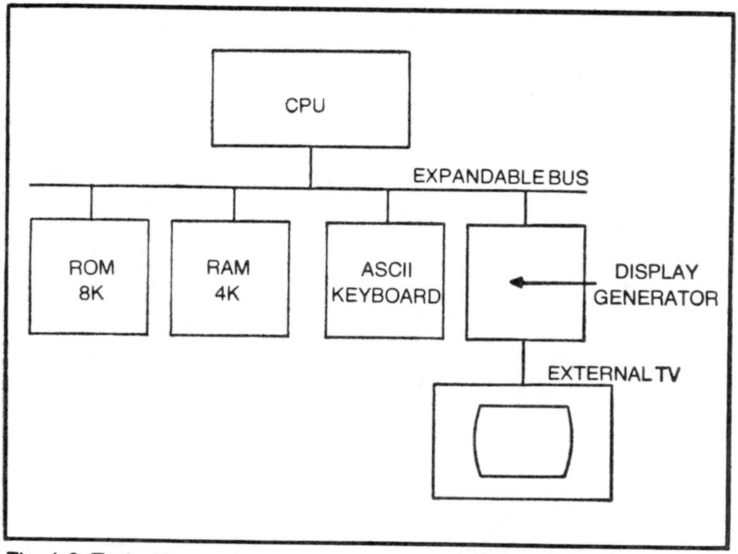

Fig. 1-6. Typical low cost computer system for keyboard input and TV screen output. Competition has brought the price of these units that it's not likely that you can build an identical unit at home for less. However, buyers should study the specifications of several kits before making a decision. Also, watch the radio interference and susceptibility: Some units are very bad in this respect.

bases of the simplicity of its instruction set, the adequacy of its other features, and the availability of a high-level language. The decision was definitely a compromise—slanted towards the beginner. For system expansion, the slant should be towards broader abilities.

Addition of a second processor will be necessary for multiprocessing, but the second chip does not need to be identical. Different capabilities can simplify multitask and multiprogram work. As an elementary example, assigning all timing elements to the SC/MP chip takes advantage of its built-in timing instruction, while freeing the second processor for other tasks.

Another decision that seems desirable is not to follow the current mainstream of hardware development too closely. The line is well developed, and there is enough competition to keep hardware prices so low that it would be difficult to equal purchase price. We want to approach improvements and expansion in a different way.

If you spend some time thinking about the system of Fig. 1-3, you will see some alternatives that are closer to the simple computer concept and go a long way toward attaining the capability of the well-developed desk-based packages. You will also see some approaches that appear to offer cost advantages to the home builder.

The most productive change is to place as much of the operating system in firmware in read only memory, ROM as you can. In principle, this can be all of the operating system (OS), subject to the constraint that you might have to assign a block of RAM to changeable elements within the operating system.

Having a firmware OS does have some marked advantages. The chance of destroying the OS during a run becomes zero, although you can lose the changeable part of a given run and have to do it over. You don't have to bootstrap the OS each time a program is changed, as in some systems, or at power up, as in others—it's always there.

And it may be cheaper. As shown in Fig. 1-7 the firmware OS certainly will be cheaper in the smaller sizes, and it may be attractive in larger sizes.

Of course, there are some disadvantages. You will use up memory space. But you can avoid some of the penalties of this by using *bank select*, assigning several elements of the OS to the same memory space arranged in banks selected on command. The same technique can be used for power saving. Some of the newer ROM or PROM turn off automatically if they are not selected.

You will still need slow memory, probably a considerable amount. However, a way of reducing access time would eliminate one of the drawbacks of cassettes, fast access being an important disk feature.

If you had tape with remote-control forward and reverse, and bidirectional read, you could approach disk convenience. Such designs have been used for a long time, for example, with the PDP-8. But most of the units cost nearly as much as disk mechanisms. There is also a cassette version, and some reel-to-reel recorders have these features.

There is another approach that takes care of most problems at a reasonable cost. This is to use cartridge rather than cassette recording, with automatic return to the starting point. Used in nonredundant channel form, as is common in cassette data, each cartridge reader provides eight channels. Zero access time is the most probable case and the worst case access time equals tape duration.

Two of these decks extend capability to 16 channels. One deck sets tape length to the most common program size, say two to 4K.

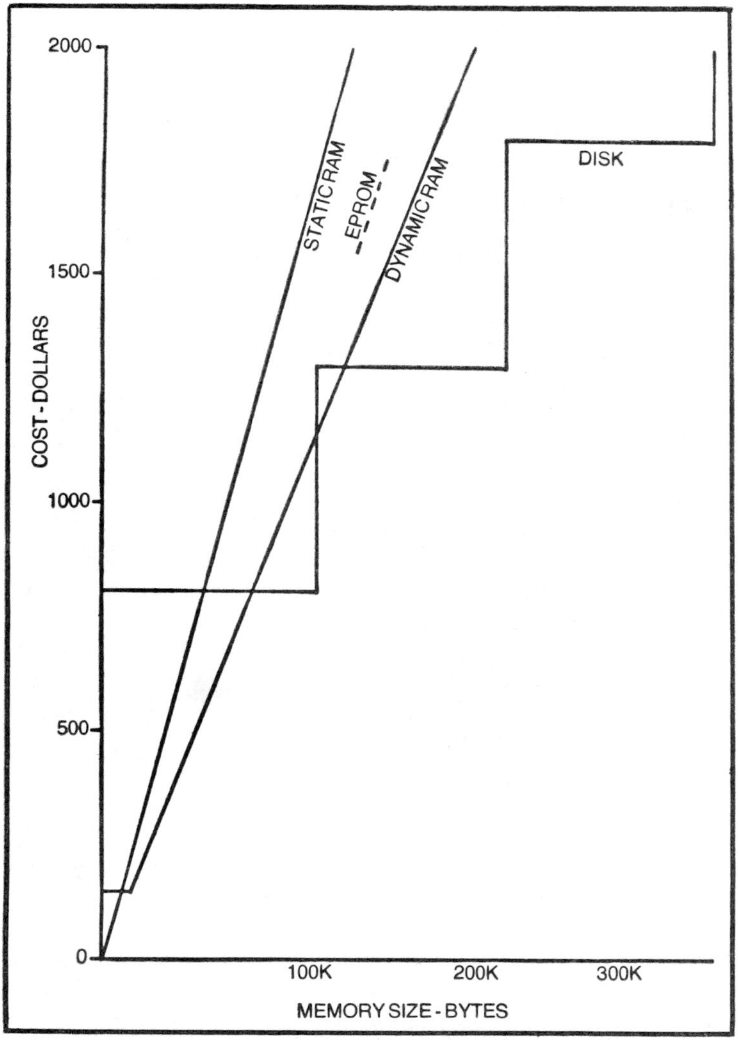

Fig. 1-7. Approximate cost of four forms of memory. Below about 32K, RAM is cheaper, but disk is cheaper in large memory systems. It's possible to get disk capability at a lower cost by sacrificing some speed.

Maximum access time is short, and not much time is lost if a load error occurs. The second deck can be used with longer tapes, perhaps up to maximum storage capability. Table 1-2 shows some of the possible arrangements for each cartridge, from very short cycle times to maximum tape capacity.

This arrangement is not exactly competitive with disk, but it gives quite a respectable performance. It does this by substituting time for complexity and cost. Actually, the time involved is surprisingly short.

Since the tape mechanism is easily made self-cycling, the CPU can proceed with its regular duties after completing a read or write block. The next block operation would involve a query about the recycled state (possibly a wait until completed, then start, track select and right/left select). The operating system could keep track of data location, perhaps by code, or perhaps by named files as is common with disk. Operating convenience would approach or equal that of disk systems.

A few points require attention. Since there is no redundancy, the possibility of read or write errors is good. An error detection element in the OS seems necessary. The usual check-sum suffices, providing the tape is kept clean. Heads need cleaning at regular intervals. And, since the tape drives always involve a rubber idler pulley, the cartridges should be released from the tape mechanism whenever it is unused for long periods. (This precaution is often neglected by cartridge users.)

In addition to expanding slow memory, you will also want to expand the fast memory. Adding boards of the original design will work, but has two drawbacks. One is the physical design, which uses small sub-panels as chip carriers to simplify board layout and construction. The design accomplishes its purpose, but requires more work than is necessary.

The original 2K board is a little small. 4K or even 8K would be better. Construction requirements are covered later with further hints on home construction of high quality printed circuit boards of all types.

In redesigning memory boards, you can take advantage of another development; the simplified graphics system now used directly or indirectly in many systems. This can be set up to interface a TV display by adding a single, relatively simple, interface board.

In addition, look at the possibilities of using one or more of the many commercial boards, such as those made for the S-100 bus. This could be done by building the simplest possible interface adapter, or by a full interbus interface.

Since the operating system will be dependent on PROM, or more properly, EPROM, a *PROM-burner* of some form is vital. Purchasing the service is a nuisance, due to the time delay

Table 1-2. Memory Capability That Can be Developed by Using Self-Cycling Cartridge Tape. Two Machines, One for Short Tapes and One for Long, Can Give a Capacity of Several Hundred Thousand Bytes With 16 Records Available at Essentially Zero Access Time.

Cartridge Tape Capability					
Length Feet	Duration*	Access	Ave Access	Bits Track	Capacity Total
6	18S	5S	9S	1K	8K
12	36S	5S	18S	2K	16K
23	72S	5S	36S	4K	32K
45	145S	5S	72S	8K	64K
94	5m	5S	2.5m	16K	128K
188	15m	5S	5m	32K	256K
375	20m	5S	10m	64K	512K

*Allows five seconds for start-stop

involved. The most practical approach seems to be a plug-in board just for the one task. While data input can be from the keyboard, it seems better to use the tape facilities, first preparing and checking a paper tape, then loading it for the actual PROM burning. (If a single-pass EPROM is used, it might be better to use the tape directly.)

When I use paper tape, I prefer to punch all initial program input tapes off line, then feed these into the computer under program control. If the program is run rarely, the tape is simply filed. However, if it is part of the OS, or used often, the data is recorded on one or more tape cartridges. The filed paper tape becomes the grandfather. It is highly immune to damage if reasonably handled. Even the occasional tear can usually be mended.

The other work to be done is the design of a second CPU board. It can be a substitute for the original design; or an extension of the design for greater capability. With improvements in slow and fast memory, and extensions to the OS, you can start on multitask and multiprogram activities. Also, you can start making use of the expansion concepts presented in the last chapter of *How to Design, Build, and Program Your Own Working Computer*.

REVIEW OF ORIGINAL DESIGN BOARDS

For the benefit of readers who do not have the first volume, the two key boards of the original design are reviewed. This review includes one change.

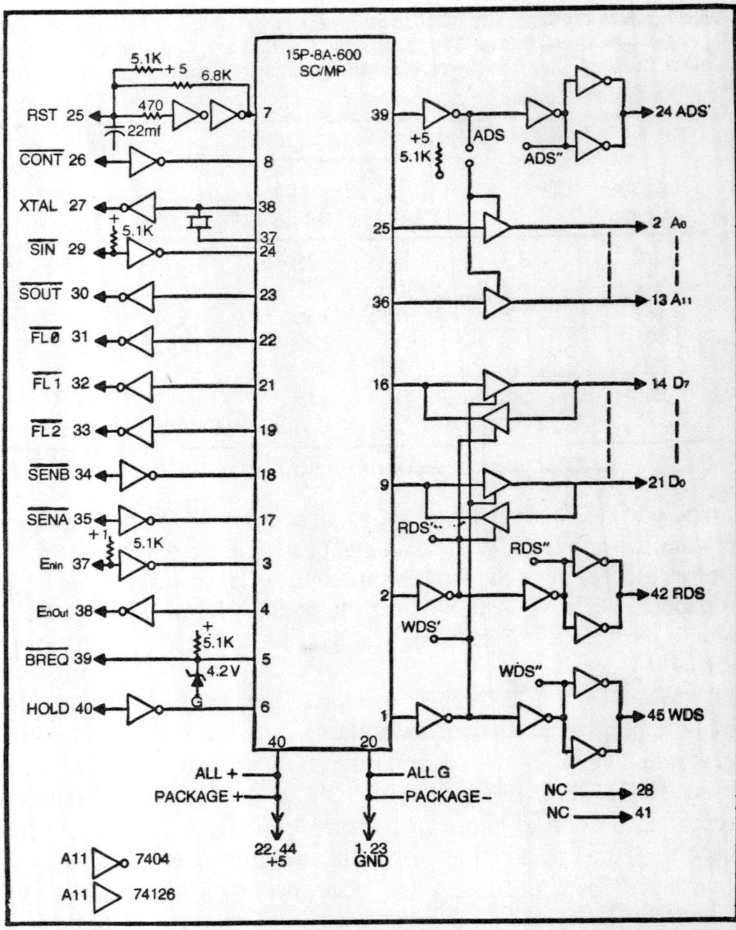

Fig. 1-8. Basic circuit of the CPU board of SCAMP:Y. Board size is 4½ x 9 inches.

The circuit of the CPU board is shown in Fig. 1-8. The major items of interest here are:

- All 1-0 leads are fully buffered except NBreq, which is diode protected
- The data and address bus lines are tristate, to allow DMA, and so on.
- The ADS, RDS and WDS lines are driven by paralleled gates, primarily to give good bus pull-down power.
- All buffers are standard TTL.

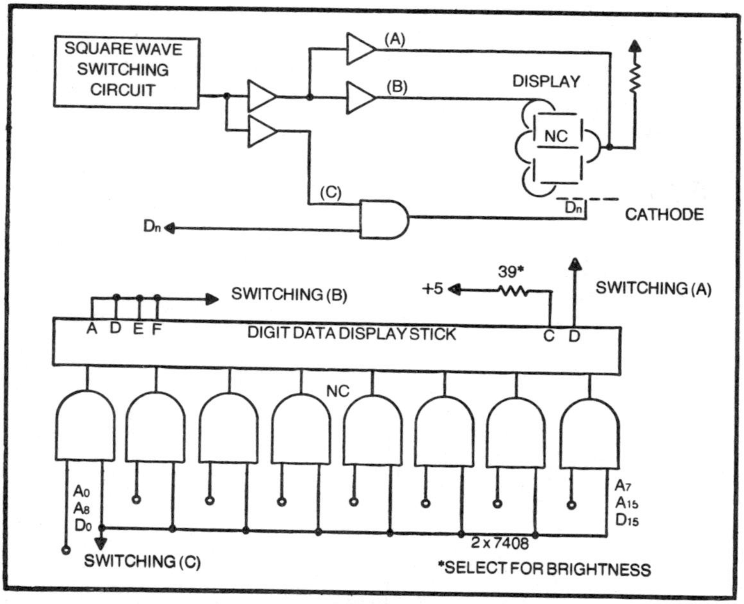

Fig. 1-9. Address and data display for the SCAMP:Y. Calculator type displays are used.

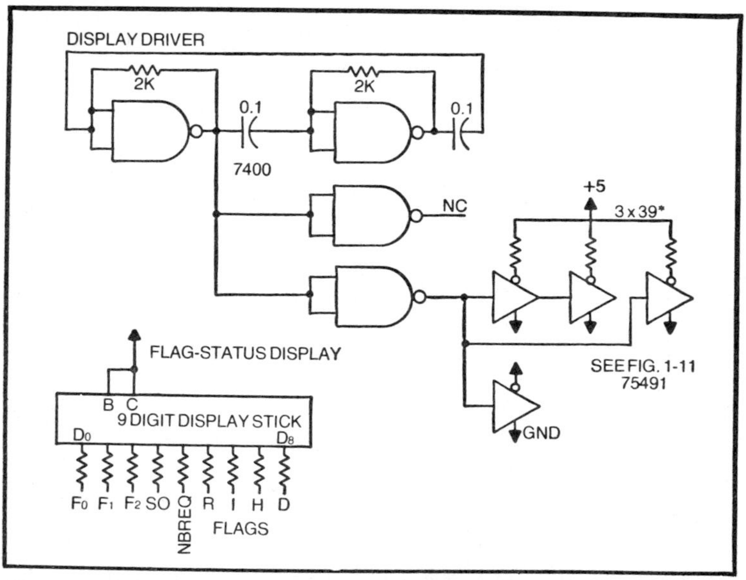

Fig. 1-10. Flag-status display for SCAMP:Y, using only segments b-c as the indicator. Also shown is the multiplex MV, the driver for the display of Fig. 1-9.

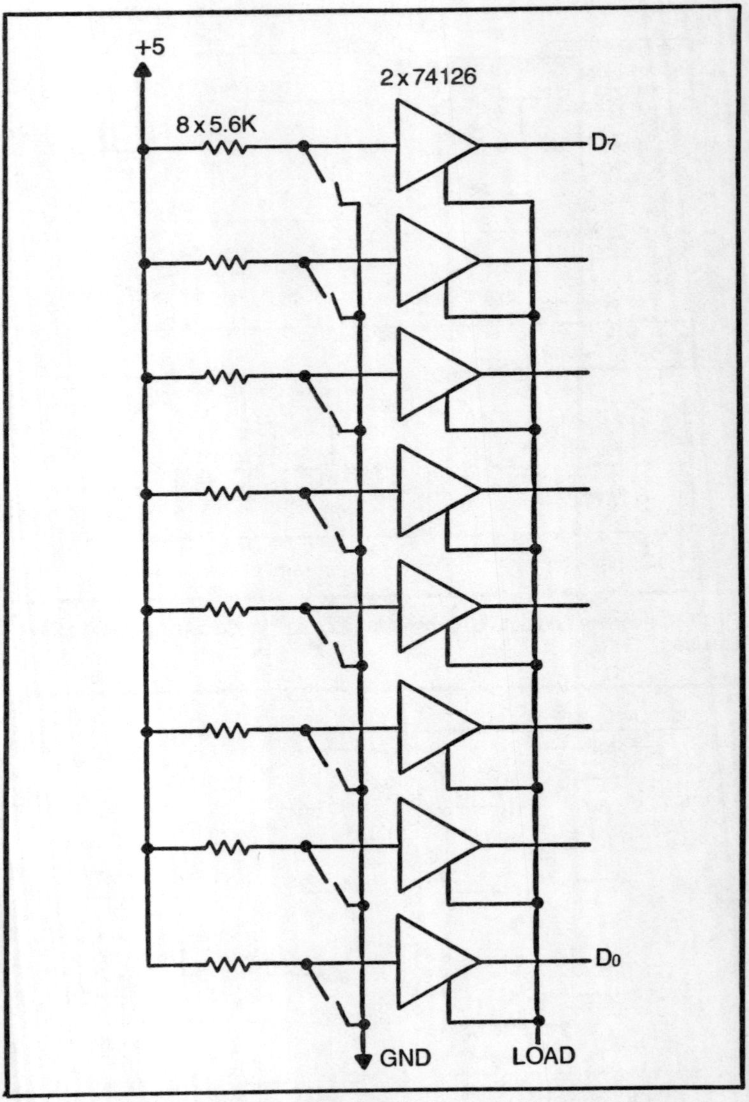

Fig. 1-11. Manual data input switch bank and two buffers for SCAMP:Y.

The circuits for the control panel are shown in Fig. 1-9 (address and data display), Fig. 1-10 (flag display and display driver), Fig. 1-11 (data bus and data switches), Fig. 1-12 (high byte address generation), Fig. 1-13 (single-step system).

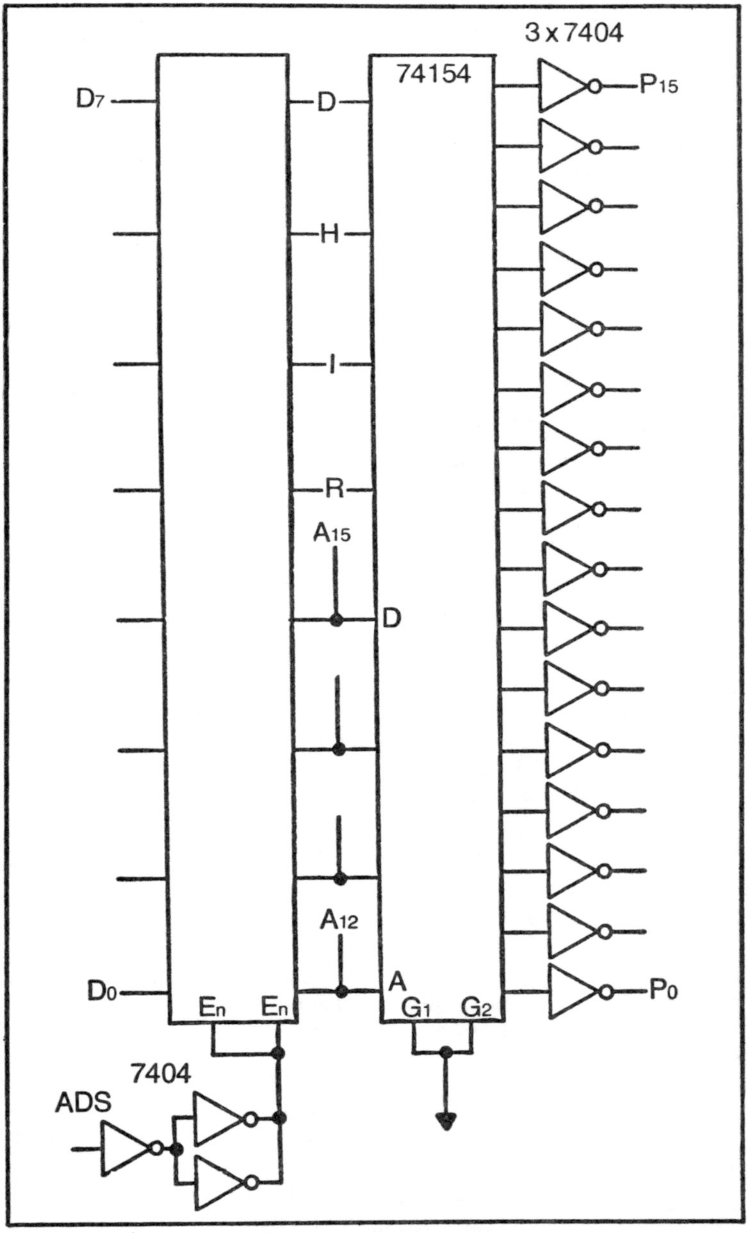

Fig. 1-12. High address or page generation for the SC/MP chip, using address information present on the data lines D_0 to D_3 at ADS time. D_4 to D_7 give status information.

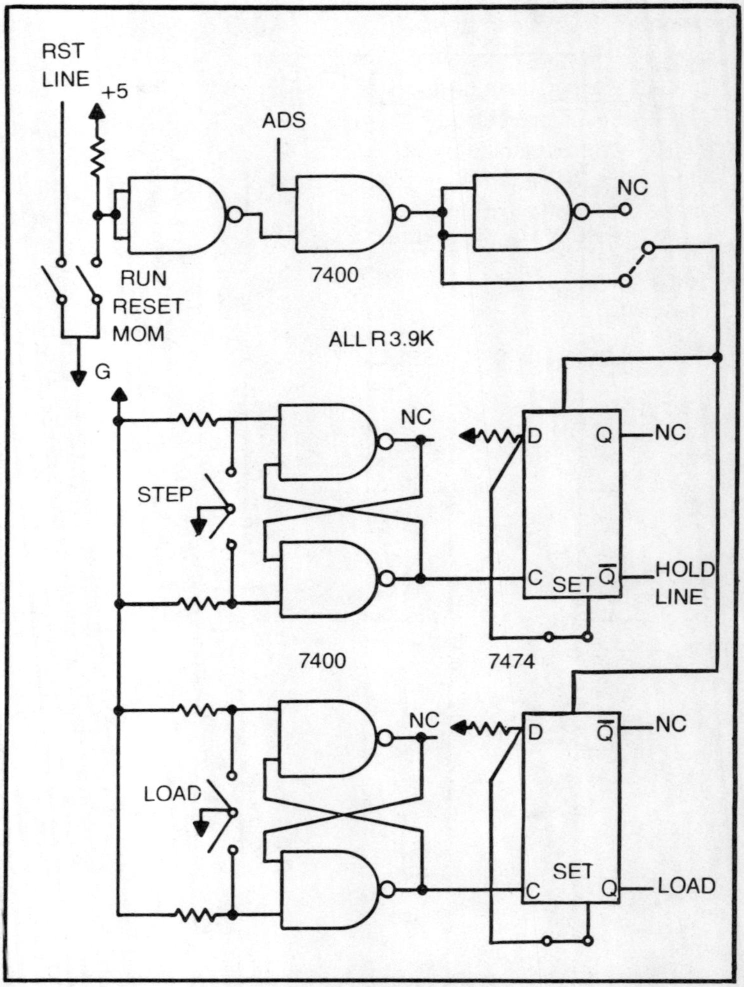

Fig. 1-13. Reset and single-step provisions for SCAMP:Y. The last is especially important for home built systems, and can be a help in program debug operations.

One change was made in the single-step system. A buffer was added at the output of the 7474 FF, changing the signal input from $\overline{Q}$ to Q to compensate. The second input to the buffer gate can be used for an externally generated hold signal.

The ADS, WDS and RDS buses are active high, and are not driven by tristate buffers, even though the corresponding chip output signals are tristate. This has a disadvantage when multiprocessing, etc., is undertaken. Active low buses are easily arranged for wired-OR operation, and would be superior. Remembered this when you start on the multiprocessor work.

Those who are starting from scratch might want to rearrange the CPU buffers to make the strobes active low, as is now common. Changes may be required on other boards. An alternative to this is to make the output portion of the buffers tristate.

With these thoughts in mind, let's move to construction projects.

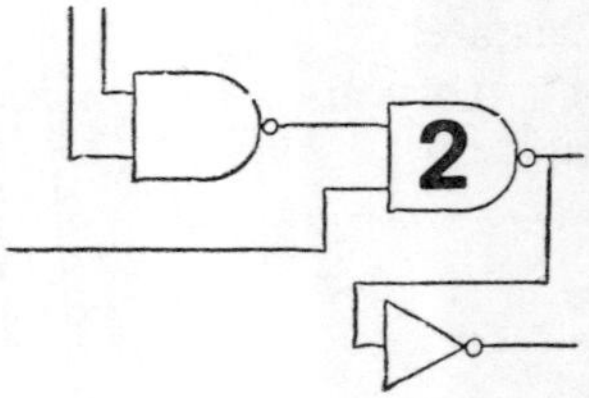

New Fast Memory Cards: Improved Construction Techniques

The design of the memory cards used with the prototype system in Chapter 1 has two disadvantages. One is the method of construction, which uses a small subpanel for each chip. See Fig. 2-1. One of these subpanels isn't at all bad, and the technique is useful. But 16 of them closely spaced on a single board, is just too much. Memory chips should go directly on the board—in sockets, yes, but directly board mounted.

In addition, the amount of memory on the board, 2K, is too small in an expanded system. Most of the time 8K or more of RAM is needed which makes more memory per board attractive. Further, the cost of auxiliary chips stays nearly constant as board capacity increases up to 16K.

However, two factors favor the small board. One is increasing construction difficulties as chip density increases; for a large number of chips per board it is probably necessary to use photo-etching.

Some applications sometimes favors small boards, in particular those involving *post boxes* where data is deposited in a memory location by one system, and read from that location by another. This is the principle of the low-cost video screens, or TV typewriters. For alphanumeric data a typical screen presentation might allow 16 lines of 80 characters each, or only 1280 bytes. Very high resolution graphics, such as 400-line resolution, reach 16K; but respectable resolution is possible with much less. By using

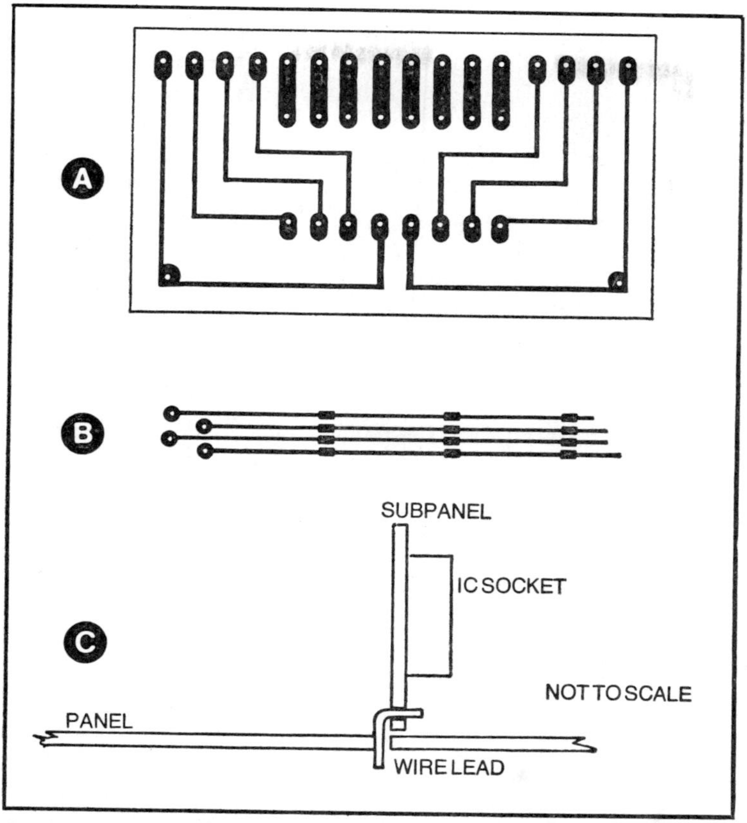

Fig. 2-1. Memory board basics for the prototype SCAMP:Y. Each memory chip and socket is mounted on a subpanel, etched as shown at the top. The center shows four of the 16 bus lines, with pads marking the mounting holes for the subpanels. The bottom part shows the subpanel mounting to the main panel. While the method works, it is tedious to assemble.

only two levels of illumination, and using the digit information within a byte to fine position the illuminated points, 160-line resolution is possible in 1 to 2K of memory.

Before settling on the size of boards, you should review construction techniques and their influence on board capacity.

HIGH CAPACITY BOARD TECHNIQUES

The memory board used in *How to Design, Build and Program You Own Working Computer System* is the small subpanel pattern shown in Fig. 2-1A. After etching, the board is trimmed to the outline shown, and drilled for the 16 holes of the socket and the

16 edge connections. Jumpers are soldered along the edge, and threaded through the mother board. Part of the etching pattern is shown in Fig. 2-1B. A cross-section of the finished board appears as in Fig. 2-1C.

This technique works, but it is tedious! The number of solder connections per socket is tripled, and the number of subboards becomes large. While the final result is adequate, rugged, and satisfactory, the labor involved is just too great.

Commercial boards use the pattern of Fig. 2-2. Figure 2-2A shows the component side of the board; and Fig. 2-2B the solder-pad side. Very close packing is possible, usually 0.2 inches larger than the pin pattern. A typical 16-pin chip can occupy only 0.5 x 1.0 inches. Thirty-two mounting spaces can be placed in an area 4 x 4 inches, although interconnections become much easier if

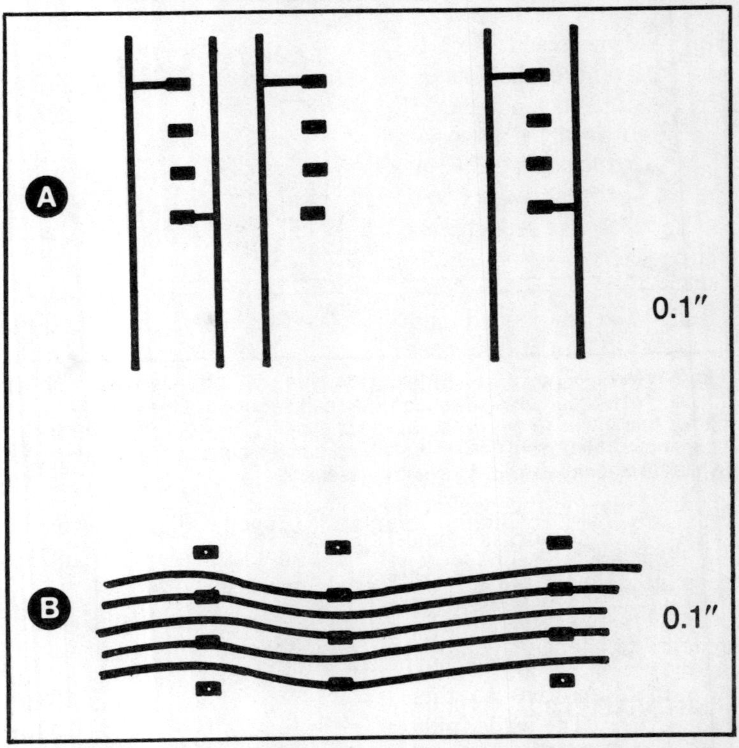

Fig. 2-2. Foil design for close-packed IC sockets. Top is the component side of the board, lower part is the foil side. Layout is easy at 2:1 or larger, scale, but very tedious at 1:1. If used, photo reduction of a 4:1 drawing is recommended, rather than a direct-etch process.

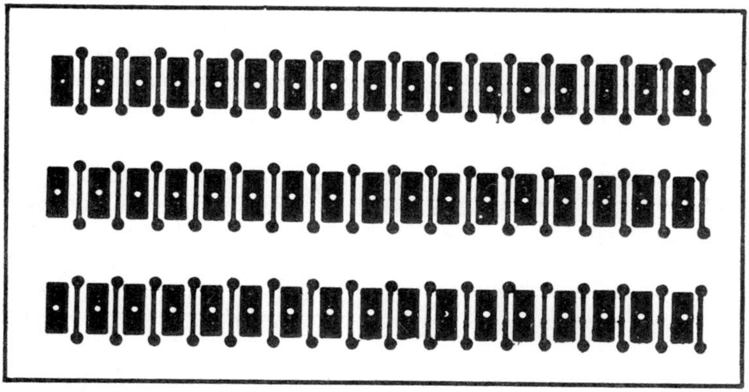

Fig. 2-3. Direct-etch material available at 1:1 scale. Layout is appreciably easier than for Fig. 2-2, but still requires care.

the area is relaxed to 4 x 6 inches. On the standard 4½ x 9½-inch board used here, this leaves ample space for bus interconnection and for other components. (Commercial boards of this type are available from at least two sources for the 4½ x 9½ size.)

Board layout is easy if photo-etching is used, and if the original art work is prepared at two or four times the final size. However, such precise size reduction is difficult or impossible for most home builders.

Drafting layout is possible at 1:1 scale. A good eye and a steady hand are needed; and an illuminated magnifying glass is a necessity. The main secret seems to be patience: Don't hesitate to erase or to remove tape, if the lead spacing is too close to a solder pad.

It is also possible to direct-etch boards of this type, but it's tedious. The problem is the threading of bus leads through the solder pads for the socket pins. Even with 1/32-inch or even 0.015-inch tape work is tedious, and poor positioning is common.

But, for work at 1:1 scale, prespaced direct-etch arrays are available in 14 and 16-pin sizes (see Fig. 2-3) and in LSI and continuous strip form. They are equally suitable for drafting. Work is still somewhat tedious, but much easier.

An alternative to this construction method uses *jogged* connections. The solder pins on one side of the socket (or IC) are bent at right angles to the normal position, the bend being as close to the socket body as possible. These bent leads contact and are soldered to pads on the upper side of the board. The remaining pins go through the board, as usual. Figure 2-4 shows the construction.

Jogged construction can be used in several ways. One is to pack the sockets as closely as possible, making all connections for half of the pins on top of the board and the other half on the bottom. The bottom pattern can be straight lines, but the top pattern must be wavy to miss the through-pins, as in Fig. 2-5. This pattern is much easier to work with than the pattern of Fig. 2-2, since there is nearly 1/16 inches of space instead of the usual 1/32 inches or less. Note the arrangement of pins here; this makes the assembly stronger than having all pins on one side jogged, and prevents foil tears.

Because of the protuding pins, a minimum of 0.25-inch spacing between one socket and the next is needed. At this spacing, it is best to mount and solder each socket before proceeding to the next one; it is very difficult to solder or resolder the top pins when all sockets are mounted. Larger spacing would be desirable, allowing more room for test prods or solder iron tips. Space needed for a 16-pin socket might be 1.0 × 0.6 inches.

Another arrangement with jogged sockets uses one side of the board for leads running up-down (the short way of the board), and the other side for leads running the long way. A typical pattern is shown in Fig. 2-6. This is a suggested pattern for the 2102 chip. It is very easy to work with if the long-way bus leads are all spaced by 0.1 inches. With this arrangement a 16-pin socket will occupy 1.5 × 0.7 inches. The above comments about test lead and soldering-iron tip space applies, of course.

Let's consider these mounting methods with respect to memory capacity of standard size boards. These have an effective

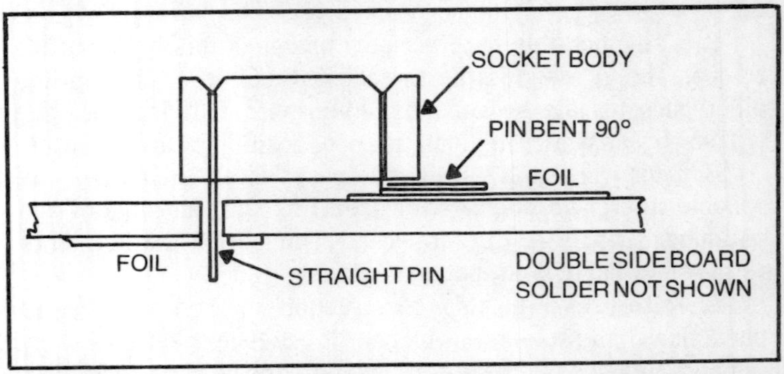

Fig. 2-4. Method of jogged-pin construction, bending some pins for easy top-of-board soldering. This is the method adopted for most construction projects shown in this volume, and is recommended for all direct-etch double-sided boards.

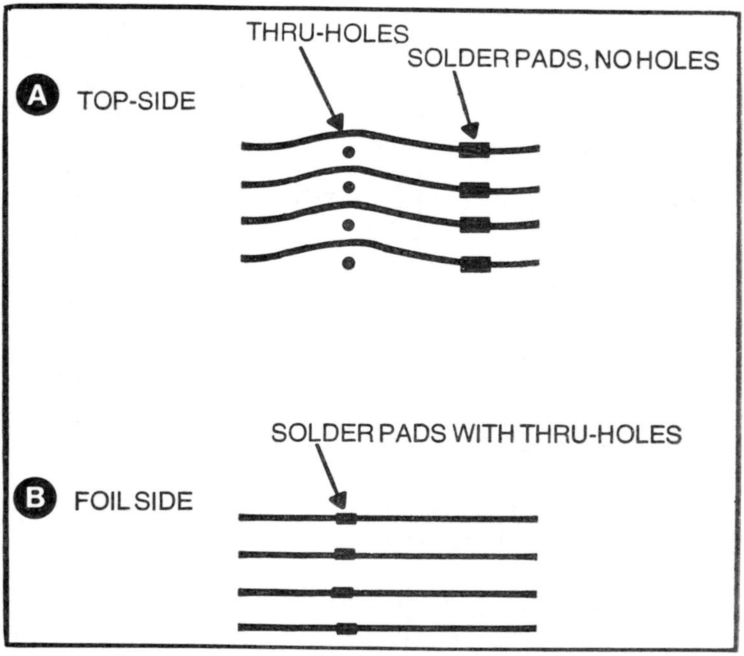

Fig. 2-5. Typical layout of four pairs of socket pins in jogged construction. However, at least two pins on the jogged side should go through the board, to prevent foil pull-off when ICs are removed from sockets.

mounting space about 4.2 inches wide, and some 6.5 to 7.0 inches long, assuming other circuit elements require 2.5 to 3.0 inches of board length.

The cross-wired, jogged circuit takes about 1.5 × 0.7 inches. Three rows could be placed the width of the board with a little crowding, with nine or 10 chips in each row. Eight chips per row would be much less crowded. This layout lends itself to a 3K board using the common 2102 chip. This is an awkward size, so it is probably best to use only two rows for 2K capacity. If larger capacity is desired, it seems best to go to a chip organized as 1K by four bits, such as the 18-pin 2114. Two rows would allow a 4K board, with a reasonable spacing and lead arrangement.

It appears that the jogged board with parallel leads on top and bottom would allow greater density. The necessary wiring space is only 1.0 × 0.6 inches, so four banks of eight each can be placed in 4 × 4.8 inches. However, the 2102, or even the 2114 do not really lend themselves to this layout, because of the relations between the input, output, chip select and write enable leads.

The board with all connections on the foil side, Figs. 2-2 and 2-3, is much easier to work with, since it is easy to run leads at right angles on the top of the board. A sketch of one cell of a 4K 2102 board is shown in Fig. 2-7.

It appears that the most reasonable size memory board for complete home construction uses no more than 16 memory chips for 2K capacity with 2102s or 8K with 2114s. Doubling these is possible by going to the foil arrangements of Figs. 2-7 and 2-8. However, the work is tedious.

These considerations plus cost add up to a simple fact. When a standard duplicative design is needed, home construction becomes unattractive. It is probably better to consider one of the well-built kits now available as an alternative to home building. Of the 4K boards available in the 4½ × 9½-inch size, there is a large assortment of boards of up to 96K available for the S-100 bus. Some are directly usable by pin wiring, others may require an interface unit for signal inversion (or perhaps a jumper to an idle pin) to function. When considering fast memory expansion, don't forget to look at the kits.

Home construction pays when a nonstandard approach is followed, for instance, if duplicate use of memory is desired, as in some forms of video presentation.

Before firming up memory design, let's look at approaches to this particular form of *direct memory access* (DMA), as well as the general DMA problem. Then decide on the memory form.

CHEAP VIDEO

The basic form of Cheap Video is shown in Fig. 2-9. Connections are made to three places in the microprocessor. One set of leads are taken to the timing bus, and another set to the data bus. Under control of the timing signals, the added circuitry generates pseudo instructions which are placed on the data bus. A form of memory read instruction in the microprocessor uses these pseudo instructions to scan the selected memory in sequence. As this is done, sync and step pulses are generated by the timing circuits.

Because the data bus is used in the sweep process, a tap is placed on the memory output to obtain stored data directly. This is combined with sync data to generate the composite TV signal. The scan microprogram can be placed in ROM and changed, if desired, to give alphanumeric or graphics capability.

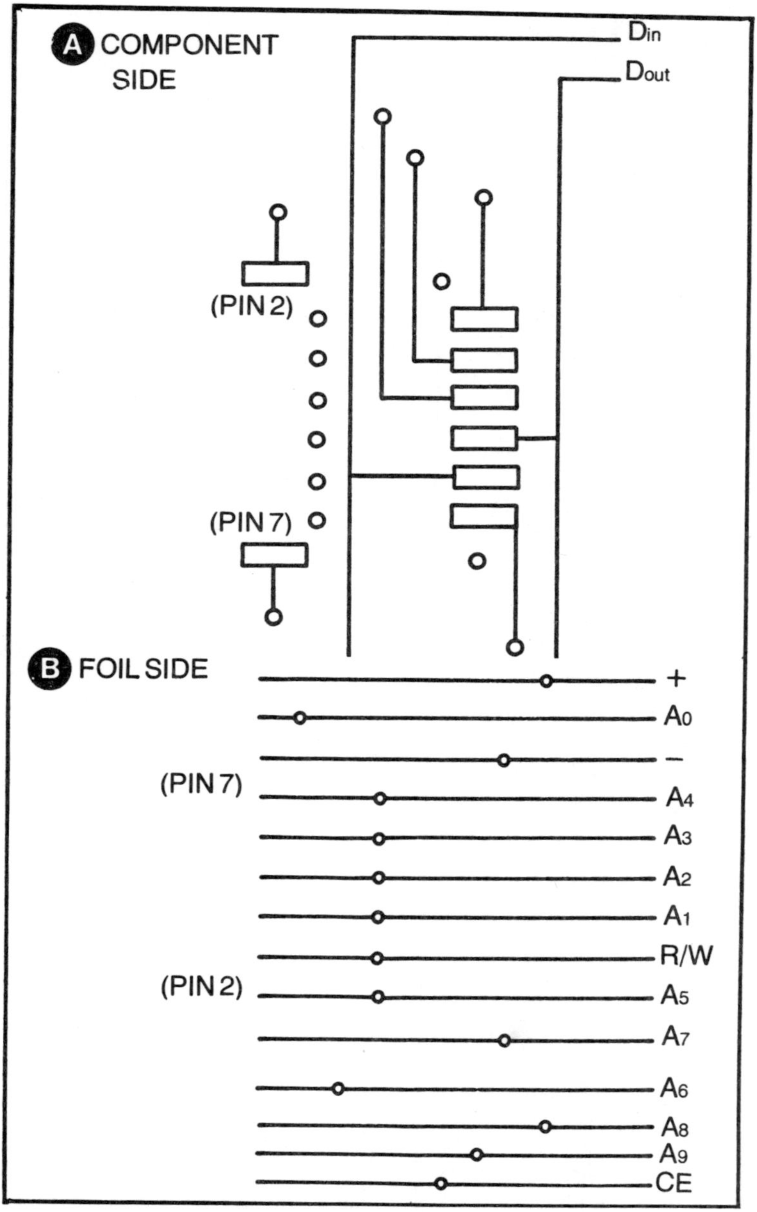

Fig. 2-6. Complete layout for one cell of a memory board using 2102 chips. Through-pins are used to connect some top-of-board leads to the parallel bus lines on the foil side. Each cell is 0.6 x 1.4 inch minimum. It is easy to place 2K on a 4½ x 9 inch board.

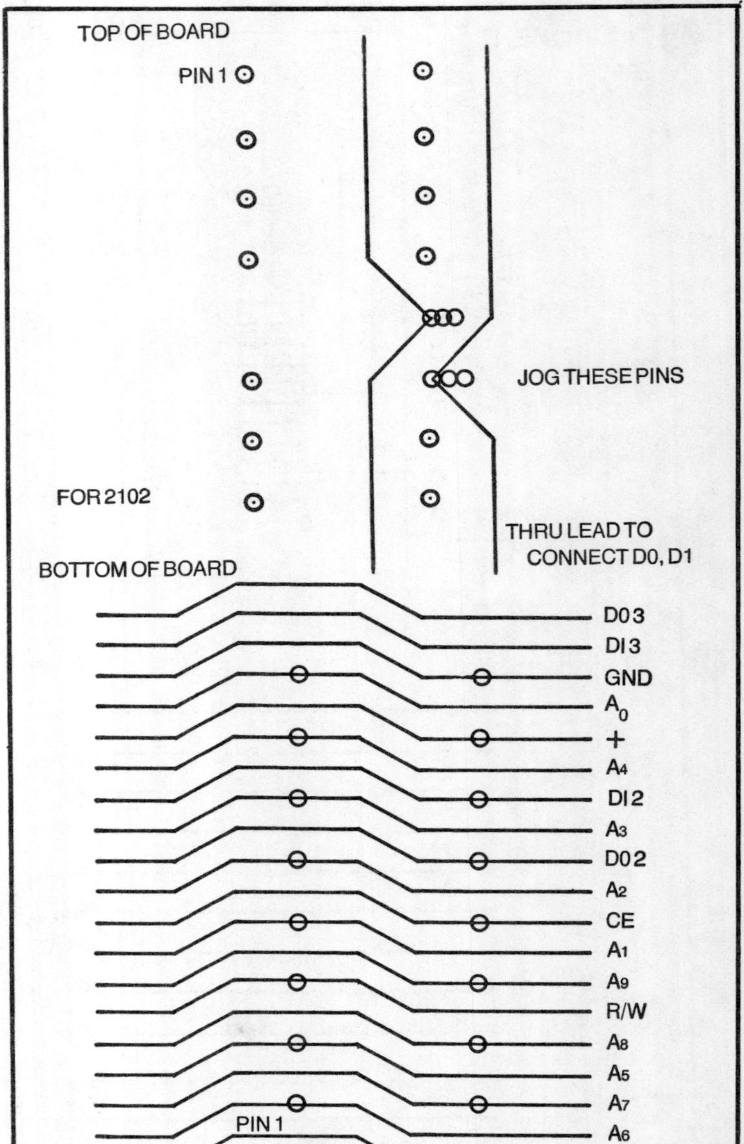

Fig. 2-7. Complete layout for one cell of a memory board for 2102 chips, using the wave connections of Figs. 2-2 and 2-3. 4K can be placed on a 4½ x 9 inch board, but direct-etch construction is tedious.

Fig. 2-8. Memory cell for an 8K board using 2114 chips. Cell size is 0.6 x 1.8 inch minimum. Layout is very easy.

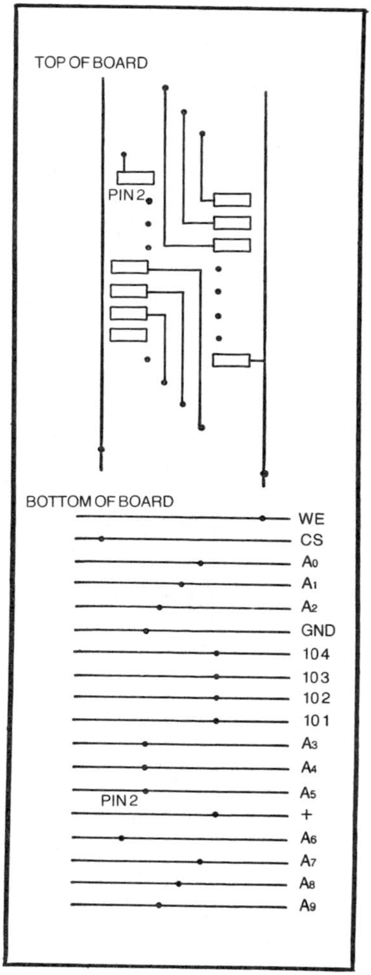

The system involves some timing tricks with some microprocessors. The 6502 and 6800 are easier than the 8080 or the Z80.

It is evident that this form of video generation uses a special form of direct memory access by an external device. It is special in that it is for a read-only process, and in that the microprocessor is used for address generation. Normally, direct memory access places the microprocessor in a hold state during the access period. However, since the data-to-video conversion does not require all that time, the scan generation and regular programming can be run on a time-sharing basis. This is often called *transparency*.

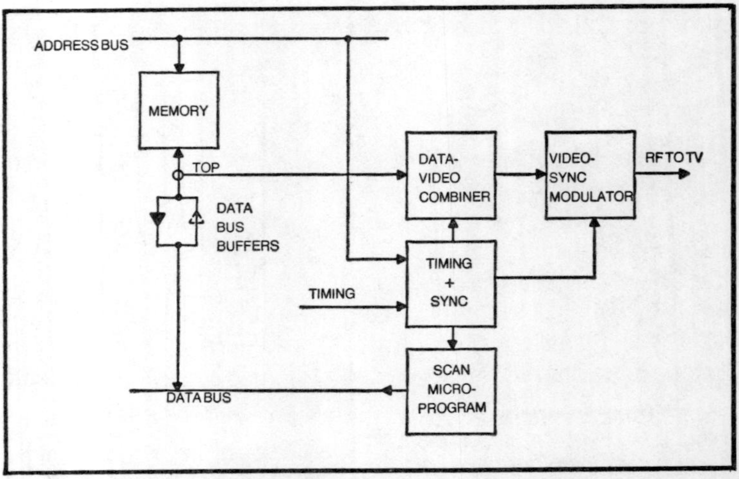

Fig. 2-9. Principles of Cheap Video. A tap is placed on the memory bus, between memory and buffer, to allow nondestructive read-out of memory data. This is combined with sync signals to form composite video. In this version, a short program is used to scan the memory; the CPU is busy when video is displayed. An alternate version placed a second tap on the address bus, between memory and its drivers, and scans by timed signals. This leaves the CPU free for regular operation. In both systems memory write is by the CPU only.

THE TRS-80 VIDEO SYSTEM

In the popular TRS-80 system, the same principles of computer write-in and video read-out are followed, but in a form which does not tie up the central processor for read. Instead, as sketched in Fig. 2-10, the job of addressing the memory for data read-out is assigned to the sync generator. This also controls the conversion of data to video form. The addresses generated during this process are routed to the memory via a multiplexer. If there is no CPU-to-RAM write in progress, the addressing is from the sync generator. The CPU has priority.

The functioning of the RAM and associated components depends on the OS. Typically, keyboard data is fed to the RAM, for display to the screen. As the RAM fills up, data is transferred to regular memory, by scrolling, or by page transfer. A cursor can be provided to allow data changes at specific locations. The visual effect is not greatly different from that of the Cheap Video system, but there is one major advantage—the CPU can be completely independent of the video screen. This is a great advantage in speed. For example, the CPU can proceed to a second part of a program while the video screen is displaying the results of the first part.

A MULTIPROCESSOR VIDEO TECHNIQUE

A straightforward approach to the video display is to combine the two systems. A sync generator of the TRS-80 type is replaced by a second CPU, operating with its own ROM. While this may be a waste of computational ability, it can be cost effective, especially for densely packed systems where chip count is important.

Figure 2-11 shows the principles of such a system. It represents both a form of direct memory access, and a form of multiprocessing. An advantage of this technique is simplification to

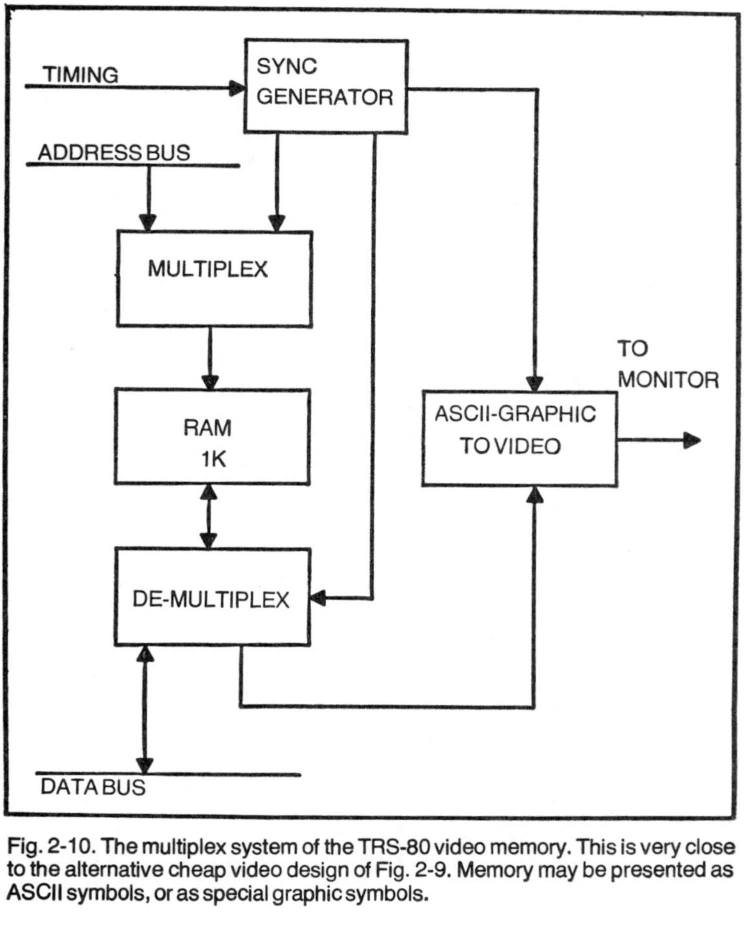

Fig. 2-10. The multiplex system of the TRS-80 video memory. This is very close to the alternative cheap video design of Fig. 2-9. Memory may be presented as ASCII symbols, or as special graphic symbols.

the OS. For example, the post office box address for depositing all data can be a single memory-mapped location, rather than a block of addresses determined by CPU computation. This method is also simpler when shifting from alphanumeric to graphics displays.

In the original Cheap Video system, the change was accomplished by changing a module, so the presentation was an either-or type. In the TRS-80 circuitry is provided for both. As noted in Fig. 2-10, the signal for graphics selection comes from data bit 6, so both alphanumeric and graphic symbols can appear on the TV screen at the same time (actually, in sequence a scan rate, but simultaneously at eye rate). The ROM for the separate microprocessor can be programmed in this way, or some RAM could be provided to allow a new form of graphics to be introduced. Many variations are possible.

PRELIMINARY DESIGN OF NEW MEMORY BOARDS

Now, let's proceed with design of improved fast memory boards. One goal is to improve and simplify construction. A second is to design the boards to be compatible with simple video systems of the types just described. The third is to increase board capacity. (However, it isn't necessary to have more than 1 to 2K capacity for the boards used with video).

Review of the home construction techniques indicates that 16 chips in the memory bank is about the limit for easy direct-etch layout. This would allow only 2K if the popular and low cost 2102 chips are used. This is suitable for a board used with video. It also wouid be a direct replacement for the original design.

For greater capacity, several approaches seem reasonable:

- Use Fig. 2-5 layout with 32 2102s for 4K.
- Use 16 2114s for 8k
- Use Fig. 2-5 layout with 32 2114s for 16K.

There is not really a great difference in degree of difficulty for the first and last, but the 8K 2114 design would be appreciably easier to fabricate. Further, expect to use considerable ROM, so there is not as great a need for fast memory. It appears that the 8K design is the best to use for the second type of memory board.

In the 2K design, you must settle the method of video pick-off. Since this is a wire-wrapped bus structure, the data, address and control buses are readily available. For video you need read-data at the output of the memory chips, before the bus buffers. You can handle this by using a WRITE/$\overline{\text{READ}}$ signal, placing the chip in the read state unless the CPU requires both board select and the write

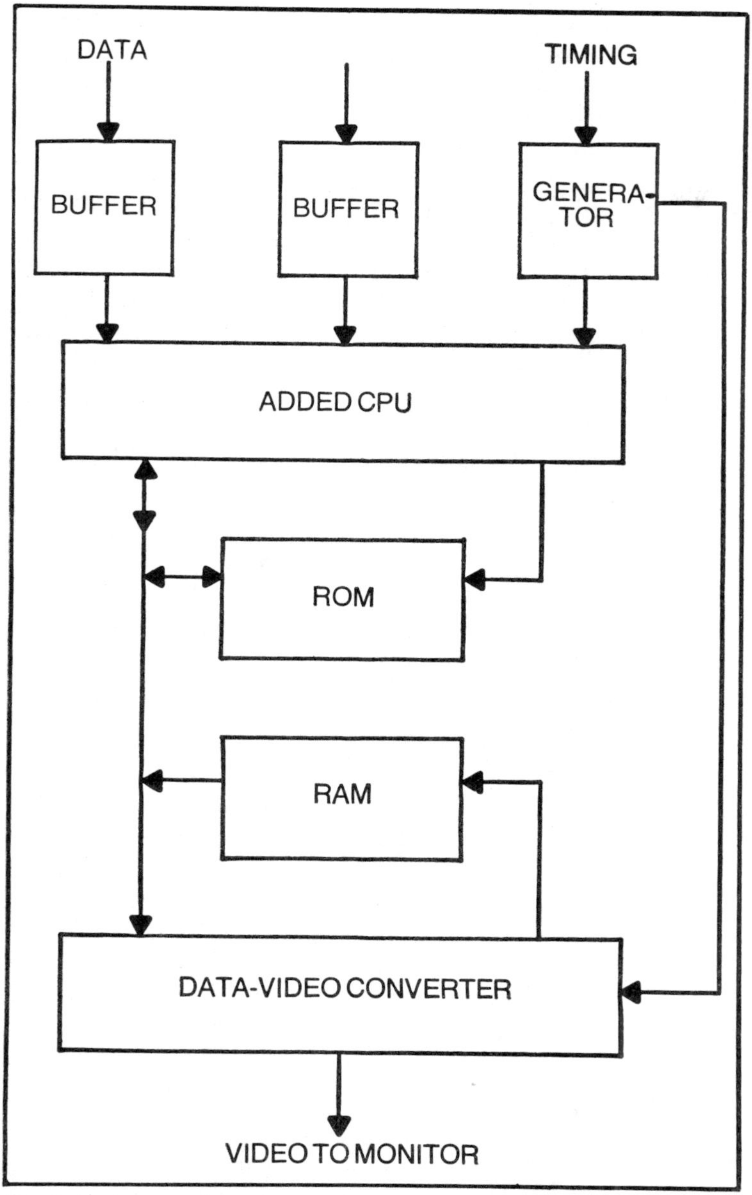

Fig. 2-11. One form of a dedicated CPU for video signal generation. Data is deposited from the main bus, the added circuits acting as post office boxes. Memory look up and deposit, and many scan functions are handled by the added CPU. Subject to some speed limitations, this system can operate independently of main CPU timing.

state. Bank select is necessary, and can be handled by a wired OR chip-select circuit. In addition to page select, the 2K board requires half-page designation. This is easily handled with a jumper at installation.

You can simplify layout somewhat by using some of the four or eight-line buffer chips, including bilaterals. The cost of these is now sufficiently low to be competitive.

While there are extra pins on the memory board connector, it seems better to have the video circuits on a separate connector. This can be done by placing a second set of finger contacts at the other end of the board. To prevent mistakes, the pins can be offset from the board center.

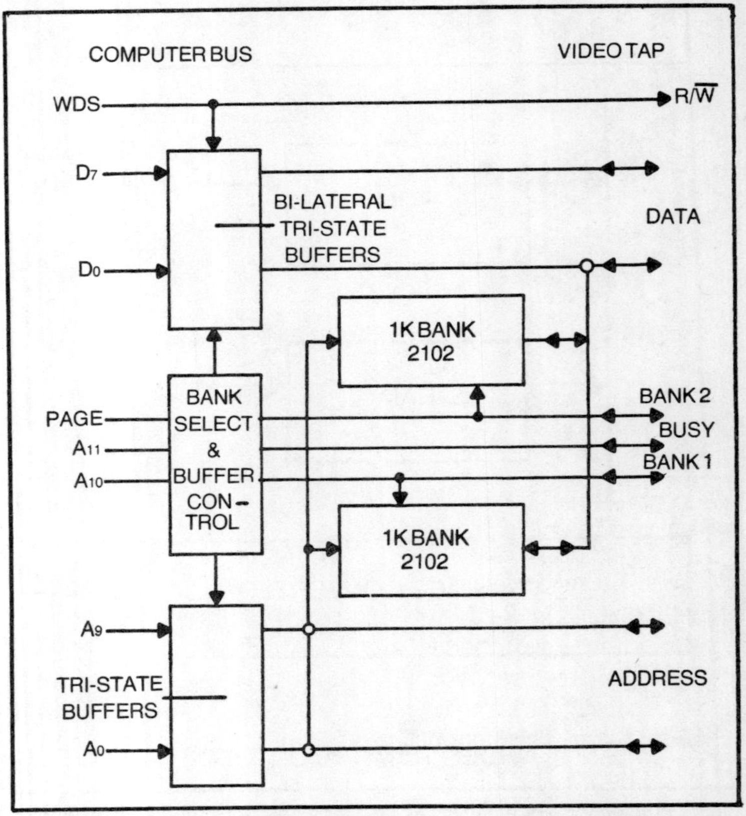

Fig. 2-12. Block diagram of the 2K RAM board, second version. All data and address lines are available on a second board connector. The board can serve in a video display system, or as a set of post office boxes in a multiprocessor system.

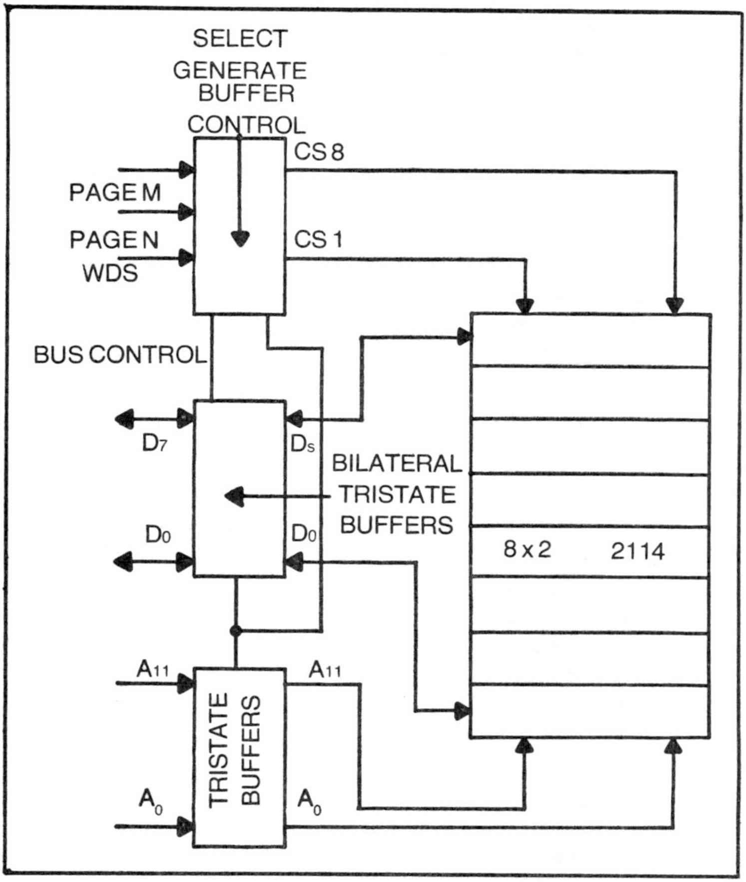

Fig. 2-13. Block diagram of the 8K RAM board. This design does not include special provisions for video or post office use.

With these decisions, the block diagram of the 2K board becomes that of Fig. 2-12. Note that in some respects the 8K board is simpler since we decided that we do not need the data and address taps. However, you will need to select chips by pairs, a total of eight signals is required. The bus buffering can be the same as the 2K board. The block diagram for this board is shown in Fig. 2-13.

CIRCUIT DESIGN FOR THE 2K BOARD

For data bus buffering, several chips are suitable. The 8216 appears a good choice. There is an even larger selection for bus buffering, with the 74367 being good. The controlled circuits of the board are shown in Fig. 2-14, which also shows the 2102 connections with only one chip in each bank.

In this drawing leads ending in circles go to the main bus; those ending in a bar go to the video connector; and those ending in arrows must be generated on the board. The circuit selected for these board-generated control signals is shown in Fig. 2-15. Note that one of the gate packages is an open-collector with pull-up resistors. This allows the wired OR connection for board and chip select. The other package provides buffered address signals as required for board and chip select operation. The board half-page designation is made by jumper.

CIRCUIT DESIGN FOR THE 8K BOARD

The bus buffering for the 8K board is essentially identical; the major difference is omission of the video taps. Also, as shown in Fig. 2-16, there is a difference in chip select, since the 2114 as organized is 1K by four bits, and, of course, a corresponding change in data pin connections.

Of the many possible ways of generating the chip select signals, the one shown in Fig. 2-17 is about the simplest. The write enable signal generation is also simple, the first gate giving a page A OR page B signal to combine with the write strobe.

The pin connections of the 2114 are different. Figure 2-18 shows these, plus a suitable wiring arrangement for each memory cell.

BOARD CONSTRUCTION

How to Design, Build and Program Your Own Working Computer System included a discussion of board layout and construction techniques. These are reviewed here, with changes and additions that have been found helpful. With just a little practice, it is possible to turn out neat boards of good quality without special tools.

The first step after deciding on the circuit design is to assemble the pin-out data you will need. It is convenient to make these on cards, one for each type of IC. Show both top and bottom views, and label these clearly. It is also helpful to show polarities of control signals.

With these and the circuit diagrams, sketch out an approximate location for each chip. Then start actual layout. I find that time is saved and errors are reduced by working in two steps. The first

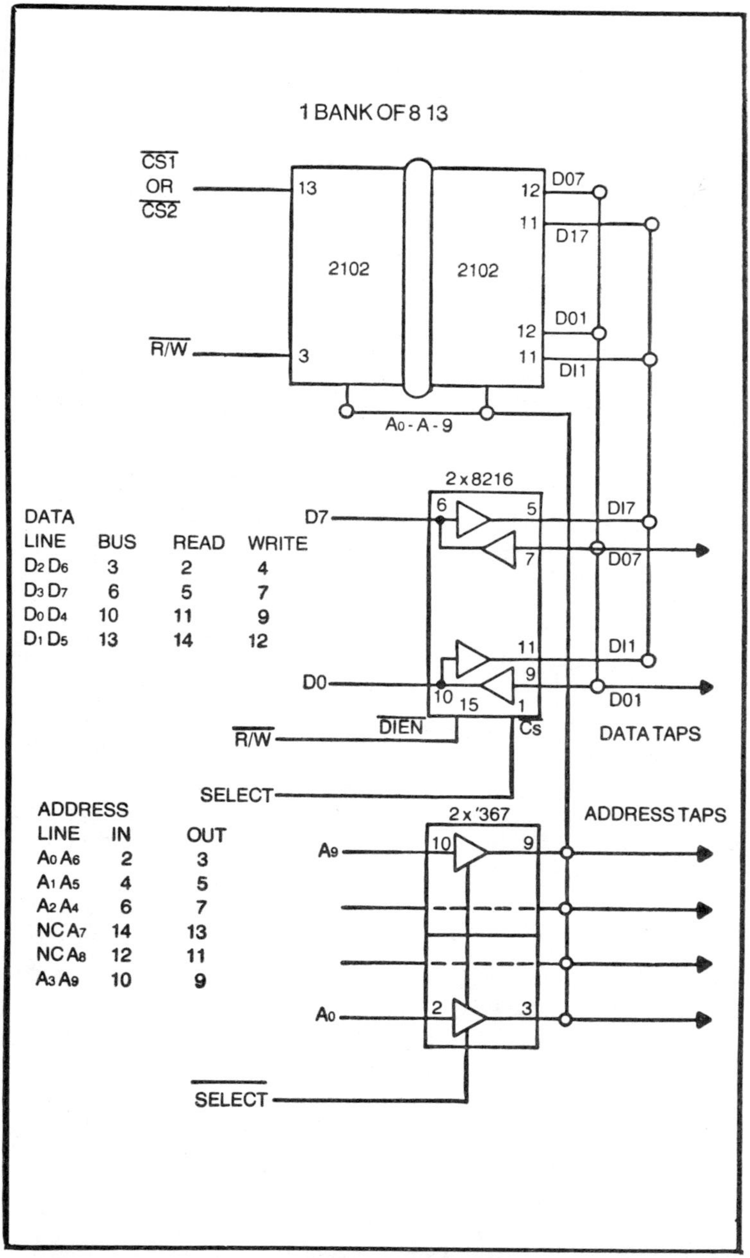

Fig. 2-14. Controlled circuits of the 2K board. Bilateral tristate buffers are used for the data lines.

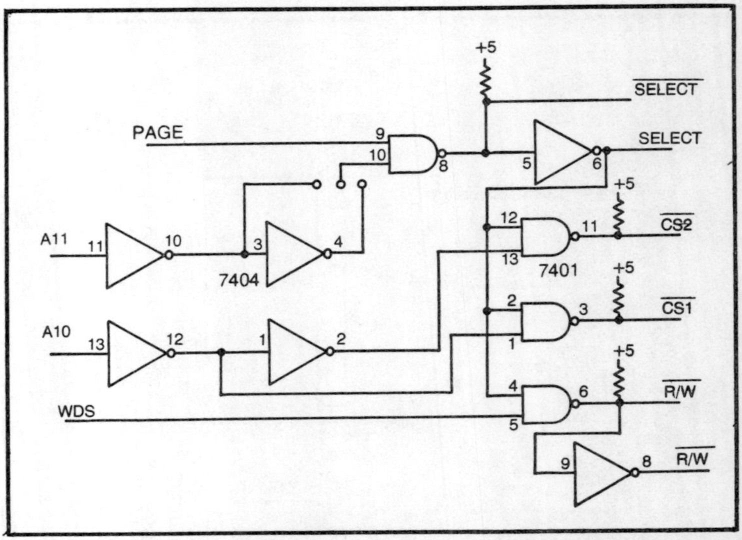

Fig. 2-15. Controlling circuits for the 2K board. Board select, chip select and read/not write signals are developed from page and address signals and the write strobe.

is a pencil sketch at two-times scale, made on inexpensive two-sided graph paper. Label the two sides, Foil and Component. The second drawing, for direct-etch use, is on two sheets of one-times scale, 0.1-inch guideline or equal, tracing paper fastened together, back to back. (In transferring from one drawing to the second, improvements in lead routing may be found: Correct both drawings if a change is made.) If photo etch is to be used, the second drawing can be one times, two times, or four times.

Layout

In layout, it is usually best to keep the long bus leads on the foil side of the board, and short connecting leads on the component side. Lead crossings, which always occur, can usually be handled by a through-hole, continuing the lead on the other side of the board. While through-plating is not possible, a satisfactory substitute requires only a doughnut pad on each side of the board with a short piece of wire soldered in. Sometimes a jumper works better than through-board connections.

I do not usually show by-pass capacitors on schematics. Add these when making the layout. It is safest to be generous: one 0.1 mylar or ceramic for each memory chip, one for each three to four other chips, plus one 10 μF electrolytic about each four inches of board. A trick sometimes used is to leave the positive power lead to the last layout item, then show these as jumpers with ferrite beads. This helps with RF noise generation and susceptibility problems.

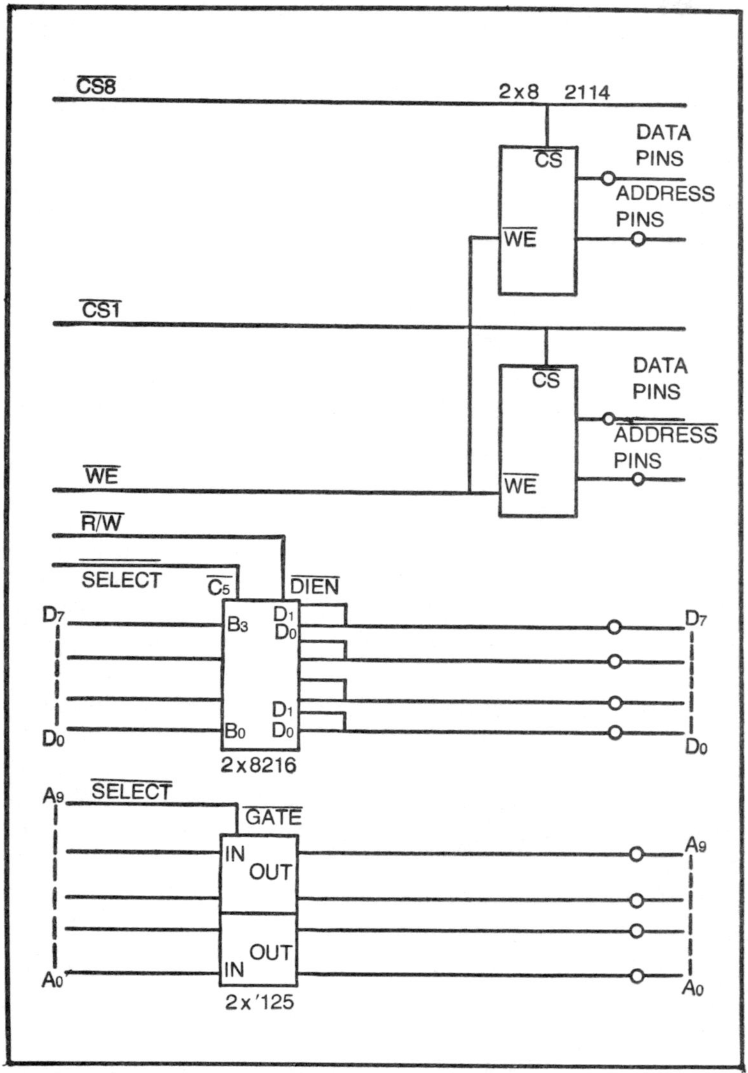

Fig. 2-16. Controlled circuits of the 8K board. Principles are the same as for Fig. 2-14.

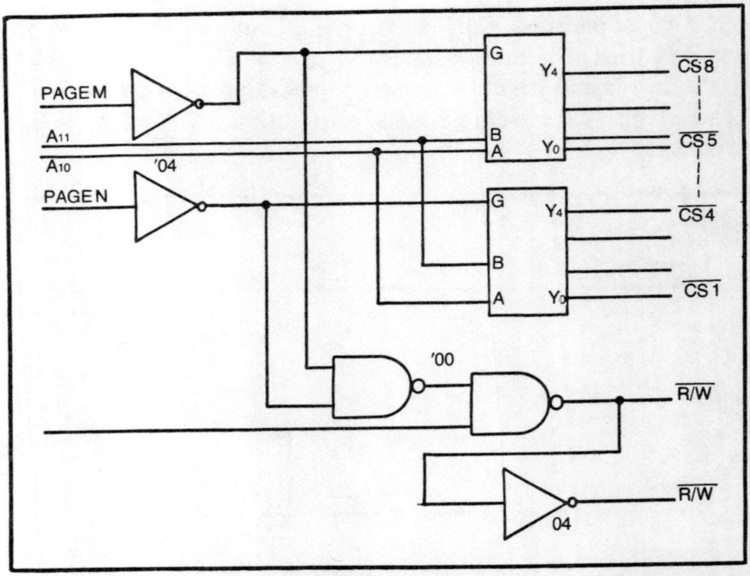

Fig. 2-17. Controlling circuits for the 8K board. Eight chip select signals are developed from two addresses plus two page signals, with either page signal plus the write strobe being used to generate the read/not write signal.

Soldering

Don't forget to allow space for soldering on the component side of the board. A card showing layout and clearance dimensions is a help. Also, remember component height. It's especially important if boards are to be rack mounted at less than about one-inch spacing.

Normally, and especially with newer chip packages, there will be a direction of signal flow that lays out easily. Any other direction gets into lead crossing problems. Sometimes, changing to a different chip of the same function will simplify layout. However, practice will let you get to an adequate layout, usually on the first attempt.

When the two-times layout is complete, check it against the circuits and also against connecting drawings. It is helpful to use the margin to show pin-out names and signal polarities. Then transfer the sketch to the one-times vellum sheets. When these are complete and have been checked, separate them, and make sure that they are correctly labeled.

Actual board marking is done by prick-punching through one of these sheets, the component sheet usually being best. In prepara-

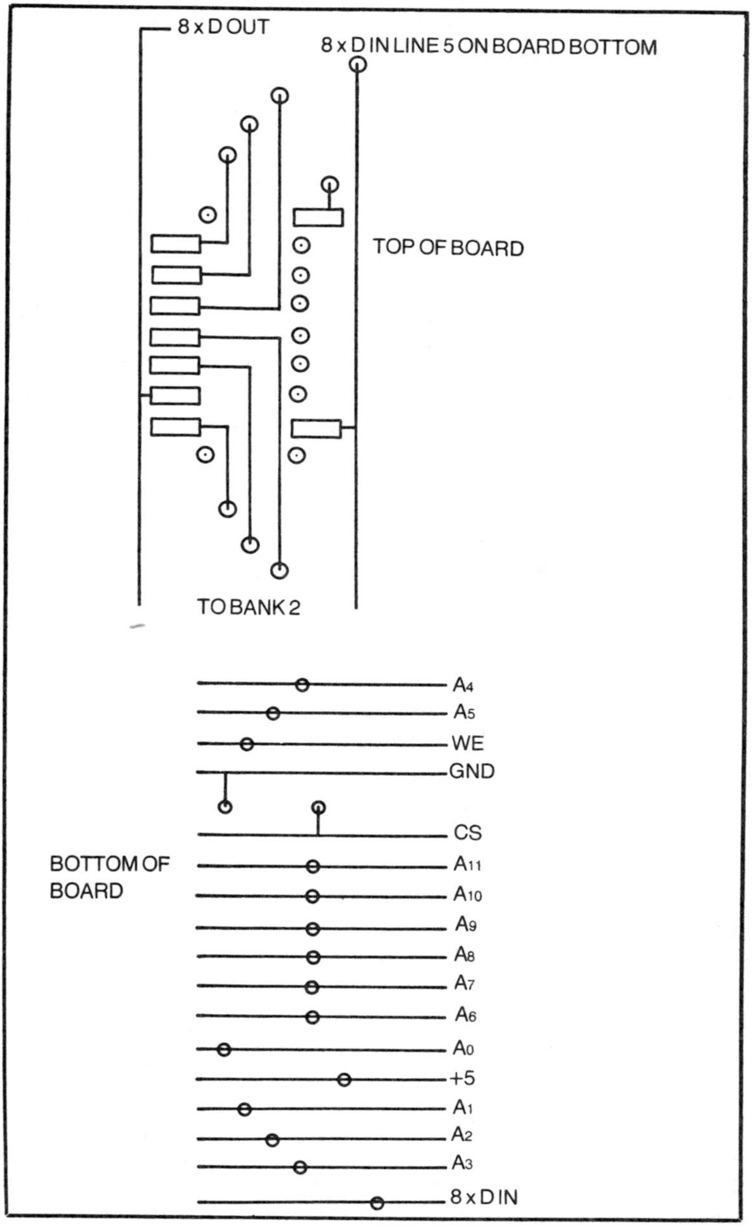

Fig. 2-18. Memory cell for an alternate chip for the 8K board, the 2147. Cell dimensions are the same as for the 2114, but connections are different due to the different organizations.

tion for this, clean the board with steel wool, and polish it until it is shiny. Make sure that it is free from grease, washing it if necessary. Then coat the board thinly with rubber cement, and position the drawing on it. Be sure that you get the right side out. Let the board dry under pressure.

When dry, use a prick punch to mark each drill hole into the foil. Be as careful as you can to get good accuracy. A sharp punch and a magnifying glass help. Also punch the board edges (if the board is to be trimmed later) plus one locating hole for finger contacts. Punch every pin to be drilled. With jogged construction, not all pins go through the board.

After checking for completeness, peel the vellum off the board. Clean it, and file it in your construction notebook. Also clean the board. Now decide if you are going to use the board directly, or keep it as a master. If you are going to need several boards, the master is the way to go. For this, fasten a second board to the back of the punched one with rubber cement. You can even drill three, or perhaps four, boards at the same time, but the drill bits must be sharp and the feed pressure light; otherwise drill wander may give excessive errors.

After the cement has set, drill. A drill press is a necessity, and a magnifying glass a big help. For IC sockets and most small component leads, a No. 63 drill bit is a fair compromise size. A range of bits from No. 50 to No. 70 would be better, especially if you are placing many chips on a board.

If two or more boards are prepared separate the boards after drilling. Use a fine file to remove burrs. Finally, polish again, with fine sandpaper. Don't use soap-filled steel wool—the soap gets in the holes, and causes problems later. However, the board must be grease free.

Etch Resist

The board is now ready for the etch resist. DATAK or equal dry-transfer direct-etch resist is by far the best. I find it easiest to place the finger stock first, then the IC pin solder pads, and finally the doughnuts for through holes, component leads and discrete IC leads. Each group should be transferred, then rubbed down through a sheet of the treated separating paper. This is to get best adhesion.

If an error occurs, the transfer is easily removed with a toothpick or fingernail. Don't forget—with jogged construction two touching pads should be used to give good solder space. There will be no pad on the opposite side of the board. Also, while through points require doughnuts on both sides of the board, component and IC-socket pin-pads are necessary only on the solder side. However, find them helpful on both sides, if there is room.

Dry transfer can be used for connections between solder pads, but the gummed crepe tape works equally well and is faster to use. About three sized are needed—1/32-inch for normal leads, and 1/16-inch for power buses. Large areas are outlined with 1/16-inch tape, and filled in with colored fingernail polish. Board appearance is better if an attempt is made to keep leads parallel and equally spaced. Curves or angled connections are optional.

After the tape is placed and checked, go over all tape to ensure that it is adhering firmly. A finger or a soft eraser can be used to give pressure. It is helpful to let the board sit for several days before etching; this improves etch resist and tape adhesion. Also, you might cement small rubber pads in three or four places on both sides of the board, over large unetched areas. These are to hold the board off the container bottom during etch. Pregummed weatherstrip is satisfactory.

Etching is conveniently done in a flat plastic tray. It is faster with the tray in sunshine or under a heat lamp. If separated from the container, the bottom of the board etches much faster than the top. Inspect it frequently, and turn the board over when the lines are cleanly etched.

After etching, clean off the tape and resist. Steel wool or fine sandpaper may be necessary. For the next step, the foil must be clean and well polished.

Some method of preventing foil corrosion is a necessity. The easiest is a heavy coat of paste wax. The dip type of silver or tin plating, sold in some radio and hobby stores also works. Follow it with a wax coat. Dipping the board into spent photograph fixing solution (hypo) gives a good coat of silver, but so far, a way of getting good adhesion has not shown up.

Still another way of getting protection, and probably the best, is to solder-coat the foil. To do this without a solder pot, cut a one-inch length of ⅛-inch woven shielding, and fasten this around the tip of your soldering iron with a spring, or a few turns of iron wire. Heat the iron and tin the shielding. Using well-fluxed, low melting-point solder, spread a thin layer on the board foil, using a wiping action to keep the layer thin. A build-up on the solder pads is all right. If necessary, melt the built-up area, and wipe it off with a natural bristle brush (an old tooth brush is convenient). With a little practice, this looks much like a solder-dipped board.

For component mounting, I prefer start with the sockets first, then resistors, then the through leads and, lastly, all components that stand up from the board. Save the trimmed lead ends for the through leads of the next board.

Figures 2-19 and 2-20 show a 2K board prepared in this fashion.

CHECKING THE FINISHED BOARDS

Checking any finished board for flaws, errors in construction and logic mistakes is easier if a routine is followed. The one I use involves both bench and in-computer operations.

Bench Checks

The first bench operation is a straight visual inspection, in good light and under a magnifying glass. Most attention is given to solder joints, looking for evidence of poor adhesion, cold joints and bridges. Back illumination is helpful in finding bridges and occasional whisker faults.

The next step is to apply power, and to make sure that it gets to the right pins on each socket. Then, the TTL ICs are inserted and power applied. All data, address and control signals can now be checked, by taking inputs HI and LO. This can be done with clip leads, but is much easier if you make a test jig.

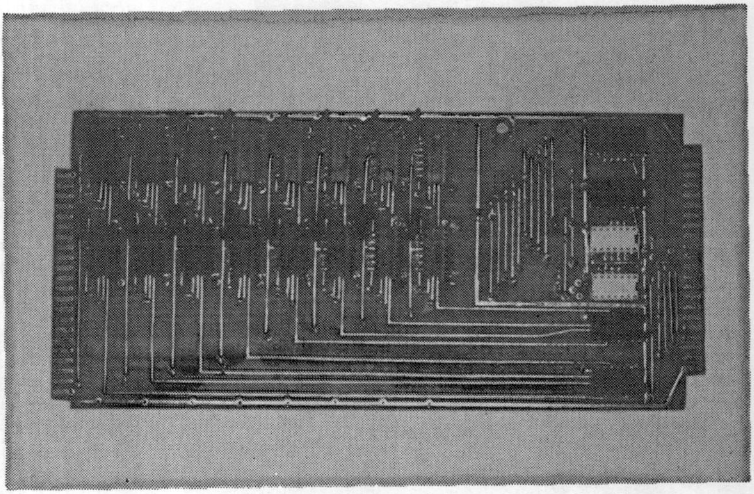

Fig. 2-19. Photo of the 2K RAM board. Because of the open spacing, construction is much easier than for the original 2K board.

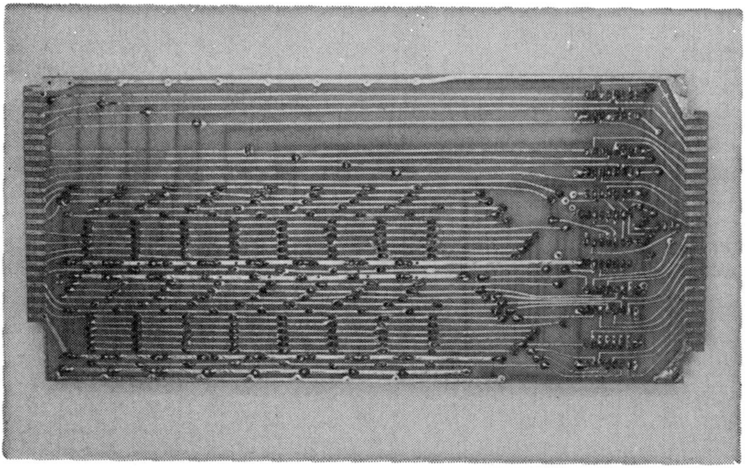

Fig. 2-20. Rear of the 2K RAM board. Simplicity of layout is evident. For this board, all conductors were tinned before sockets or components were added.

When these signals are checked, the MOS chips, in this case, memory, can be inserted—taking the usual precautions against static electricity damage. With all in place, power is again applied, and a finger test made of chip temperature, a warmer or colder chip indicating a problem. Note that MOS dynamic memory chips which contain a flaw can get hot enough to burn; this is rarely true with other types of chips.

Operating Checks

Further checking is done with the board in the computer rack, under operating conditions. For memory, a good first check is to deposit and read one location in each memory bank, using manual input. For a complete check, a memory test program should be run which should be an occasionally-used part of the operating system. Several runs are required to test for the possible types of memory failure. The most common are that location stays low, location stays high, and location is set by an adjacent location. One short memory test program is given in the Appendix.

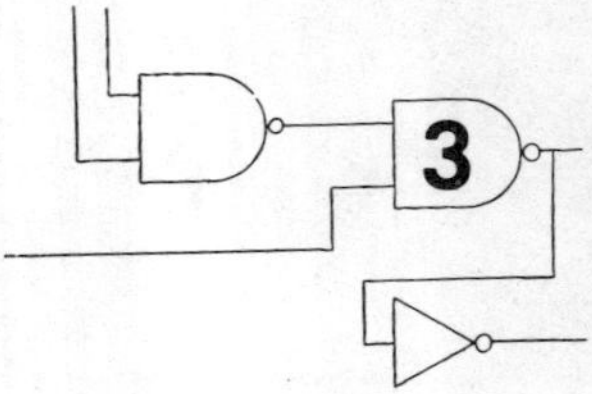

A NIBL-Based Operating System

Traditionally, the programs of the operating system are written in machine language. There are two very good reasons for this. First, the simpler forms of the operating system do not include a high-level language, and usually do not include an assembler. Programs must be in machine language. Additionally, properly written machine code is much faster in execution than the higher languages. Thus, there is no incentive to change.

For the home builder this is also the traditional approach. But here another factor may enter. If a high-level language is available in firmware, it may be easier to use this as the basis for the operating system. This is especially true if the language contains features which make operating control easy. NIBL is a BASIC available on a chip and contains a number of these features. But, before considering them, let's look at what the OS must do.

DATA FLOW ROUTING

One of the things the OS must take care of is the routing of data to the hardware elements of the system. Input must be accepted from the various sources, and outputs directed to them; there may be data interchange between devices. Routing becomes especially important if slow memory is to be written, then read, during program execution.

For the system used in this book, the major elements that use data are shown in Fig. 3-1. Here solid lines indicate wired data

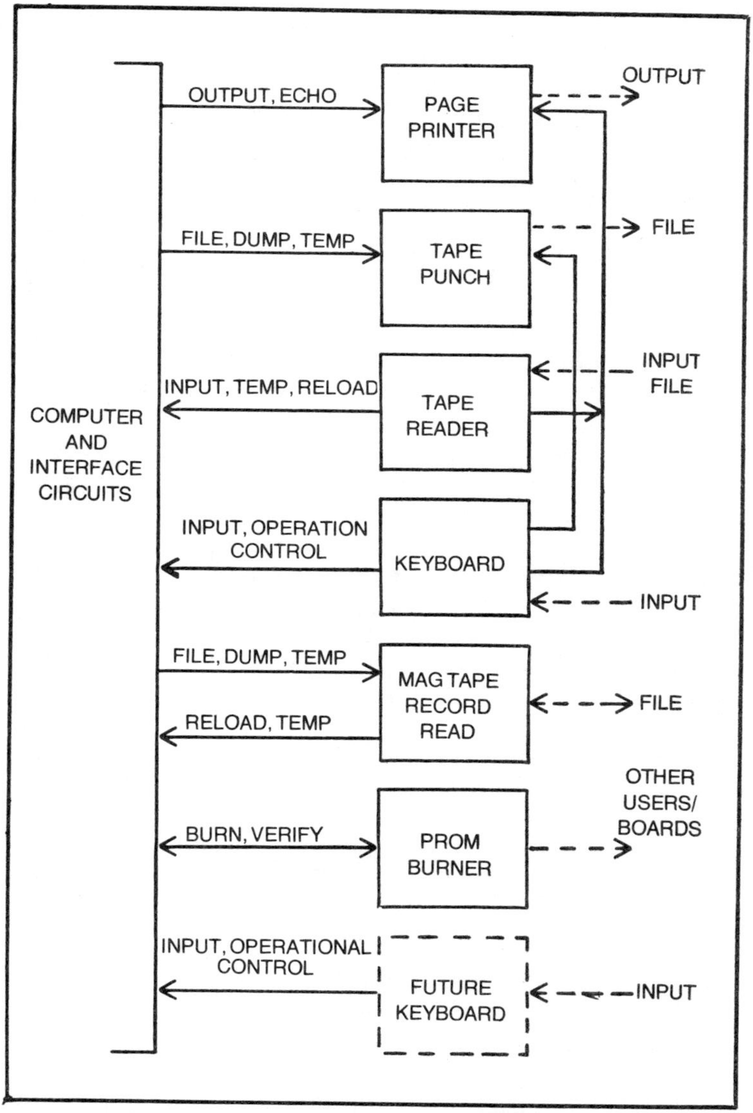

Fig. 3-1. Information flow in the advanced computer system. Dotted lines represent a manual operation—handling tape, interchanging a board, or looking up a file. One possible expansion is shown, but others can be added.

flow, and dotted lines indicate a mechanical operation, such as loading a tape. Not all peripherals are shown, just those that are typical.

Each of the data flow lines is labeled with the major reason for data existing on that line. For example, the line for the page printer carries the notations *output* and *echo*. This indicates that the line is used for human associated copy, and that it is also used to echo an input, specifically, the tape reader and the keyboard.

Another wire line also goes to the page printer, coming from the keyboard and tape reader. This line bears the notation *"local mode"*, indicating that these elements can operate independently of the CPU. This independence allows preparation of input off-line, freeing the CPU for other tasks.

At the right of the figure is an indication of the uses made of the data at that device. For example, the page printer sheets may be used as output, or for file maintenance, or for other uses (such as scratch paper!).

Some of the data forms may also become inputs. For example, a paper tape can be the file copy of a program, later serving as input to the tape reader. It may become input to another computer.

Some of the data lines are bidirectional, others one-directional. This difference is encountered in many devices. Some lines carry slow data, others operate at computer speed. Obviously, these differences must be remembered in both software and hardware design. While not all planned devices are shown, these give good representation of the data flow expected.

HARDWARE CONTROL REQUIREMENTS

Another item the OS must take care of is the control of system hardware. The control elements for the hardware just considered are shown in Fig. 3-2. Typically, there will be a check to see that a unit is ready; a device start signal, and perhaps one or more condition-select signals.

In simpler systems the device-start signal may be manual, and this is useful in all systems. However, it is usually desirable to place the device under computer control, or at least to control it at the keyboard (a form of remote control).

One of the common ways of handling these control signals is to assign each device a name or a symbol. This may be combined with a quantifier, as in READ TAPE 1, 1000, to indicate that 1000 bytes are to be read from tape unit one. Alternatively, dependence can be placed on a standard format, where, say, READ TAPE 1 always means read 256 bytes. Sometimes, a variable control may be used, perhaps a unique character, a succession of blanks, or a code.

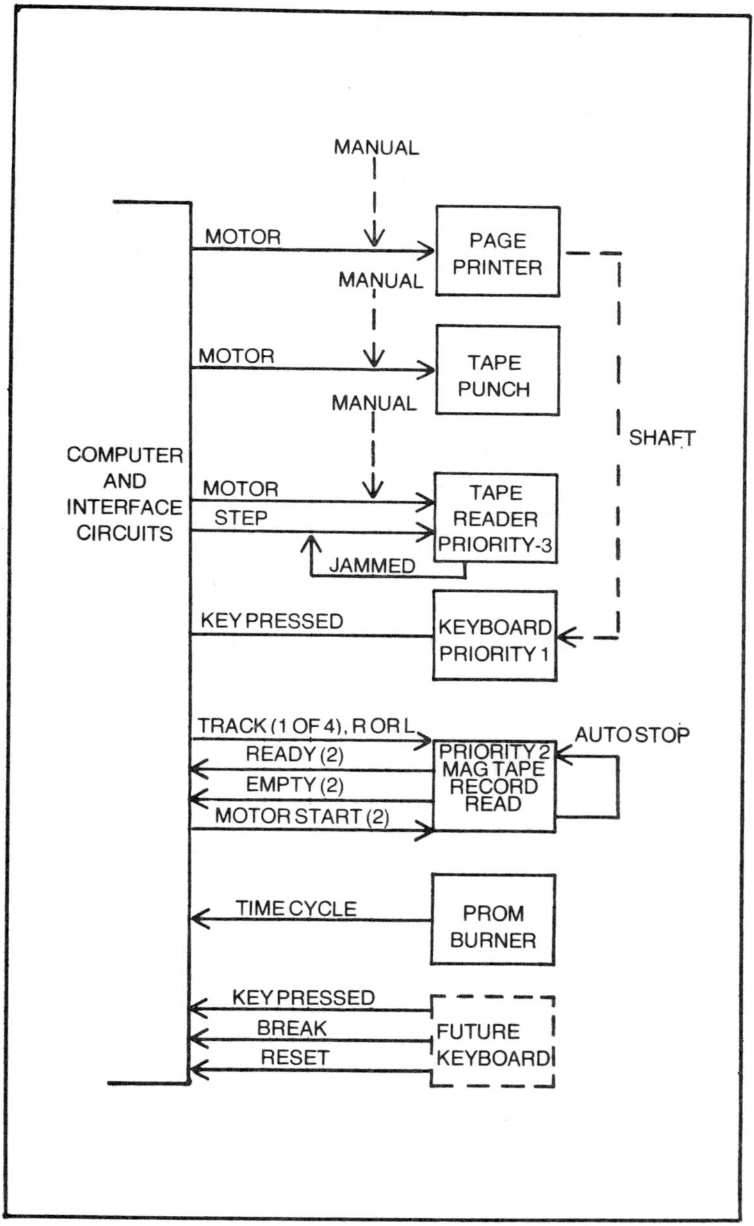

Fig. 3-2. Control flow for the system shown in Fig. 3-1. Solid leads are wired, dotted lines indicate a manual operation or a mechanical element. Data, address and CPU control lines are not shown.

It would seem that, ideally, the physical elements of the system should be set up to be independent of the software elements. However, this places limits on design, so practically, some cross-feed is necessary. For example, we have considered use of cartridge tape in a "start and run to tape end" format, as a way of eliminating CPU tie-up after reading a block of data.

Remember that some devices can operate intermittently—the mechanical paper tape reader, for example—whereas others, when started, must continue or errors will result. The need to continue is not invariable. For example, if output suddenly turns into garbage, it should be possible to stop immediately, without wasting time and paper.

A way of handling these format and sequence limits on input is to set up a priority system. This may be by hardware, through an interrupt priority decoder, or may be by software. Such a priority list is also shown in Fig. 3-2. This places slow items that can be interrupted at low priority, and fast, error prone devices near the top. It is usual for the keyboard to be the highest priority, to give complete control to the system operator.

Output priority systems are equally possible, but are not needed at all for devices that can be interrupted, such as the start-stop page printer. Those which do have problems, such as magnetic tape output, can be assigned a position in the input priority list. (Both output and program-execution priority are important in a multitask computer, such as a large time-shared system.)

We can eliminate many of these control conflicts by taking a little care in program writing, and a little care in operating. For example, we should not call for mixed magnetic and paper tape output, nor should we attempt to start a magnetic tape reader until it has cycled. Keyboard interrupt would be no problem during a page-printer output, but should be reserved for a tape jam, or a similar event, during magnetic tape operations.

Let's approach the control problems in this fashion. That is, by establishing operating and control practices. Perhaps the best way to do this is a check-list of "dos and don'ts" at the programming place and at the keyboard. Later, we can think about extensions, such as priority interrupt schemes.

OS PROGRAMS

In *How to Design, Build and Program Your Own Working Computer System*, I noted that the OS was composed of two

Table 3-1. Elements of an Operating System. The Two Major Divisions are Control Programs that Control Events, and Processing Programs that Create, Accept or Modify Data. Note the One Major Objective and Data Management is to Prevent Data from being Modified Accidentally.

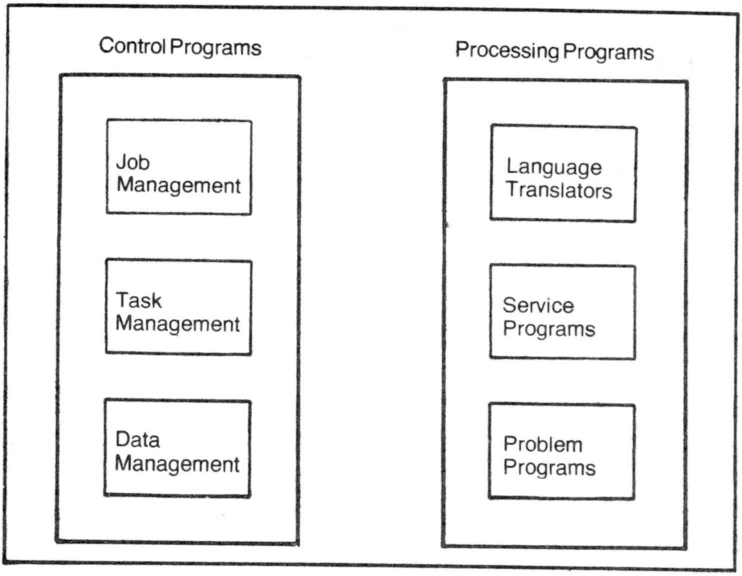

elements, as shown in Table 3-1. One set has to do with control of what the computer and its peripheral devices are doing. The second set involves the actual processing; this includes the problem program being run, any language translator needed to get this into machine language, and any service programs called to aid in finishing the problem program.

In small computers, and especially with the Tiny language versions, the distinction between these programs is usually not clear. Many of the control or management elements must be omitted completely, and others may be buried in a small segment of what is nominally a language translator. These steps are forced by the limits on memory to be devoted to the language. For small computers a different way of looking at the operating system is helpful. One such way is shown in Table 3-2.

The primary separation is by elements that get a correct program into the computer, elements that operate on the program, and elements that get results out of the computer. Also shown are some of the elements the computer or programmer can call, such as calculation routines and utility programs. Finally, distinction is made between commands that cause the computer to operate in a certain way and instructions that are in a program and cause data or files to be operated on.

It must be emphasized that the distinction between these elements is still arbitrary, and that any or all may exist in a program nominally intended for one purpose. NIBL, a Tiny BASIC, contains on-line translator and output elements, and includes an effective utility, COPY. Commonly, as the translator elements grow from the 4K of NIBL to 8K, 16K or even 32K, they offer larger and larger ranges of input, output and utilities.

Since NIBL, in particular, already has some of the operating elements, let's look at how this is done, and how you can extend it to handle the elements of the operating system you need. The results will apply completely only to NIBL, but parts may apply to any other language.

If you are using another Tiny BASIC or another language, you will need to go through an equivalent analysis. This gives you a better understanding of how your language works and how it can be used to extend your own operating system.

NIBL CAPABILITIES IN THE DIRECT EXECUTION

Look at the capabilities of NIBL with respect to OS needs. A way to do this is to look at the commands and instructions usually used, then to consider their use for operating systems.

Figure 3-3 shows NIBL elements organized by general function. The usual indication of capitalized letters for actual symbols is followed. Brackets indicate a line number, and an arrow indicates an automatic return to the starting point. Parentheses indicate an element following a particular item, such as a page number.

At the top is the keyboard-monitor routine, which causes a prompt sign to be printed as a request for input. Subsequent action depends on presence or absence of a line number. If the number is present, the line is entered. A blank (b) deletes lines.

In NIBL if no line number is entered with a statement, the statement is still executed (except for four instructions discussed later). This is true for commands, instructions and expressions. All high-level languages follow this principle to some extent, but NIBL appears to be unique in the fact that nearly all entries that are legal in a program can be executed without a line number. For example, the arithmetic statement X = Y + Z is executed using the current values of Y and Z, and the results are stored in X. (This occurs

Table 3-2. A Different Way of Looking at the Operating System; by Major Functions. Typical Programs Are Shown for Each. Some Functions Include Default Routines; for Example, PRINT ( ), With a Null in the Parentheses May Invoke a Tabular Format, Typically of 15 Spaces Each.

COMMANDS	TRANSLATERS	UTILITIES
LOAD	ASSEMBLY	APPEND
LIST	BASIC	AUTO START
RUN	FORTRAN	COPY
		TRACE
		RENAME
		NAME
INPUT	**OUTPUT**	**ROUTINES**
READ ()	PRINT ()	FLOATING POINT
GET ()	OUTPUT ()	TRIG
	DUMP ()	START ()
	PLOT	
MONITORS	**CORRECTIONS**	
BREAK	EDIT	
RESTART	DEBUG	

because LET is not required in a NIBL expression). Note that the current values may be zero, due to operations discussed later.

The executable statements are divided into groups in the lower part of Fig. 3-3. To some extent, the divisions here is arbitrary. The first group, called the *information group*, has the command TOP that causes a printout of the first free byte in the current memory page. STAT returns the current value of the status register. If it is in an expression such as STAT = N or STAT = STAT XOR #N, the status register is set to the calculated value. Note that five bits of the status register are connected to 1/0 lines.

The prepare group includes PAGE ( ), which transfers from the current page to the indicated page NIBL automatically attempts to execute any program starting at page two on power up. The prepare group also has CLEAR, which causes all variables and the FOR, DO and GOSUB stacks to be cleared. NEW ( ) clears the program in the page indicated within the parentheses. A null entry causes clearance of page one.

The execute group includes the command RUN, which first performs the same action as CLEAR for the current page, then starts execution at the lowest line number of that page. To execute a program with the current values of variables, enter GOTO ( ). The line number indicated inside the parentheses can be the start of a program, or any place within it. Execution continues until END is reached. GOSUB ( ), which is handled as a subroutine, is an alternative that returns to the monitor mode if a RETURN is encountered.

The two feedback items include LIST ( ), which causes the program of the current page to be printed, starting at the specified line number. If no number is specified, the listing starts at the lowest line number.

The other feedback item is PRINT ( ), ( ) - - - which echoes a string in quotes, or evaluates and prints output of an expression. It prints memory content in ASCII starting at the specified address and continuing until a CR is found. There is no automatic CR/FL if a CR is not found.

The linkage group includes the unusual instruction LINK ( ), which transfers program control the machine language program at the address given within the parentheses. This is accomplished by an XPPC P3 instruction, and requires another XPPC P3 to return. P2 still points to the variable stack when this is done. The action of LINK is very different from GOSUB ( ), where the parentheses contain a line number.

The linkage group also includes END, which, in effect, links any program to the keyboard monitor. END can occur many times in a program, always returning control when encountered—it is not required as the last statement in a NIBL program.

Finally, the expression group has the instruction LET ( ) where the parentheses indicate an expression. However, LET is not necessary, and the expression will be evaluated without it, possibly in less time. One of the powerful elements of this expression evaluation is the indirect operator@( ), where the value in parentheses is interpreted as a memory address. For example @245 = 60 assigns the value 60 to location 245. While this is restricted to single byte operation subscript variables can be used to handle blocks of memory.

This group also includes string expressions, indicated by $( ), where the value in parentheses is interpreted as an address. The expression form $( ) = $( ) is valid, but all operations are confined to the current page.

The NIBL instructions that are not directly executed include the two statements INPUT ( ) and 'INPUT' $( ), the parentheses in the first indicating one or more variables, and in the second an

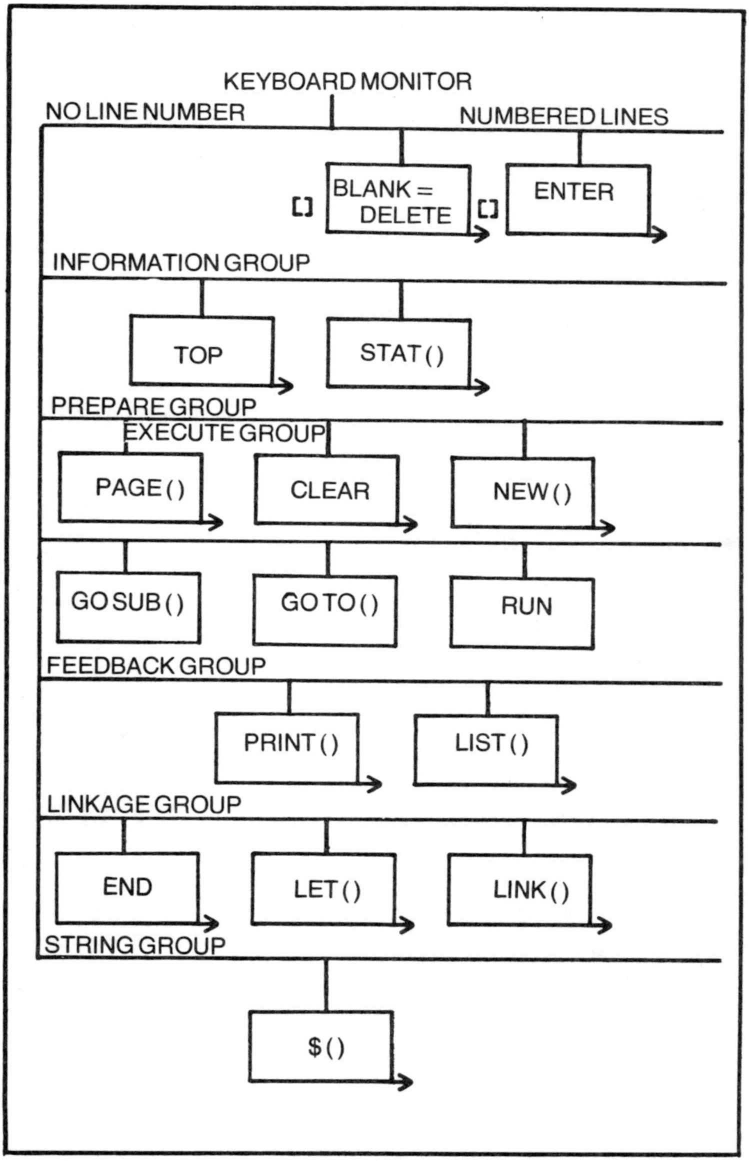

Fig. 3-3. The structure of NIBL. At start-up the system rests in the keyboard monitor routine. A numbered line is assumed to be a program line and is entered in memory. Unnumbered lines are executed immediately, if possible, with return to the monitor. After the RUN command, execution of lines in memory proceeds from the lowest to the highest line number. Error and default routines are not shown.

address. FOR-NEXT and DO-UNTIL statements in any form are not executed. And, of course, any string preceded by REM is always ignored in execution.

While NIBL allows multiple statements on a line, only the first is directly executed. Subsequent statements that precede a CR are ignored.

NIBL AS AN OPERATING SYSTEM

It seems clear from the preceding review that NIBL includes a surprisingly capable operating system. Part of this is due to the keyboard monitor provided. But the impressive OS capability is a result of the immediate execution of statements without line numbers.

The keyboard monitor is actually quite limited. For example, only the last character or a whole line can be deleted. Editing is confined to delete and insert on a line basis. However, lines may be added, if lines are numbered accordingly, say by tens. List capability can start with any line, but can only be halted by a key command, BREAK being recommended.

There is an editing aid in the @operator. An expression of the form PRINT @ #N displays the content of memory location #N, which is the operation served by the usual PEEK. Similarly, the expression @ #N = ( ) assigns the value ( ) to memory location #N, the POKE function. The symbol $ may be used for ASCII output.

While deleting capability is nominally limited to run time error messages, one of 13 messages that also indicate line location, there is another characteristic that is a great aid in debug. An END statement can be inserted anywhere in the program, causing return to the monitor routine printing the line number where END occurred. Any variable, stack, memory location or CPU register can be examined. Some are modified by the return, but most are not. The END functions as a break point, except that reading is manual instead of automatic.

NIBL also includes a way of saving programs on paper tape. For this, the command LIST is given, but before hitting CR, the tape punch of the teletype machine is turned on. Hitting CR initiates the LIST operation, also punching a tape. The last character on the tape is a prompt >, which should be deleted by a shift/0, the back arrow. The tape may be used as the file copy, or as input.

EXTENDING NIBL OS CAPABILITY

Let's look at ways to extend the operating system capability, still keeping NIBL as the base. Assume there are two primary goals:

1) To extend the number of commands available while in the keyboard monitor mode.

2) To extend the number of instructions available while the in the run mode.

Also, for now, let's assume that you want to do this without rewriting NIBL. Later you can consider ways to accomplish some of the functions by rewriting the language.

The first point to note is that you can add OS programs, very simply if you write them in NIBL, but there are some limitations. The NIBL interpreter occupies page Ø of memory, uses a part of page 1 for stacks and variables, and normally has the rest of page 1 available for program storage. However, on power up, it first checks page 2, and will immediately execute a program it finds there.

Incidentally, it should be noted that there is a peculiarity here. On reset the initialization routine writes a dummy end-of-line signal at the start of each page. NEW does the same thing for the current page only, but does not completely clear the page. This action does not affect ROM; but for RAM, after reset on an accidental NEW, the following should be entered:

For page 1: @#1120 = High Byte of first line number
@#1121 = Lo Byte

For other pages: @N002 = as above
@N003 = as above

Even with an accidental NEW, this will retrieve the program. Note that NEW is not legal as a line in a NIBL program.

In addition to pages 0, 1 and 2, NIBL allows programs in pages 3 through 7. Each page will accommodate about 200 typical program lines (160 for page 1, since the first part of the page is needed for variables, stacks and so on). Since pages can be linked, quite long programs can be handled, up to about 1200 lines total.

Of course, allocation of some of this capacity to the operating system reduces programs handling capacity. You can get an idea of how serious this reduction might be by considering an OS program already available in NIBL. For example the SC/MP Dissembler requires 218 program lines, and while some space could be saved by using multiple statements per line, this program alone requires essentially a full page.

Just this one example makes it clear that you must be careful in allocating NIBL page space or you will be short for program execution. Of course, this is where the great advantage of disk systems shows, since rapid exchange of OS and problem programs is possible. And you want to approach this capability with your cartridge tape system.

Ths space problem becomes much less if the OS programs are placed in machine language, in fact, it essentially disappears. First, note that the instruction LINK ( ) allows access to any addressed memory space; this, of course, includes the various machine language routines of NIBL, and all of the eight pages of memory not allowed in NIBL. Also, many of the NIBL routines include an automatic return to the keyboard monitor mode, and this is easily arrayed in new routines.

The space available seems entirely adequate. For example, the machine language OS programs in *Compute* include:

Kitbug Monitor	508 bytes
Interrupt	64 bytes
Double Precision Math	875 bytes
Punch Tape	63 bytes
Hex Load	64 bytes
Editor	364 bytes
Plot	574 bytes

These programs total only 2514 Bytes. It is evident that you can build quite an operating system within a page of machine language capability.

The following method of operation seems reasonable:

- Put OS programs in NIBL that are not regularly needed on tape, and call them by a LOAD TAPE () command equivalent.
- Locate a commonly used OS program in NIBL at page 2 in ROM. If several programs are needed, they can be on pages 2A, 2B---, operating in a power-down mode. Alternatively, all but the least common can be rewritten to machine language.
- Place machine-language programs commonly used in ROM on pages 8 - 15. In the unlikely event that more memory is needed, page 8A, 8B---, can be added.
- Less commonly used machine language programs can be kept on tape, and loaded if needed. It might be worthwhile to set up one page of RAM in machine language for loading and running.

●Keep rarely needed OS programs in NIBL or machine language on paper tape; the reader can be manually loaded as needed. (The penalty can be reduced with a high-speed paper tape reader.)

With this in mind, you can revise Fig. 3-3 to show just the direct OS capability (as shown in Table 3-3).

NIBL AND PERIPHERAL CONTROL

So far we have concentrated on the CPU-memory relation, with only minor attention to other parts of the computer. Let's now look at peripheral-device control, paying whatever attention is needed to data flow. In other words, let's look at the generation of the control signals and monitor functions shown in Fig. 3-2.

NIBL has bare-bones peripheral control built in. Sense B detects the presence of a Teletype character bit, and flag Ø generates a printer bit. Flag 1 controls the relay of the tape reader associated with the Teletype. While these elements could be changed by rewriting the program (at great trouble if it is firmware), it would seem best to let these assignments remain.

Table 3-3. Some of the Operating Commands of NIBL. The @ Commands Can Also Poke or Peak at Any Accessible Location, and Are Used in Such Operations as Variable Extension; Regular Variables are Used as Pointers. LINK Allows Almost Unlimited Expansion of NIBL Through Special Machine Language Routines.

NIBL INSTRUCTION	OS FUNCTION
TOP	Unused Memory
STAT	Set Flags
PRINT STAT	Read Flags; Read Sense
CLEAR	Set Initial Conditions to Zero
RUN	CLEAR, then Execute
GOTO	Execute with Present Initial Conditions
LIST	Tape Save
@(b) = (a)	POKE (a) to (b): Set Stack Value
PRINT@(b)	PEEK at (b): Read Stack Value
PRINT $ (b)	DUMP, Start at b, end at CR
LINK ( )	Useful Routines Available
	Get Character
	Put Character
	Find Line Numbered ( )
	Skip to Next NIBL Line
	Cursor save, exchange

However, you do have four powerful means to initiate control from the keyboard, or to include it in programs. These are

```
@() = ()
LINK()
GOSUB()
STAT = () / PRINT STAT
```

The indirect operator @ is the most immediately usable control function, and can address any memory location. Suppose you reserve a location for a peripheral device. Then, by INPUT@() you can read whatever byte the device puts on the data bus in response to the read strobe. Similarly, @ ( ) = ( ) can write a byte into the device. This can be interpreted as any of 256 instructions. Alternatively, it could be interpreted as up to eight simultaneous instructions (microcoded).

Instead of obtaining a printout, or an external data read, the same construct can be used to place a byte into memory or to transfer from memory to output. You can also use various routines to do such conversion as half-byte, hex, BCD and so on. The GOSUB and LINK constructs also lend themselves to these routines. Both must have a form of return, as discussed earlier.

STAT gives a way to control a device or read a device request. It can be combined with hardware elements to allow an extensive list of devices or tests. Initially, it can be left for special uses, one of which will be developed in a later chapter.

All of these constructs can be placed within a NIBL program. The OS can be started with manual operations and later changed to have its elements under program control.

Since it does not appear that machine language memory space will be critical, you can assign a full page of memory to peripherals and control functions. This allows up to 4096 devices, but seems unnecessarily large. Suppose, instead, each device is assigned eight memory locations, say those designated by bits 0, 1 and 2 of the address. You could still handle 512 devices—more than will ever be needed. The combination of address and data bits allows simultaneous reading and control when needed. This method of operation handles most peripheral problems with one addressing operation, and, at most, a few in succession.

There are two disadvantages. First, the device must be addressed by location, rather than by an English name and instruction. Also, you must remember the sequence for each device, using program lines.

The sequencing is relatively easy to handle by using the LINK ( ) instruction. You can write a machine language series that takes care of the sequence. However, this works only if the sequence is invariable. If it must change, transfer must include the change condition. This is called *passing a parameter*, and is useful generally for subroutines and called programs. It is not easy to accomplish in NIBL with its limited number of variables. However, you can add the capability by reserving more memory.

In NIBL, the register P2 always points to the variables and stacks, specifically to location 101C. RAM variables are addressed by displacement, down to location 1000 and up to location 111F. Program storage starts at location 1120.

By taking advantage of wrap-around, some of the memory at the top of a page can also be addressed by a displacement from P2. This can include all displacements from E3 to 81, a total of 99 locations (displacement of 80 location calls the extension register).

You can use these locations for programs as intended, or in other ways by addressing through P2. For example, you could allot 52 locations to an additional set of variables. A number of locations, say eight, could be assigned to parameters to pass these variables from one page to another or to pass them to subprograms and machine language routines. For example, a "last case" memory is useful when working with Baudot I/O. Pointers to subroutines, buffer contents, and parameter content are always useful. Additional temporary storage may be helpful.

As an example of parameter passing to an I/O device, consider what is needed for a magnetic tape read. The sub-routine must be told that this is an input, the tape unit and track from which it is coming (1 of 16, say), and the format, whether Binary, Hex or ASCII, or perhaps Baudot and half-hex. It needs at least a start address for storage. It might be desirable to indicate the speed, to allow for future changes.

Instead of passing parameters, you might pass pointers. For example, if you wish to use an arithmetic board, we you could load its routine address in the top memory area, and cause an indirect jump via P2 to the routine. This could, for example, read the next line of the NIBL program, interpreting this as calculator operations. After storing the results (using variables as pointers to allow necessary precision) the routine could increment the line buffer, then return. Many possibilities can be worked out.

If this P2-accessible memory area is used for such purposes, there is a possibility of interference with programs. During entry of

BASIC into page 1, this can be checked from time to time by the PRINT TOP instruction. Overwriting can be prevented through circuitry as well, but this involves address decoding and is best avoided unless absolutely necessary.

Let's assume that you follow this approach, and look at both the effect on peripheral design, and, especially, on the influence on a NIBL-based OS. To do this, let's consider a two-unit tape deck, about as complex a problem as you will want to undertake for some time.

Recall that you want to select either of the tape decks, any of the four tracks, and the right or left channel. You want to read or record data. Start is a controlled operation, but the cycle is self-completing. You also want to check that recycling is complete and that there is a tape in place.

Without going into design details, you could use the arrangement shown in Fig. 3-4. Here eight memory locations are assigned for device control, with memory lines A_4 through A_{11} plus page being the device select. Data read-in can also be used for control. Data read-out gives the status. Data to be recorded or read out proceeds over a separate path, the most likely being a two-speed serial input, one at teleprinter speed, the other around 1200 baud.

The operation of this system could be handled in several ways. One it uses LINK () to test, loop if not ready, and LINK () to select and start operations and to proceed with normal input. An alternative to this is LET ( ) and LINK ( ). The assign statement sets up the parameters, and the LINK statement calls a routine (this could call the teleprinter routine in NIBL if desired).

There is quite a range of possibilities. An attractive feature of the two above is that they can function within a program, or under keyboard control. However, they place a load on the programmer who will need a page of the correct instructions to use.

With more complex elements in the OS, it is possible to get to the capability of some of the higher language instructions, such as:

READ INPUT TAPE 17,6,K(B(K)),J,(C,(I),I=1,J)

which is a valid Fortran II statement, specifying the peripheral, the data segment and the storage locations. However, to really use this class of instructions you must extend NIBL far beyond its original intent, and we are trying to avoid this. For now, let's stick to the simplest approach—the two successive LINK statements.

Let's summarize this NIBL extension to give greater peripheral control. The key elements are:

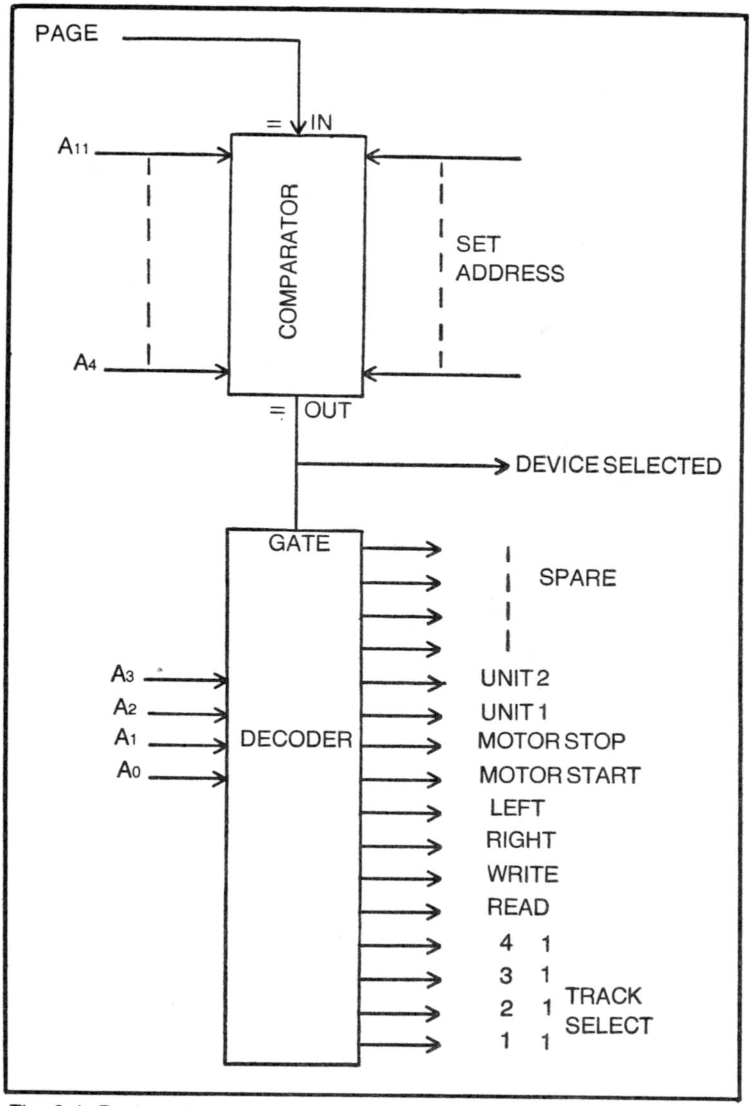

Fig. 3-4. Basics of a tape deck used to illustrate one method of generating device-control signals by assigning a block of addresses. The addresses are decoded to produce control signals. Alternately, a single address may be used with a data word to generate the control functions.

●Design the peripheral control circuits to occupy a block of memory, say eight locations, and use the individual locations as microcoded control signals.

●Provide status on a LINK read-out instruction.

●Provide detail control on a LINK read-in command.

●Transfer data by the same route or by separate I/O channels, including the built-in teleprinter channels.

●Later, consider adding OS routines for the entire operation called by an appropriate LINK instruction.

●Still later, rewrite or extend NIBL (or use another BASIC) to give control by direct instructions.

We can summarize the entire findings for use of NIBL as an OS base in chart form (Fig. 3-5). Here's a suggestion—make and keep current the chart for your system.

TYPICAL OS PROGRAM FEATURES

Look at some typical OS programs currently available for the SC/MP based computers. (See Appendix.) For the first, choose the SC/MP Machine Language Editor by Jamieson Rowe.

While the program is relatively short, only 864 bytes, it is fairly complex. The major functions provided are main command loop, line count routine, text input routine, text delete routine, list and punch routine, insert routine, change routine, and read and append routine. These operations are normally performed with respect to a line or lines in the NIBL program space, but a larger memory space can be used if it is considered necessary.

The editor makes use of the following commands:

A)	Append line to buffer
R)	Read line into buffer
L),ML),M, ML),	List all lines, or line M, or lines M-N
MC),M,NC	Change line M or lines M-N
MI)	Insert before line M
MD),M,ND)	Delete line M or lines M-N
K	Kill text in buffer
P),MP),M,NP)	Punch all lines, or line M or lines M-N
B),MB),M,NB)	Same, add bell at end
/	Print number of lines currently in buffer
Bell	If in text mode, return to command mode

While these do not accomplish more than can be done with the edit function in NIBL, it is far easier to control blocks of lines, and

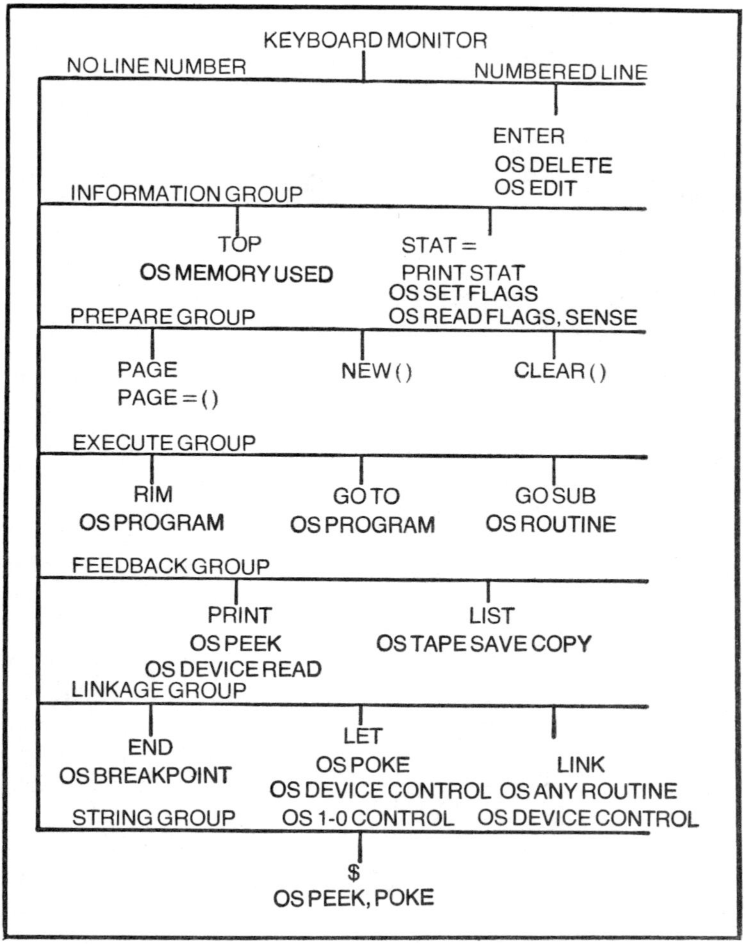

Fig. 3-5. Extension of the NIBL chart to show possible operating system. For example, in standard NIBL, LIST is used to generate a paper tape or to copy a tape.

generally, there is more specific control. The program is a worthwhile OS extension.

For the second program, choose the NIBL language disassembler by Jim Derosa written in NIBL.

In the program, a starting address is first declared. The operation codes are divided into blocks, basically those that require no operand, those that refer to an immediate instruction and those that refer to memory including address determination.

Jumps give the displacement, and arithmetic and logic operators the operands. As is common in disassemblers, no labels are generated.

The utility of this program depends on the amount of disassembling to be done.

The program is worth entering on magnetic tape if disassembling is rare, or in PROM if it is common.

There is one extension to the program that can be very helpful. This addition generates a table of reference lines; a cross reference table that gives the machine code address, the line where the definition exists, and the line or lines referring to this address.

For the final example, consider the Plot program. This 654 byte program generates a labeled set of coordinates on the teleprinter, and places on this a series of + signs to mark the variable. Other plus signs mark the tens coordinate points on the horizontal axis; tens vertical axis points are indicated by dashes. There are 65 horizontal plotting positions, assuming that the input values are normalized to give 100 percent of the normal value at the fiftieth point. The vertical axis will continue as long as data is fed in.

This type of OS element is useful to people who are specializing in math, engineering and certain classes of financial problems. It is of little or no use for many other purposes, for example, house finance, or even for games. Tic-tac-toe is much more interesting when played on a video screen than on a page printer. It also saves paper.

This example is included to emphasize that the operating system for your computer must be adjusted to reflect the type of problems you encounter. In this connection, if you are planning to purchase an OS package, be certain that adding or deleting elements is not only possible, but easy. There is no benefit in memory filled with useless program material. And there is real trouble with an OS that cannot be modified because the vendor won't supply a usable listing. See the Appendix for other SC/MP programs.

RELOCATING NIBL

If the suggested keyboard monitor system is used, it might be convenient to place it at Page Ø, moving NIBL to a higher page. Rewriting the machine language program for this is not difficult. In 22 locations, the high byte of the address must be changed to the new page. If the new location is page 7, two locations (003C and 0040) must be modified to limit page count to six. Addresses 0088, 04FB and 06C4 require modification.

The complete list of changes is given in Table 3-4. These are for new location at page 7, and were prepared by John Kennewell.

MODIFYING NIBL

For a number of reasons, it may be desirable to modify NIBL. For example, it might be desirable to run at faster I/O speed, or

Table 3-4. Relocation of NIBL.

OLD ADDRESS	NEW ADDRESS	OLD CONTENT	NEW CONTENT	ROUTINE
003C	703C	80	70	INITIALIZATION
0040	7040	80	70	
0043	704E	0C	7C	
0088	7089	00	70	IL EXECUTOR
0177	7177	0F	7F	PRINT STRING
01DE	71DE	0F	7F	PRINT NUMBER
01E8	71E8	0F	7F	
020E	720E	0F	7F	CR/LF
0224	7224	0F	7F	ERROR ROUTINE
0230	7230	0F	7F	
024D	724D	0F	7F	
027D	727D	0E	7E	
02A2	72AZ	0C	7C	BREAK, etc
02B7	72B7	0C	7C	
02FC	72FC	02	72	LIST
0307	7307	0F	7F	
0328	7328	0C	7C	
04DB	74DB	00	70	STORE VARIABLE
04FB	74FB	C3	DB	TEST, VARIABLE
05C9	75C9	00	70	OPERATORS
05E0	75E0	02	72	IF TEST
0646	7646	02	72	CHECK MODE
06C4	76C4	C3	DB	TEST, NUMBER
0762	7762	0F	7F	
0777	7777	0F	7F	GET LINE
090E	790E	00	70	EDITOR
0B11	7B11	0F	7F	PRINT "$"

Table 3-5. Table of Addresses and Data to Change From 110 Baud to Another Speed for Output and Input. NIBL Has Been Tested to 1200 Baud.

ADDRESS	110 BAUD	300 BAUD	600 BAUD	1200 BAUD
0F85	C3	29	8A	BB
0F87	08	03	01	00
0F94	45	11	D4	34
0F96	11	06	02	01
0FB9	11	06	03	01
0FC4	BB	66	2D	99
0FC6	2F	06	03	01
0FD0	54	21	E5	44
0FD2	11	06	02	01

with a Baudot or IBM printer instead of an ASCII device. Added instructions may be desired.

Some changes are relatively easy to make. The I/O routines are at the end of the program and can be rewritten without affecting the rest of the program. They can be extended. For example, a series of I/O routines could be placed on another page, with a selection of the location being determined by parameter passing.

Extension of the instruction set is more complex, requiring IL routine additions. These must be placed just before the syntax error routine, at least, and ideally should be placed in the routine table at a position corresponding to their relative use: a heavily used instruction at the end of the table slows operations. Much extension of the instruction set could result in a marked increase in execution time. If this must be done, the sequential search should probably be abandoned in favor of a hash-function approach. But then, it wouldn't be the same NIBL. In this connection see the extended NIBL for the Z80 discussed later.

A large fraction of computer time is taken up with I/O, at speeds which are determined by the peripheral devices. In NIBL, with only one I/O channel, the teleprinter may be the controlling factor in executional speed.

For some purposes, a large speed change is possible, and requires relatively few changes in NIBL, only nine locations are affected, as shown in Table 3-5.

If NIBL is in ROM, some of the changes may be handled by a hardware decode, turning off the NIBL chips for all addresses above a selected value, and activating a new ROM that contains the revised program. A design to do this is shown in the next chapter.

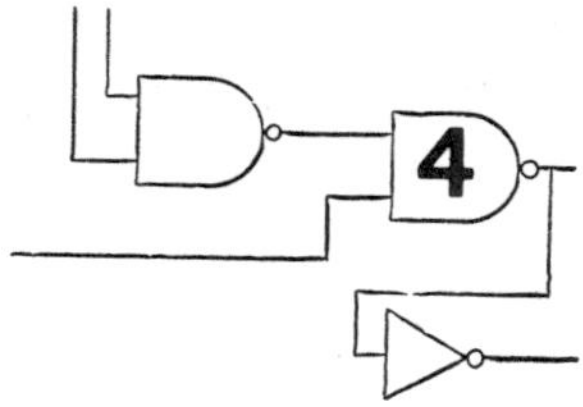

ROM Cards, NIBL in ROM and NIBL Extension

Putting the OS into ROM means that you will have quite a demand for ROM cards. If you should fully expand the system, you can expect up to 32K of machine language, and perhaps 4 to 8K of NIBL programs in BASIC, in addition to 4K for NIBL itself. Super expansion could use the block deactivation of boards. While full expansion is unlikely, you want the option.

At the present time there is only one approach to ROM that seems reasonable. This is EPROM. Except for the few instances where a preprogrammed unit is usable, the mask type program (used, for example, in the later TRS-80) is impossible, because of its extremely high cost. The fusible link type (PROM) is not out of reach, but cost is still higher than EPROM, and a programming mistake or change means scrapping a unit. The advanced form of electrically alternate ROM is still relatively scarce and expensive.

Of the various EPROM available, the common 2708 is the best choice. It is widely used, available from several sources, and low in cost. True, its 8K size is undesirably small in some uses, and the need for three power voltages is a nuisance. Although the 2716 offers 16K and requires only a 5V power supply, cost is appreciably higher per storage bit. Keep in mind that the design includes power supply capability.

As the home built system develops, it is likely that the OS programs in EPROM will need changes. Memory chips that are too large increase the chance that the whole system must be shut down to reprogram a single chip.

Chip capacity and construction simplicity are equally important in board design. A few sketches of the memory chip board design of Fig. 2-19 will quickly show that placing eight of the 24-pin EPROM chips on the board is no great problem, but that 16 would be very difficult unless photo etching were used. The eight are sufficient for 8K of 2708 memory, or 16K of 2716—or even 32K of 2732. Two of the smaller boards allow four pages of operating system—enough to occupy a home builder and programmer for a long time.

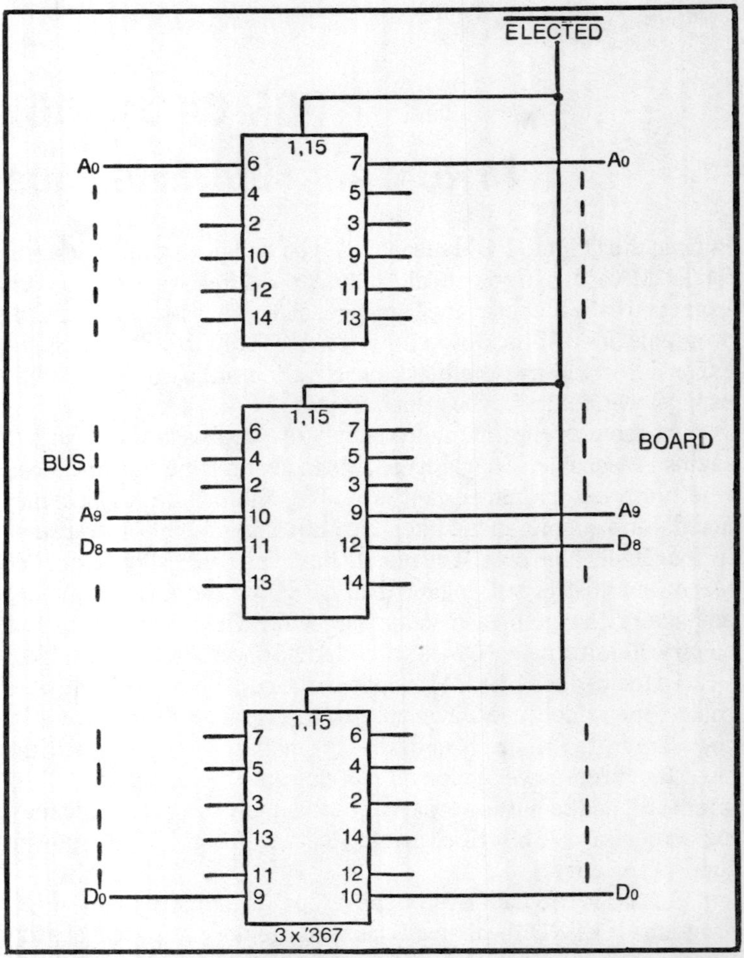

Fig. 4-1. Address and data buffers for an 8K ROM board. The same elements can be used for any output-only board.

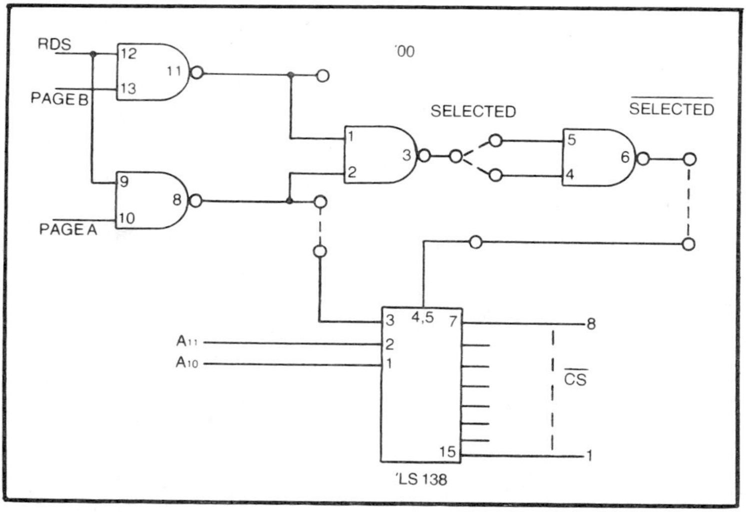

Fig. 4-2. Chip select generation for the ROM board. Two page lines are needed. These combine with A_{10} and A_{11} to produce one of eight chip selects at read strobe time.

The following designs are based on the 2708 memory (or the 2708-1 if 350 ns access is needed). Changing to the 2716 should be no problem, if greater capacity seems necessary.

CIRCUIT DESIGN

Being read only, the circuit for ROM is simpler than for other memory. Only three functions are needed: address in, chip select, and data out. Since the sketches show that space is no great problem at 8K, the choice in chip selection is good.

Keep the number of types of chips small. You could use the same buffer as the RAM board; however, the bilateral data buffers are unnecessary and can be avoided. Instead, the unidirectional 74367 can be used for both low address (A_0-A_9) and data bus (D_0-D_7) buffering. The circuit is shown in Fig. 4-1 and is as simple as possible.

The upper addresses and page select must function in a chip select mode. It's easiest to combine the read strobe and two-page select signals to give two bank selects. The OR activates the buffers and is combined with A_{10}-A_{11} address data to give the actual chip select signals. These are formed by the three-line to eight-line decoder. The complete circuit is shown in Fig. 4-2. As usual in these designs some freedom in signal use is allowed by jumpers.

There is still one detail to take care of. The main frame power supplies are +5, +12 and – 12 V. The 2708 requires +5, +12 and – 5 V. (The programming pulse is not needed on a read-only board, of course.) You need to add a small – 5 V regulator, working from the – 12 V supply. This is easy. Use the TO-220 type as in the circuit in Fig. 4-3. The only precaution is the use of high quality capacitors for stability.

In the prototype board construction uses the jogged construction as for the RAM board. The layout is simpler because of the larger size of the 24-pin ICs, and because it is organized 1K by eight. The pin-out arrangement is good for jogged construction. See Fig. 4-4.

There are several good designs available for zero insertion force sockets, but none are inexpensive—yet. Only one is readily available through home builders' channels. Unfortunately, it is quite large and not adaptable to jogged construction. Therefore, standard low-profile solder tail sockets are used for the prototype. The completed board is shown in Fig. 4-5.

NIBL IN ROM

National Company produces a firmware version of NIBL. It is contained in two chips called the 16K MM2316A, or MAXI-ROM.

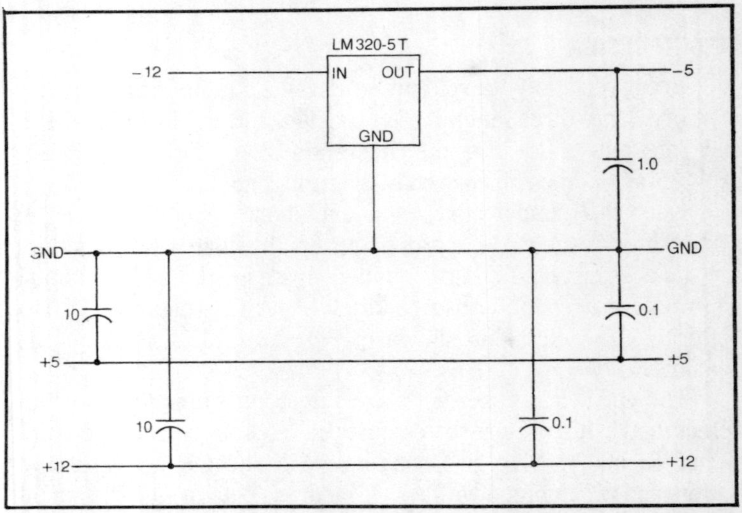

Fig. 4-3. Basic power circuitry for the ROM board. The – 5 V line is needed for the PROMs. Not all of the line bypasses are shown. There should be one set of 0.1 capacitors for each chip.

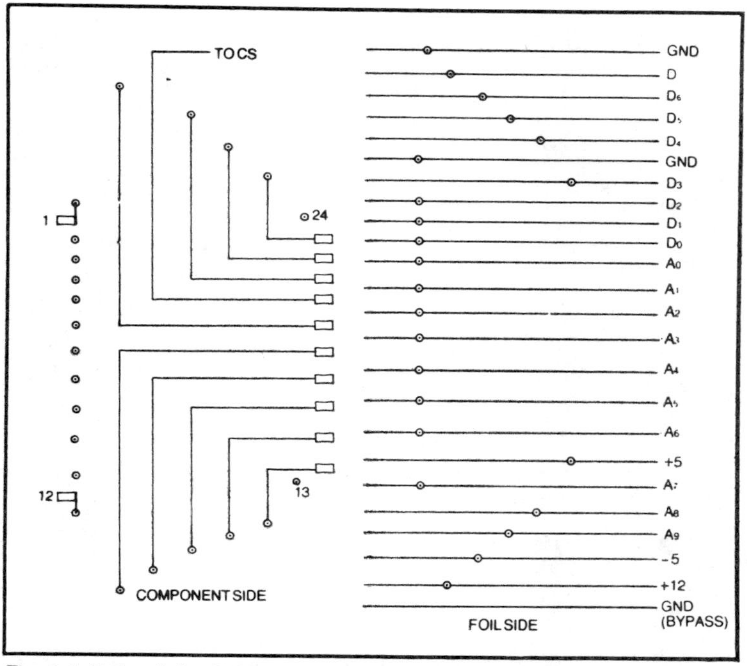

Fig. 4-4. Foil cell for 2708 chips. Each cell should be 1.0 x 2.3 inches or larger.

The NIBL programmed version is designated as ISP-8F/352, and is currently available from at least one organization catering to the home builder. It is also available on special order through the distributor chain. A single-chip version, INS8295 (See Fig. 4-6) and an eight-chip version are also sold. Further, there is an even larger version available that includes an editor, a one-pass assembler, and a tape punch. Called "SUPAK" this version is proprietary, and relatively expensive.

An altenative to burning your own EPROMs with NIBL is to use one of the above firmware versions. The eight-chip set could be placed on its own board, requiring a different memory cell arrangement, plus some work on chip select. The two-chip and one-chip versions could be placed on their own board, but the space needed for them is small, and a composite board of 4K NIBL plus 4K of EPROM is easily possible.

It seems that provision for modification of the NIBL program is desirable even though it is in firmware. This is not too difficult if modifications are restricted to the area above a particular address. Since the I/O elements are at the end of NIBL, changes are not

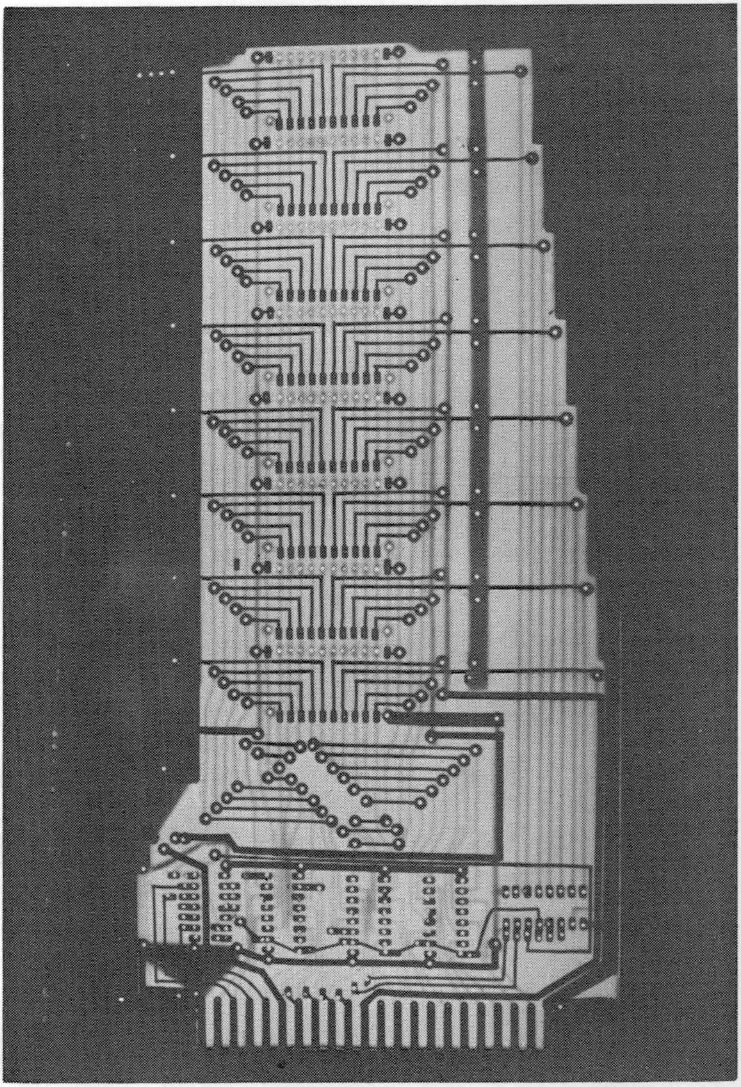

Fig. 4-5. The 8K EPROM board at an early stage of production. The pads for the jogged sockets are particularly noticeable in the memory area. Note the use of through pins to pick up bus leads on the back of the board. This board was dip solder coated, and the excess needs to be removed.

difficult. It is not even too difficult to leave the intermediate language routine open-ended to permit adding new instructions. Let's consider the hardware for this.

First assume that you can accomplish changes by rewriting no more than the last 1K of NIBL. Actually, you can expect to confine the changes to a smaller range; the starting address of the I/O routine is OF77, and the last intermediate language routine for instructions is OE26. Expect changes to lie above these addresses.

To make the changes you need a memory chip with the new routines and a means to switch over to the new chip when the address is ready.

For the switch over use the 7485 comparator. This chip provides a signal if one input is equal to the value of a second input, a second signal if it is less, and a third signal if it is greater. The last signal is the one you want. These chips can be cascaded to give 4-8-12-16 select capability.

A circuit for generating the switch-over signal is shown in Fig. 4-7. Here the incoming address signals are first buffered, to give

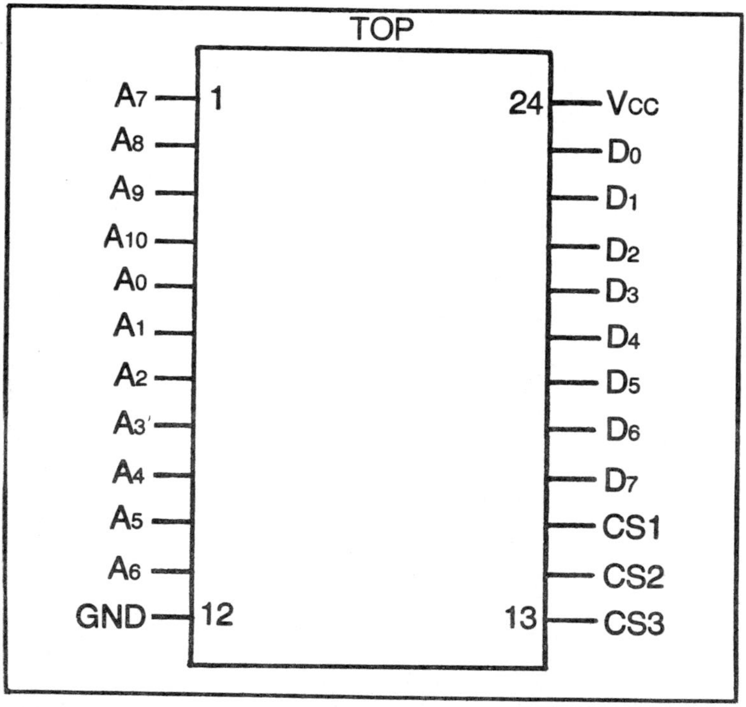

Fig. 4-6. Pin connections for the INS 8295 chip. Two of these contain the 4K of NIBL. Note that there are several versions of NIBL in ROM, one containing a line-by-line assembler.

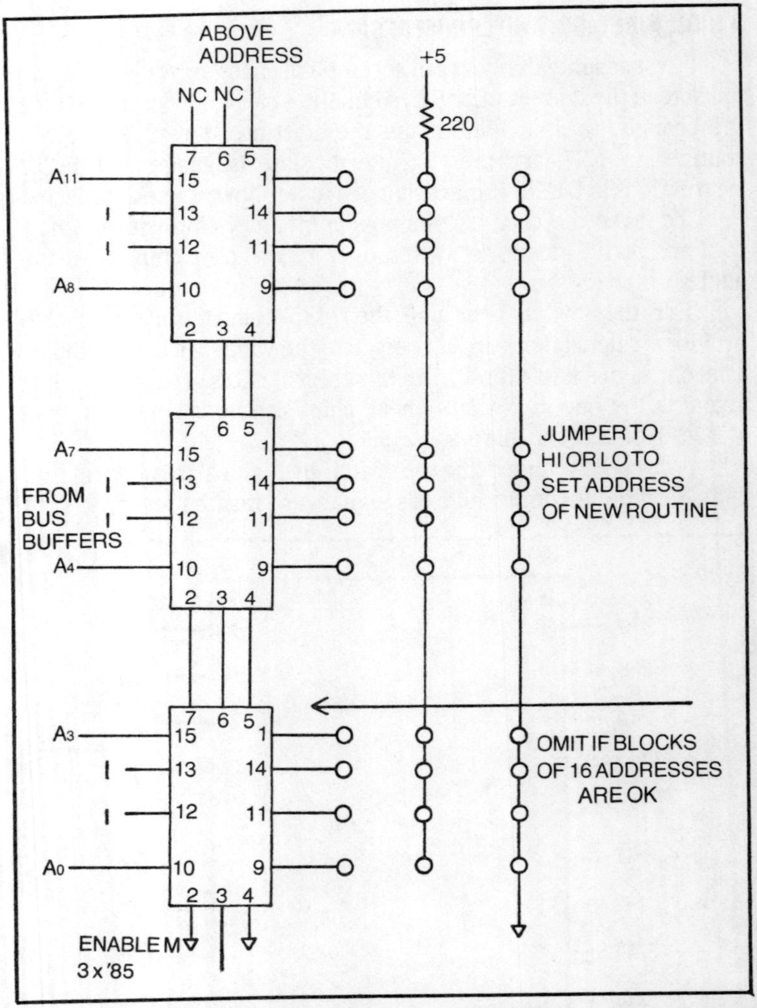

Fig. 4-7. Method of generation of a switch-over signal at a particular point in page space to allow use of alternate I/O routines with NIBL. One of the 7485 comparators may be omitted if setting blocks of 16 is acceptable. A second (or below) address signal is also available.

adequate drive to the board chips. The upper eight address signals, from BA_4 to BA_{11} are fed to two comparators. The other inputs to the comparators are by jumpers to high or low busses to set the switch-over address. The "equals" input to the lower comparator is used as an enable, and the switch-over output signal is taken from the "greater than" output of the upper comparator.

This arrangement does not allow full freedom of address choice, since only every sixteenth address is decoded. If it is considered necessary to have the switch-over occur at an exact address, a third 7485 can be added, cascaded like these two.

These comparators are very useful devices. For example, two banks can be used in parallel, with AND combinations of outputs to give block rejection or block acceptance. Various memory protect, trap, and special situation switches can be set up in hardware with them.

To use the generated switch-over signal a "device disable/ sound device enable" is necessary. About the simplest way to do this is to work on the chip enable signals. Since board space is not critical, let's see how this can also handle 4K of separate PROM.

A circuit to accomplish these objectives is shown in Fig. 4-8. The two page and the read strobe signals are first combined to secure enables, and further ORed to give a board select signal. This activates the buffers. The page N/read signal activates one-half of a dual two-line to four-line decoder, which decodes address BA_{10} and BA_{11} to give four chip selects for the four PROMs that occupy page N. These may or may not be available to NIBL.

The second half of the two-line to four-line decoder is used to select the two chips of the ISP-8F/352 version of NIBL (since ISP-8F/352 is the easiest NIBL to obtain). Selection is determined

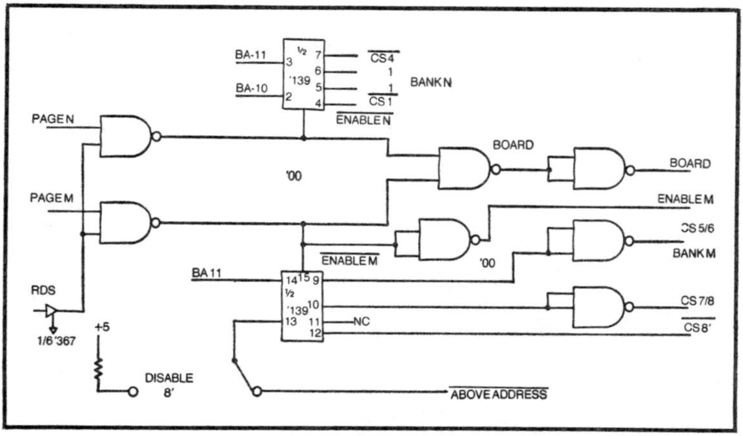

Fig. 4-8. Circuits for generating chip select signals for a board containing 4K of EPROM, 4K of NIBL, plus up to 1K of contiguous revised routines. Page N is for the EPROM, page M for NIBL. The address signal deselects the NIBL ROM, and selects the EPROM containing the alternate routine. Alternately, the change-over can be manual.

by feeding the buffered A_{11} signal to the A input line of the decoder; the two lowest outputs being the NIBL chip selects.

The same decoder also causes the switch over. When the "greater than" signal is received and fed to the higher input, the two chip select signals disappear, isolating the NIBL chips. At the same time, the highest output appears and forms the select signal for the PROM that contains the substitute programming. A jumper is provided so the switch-over signal can be disabled; this allows NIBL to be used in its original form.

The remainder of the board, the data buffers and −5 V supply, can be identical with Figs. 4-1 and 4-3. A board using these circuits is shown in Fig. 4-9.

ALTERNATE CIRCUIT SELECTION

The jumper in the NIBL board chip-select circuit is fine for preset operation, but does not allow fast change, say from ASCII to BAUDOT I/O. Replacement of a jumper by a switch would permit manual change over.

For easiest operation, the choice between the two routines should be placed under program control. A cheap and dirty way to do this is to use the remaining flag, F2, as an energize signal, feeding this through a simple AND circuit, as shown in Fig. 4-10. Since the flag can be set or removed by a CAS instruction, the change can be made during program execution, as an OS function. Note that this simple form can also transfer OS manual control from the ASCII keyboard to the BAUDOT keyboard when the change is made.

Fig. 4-9. A board using the circuits shown in Figs. 4-1 and 4-3.

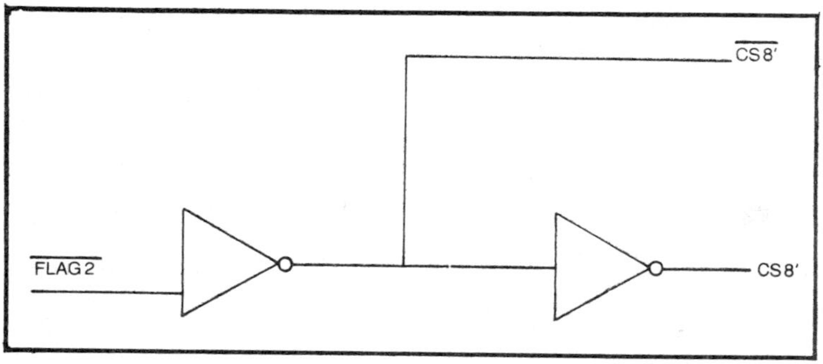

Fig. 4-10. Simple circuit for placing the change-over to an alternate I/O routine under program control. This method is useful when widely different I/O devices are used, such as a mixture of Baudot and ASCII devices.

EXTENDING MEMORY ADDRESS SPACE

There may be occasions when an extension in memory capacity is needed. Probably the most direct way of accomplishing this is to modify the page selection process in one of several ways. Figure 4-11 shows use of a two-line to one-line data selector. This allows replacement of any four pages with a second (or third or

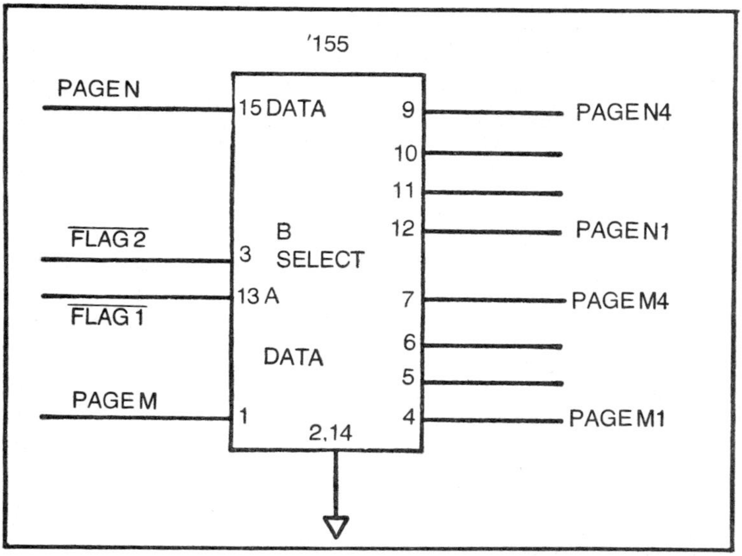

Fig. 4-11. One method of expanding memory space in a paged system. A 74155 dual two-line to four-line decoder is used to convert a page line signal to four signals selected by flags. The two sets of alternate page lines are not independently selectable in this version.

fourth) set of memory. The scheme can be extended as far as desired.

This simple memory change system has the disadvantage that power is still consumed by the inactivated banks. Much of this can be avoided if some of the newer memory chips are used, those that automatically go to a low power state on deselect.

The 2716 EPROM is an example of this automatic power-down demand in 525 mW, but when inactive, it is only 132 mW. For the 4K x 8 equivalent chip, the 2732, the two values are 788 and 158 mW. Low standby designs are also available in RAM. For example, the 2102 AL requires 165 mW in the read-write cycle, and only 35 mW when in standby. Normal read/write times are sufficient for the state change.

The total change in power demand is appreciable in large systems. For example, in a system with 32K of RAM, the usual 2102 type would require a total of 73 watts for memory, about 14.8 A from the 5 V supply. In contrast, the same memory of the 2102 AL type would require only some 10 W, about 2 A. At present, however, cost continues to favor use of the higher demand chips, even when power supply costs are included. The cost of the several chip types appear to be converging, and it is advisable to consider the low power design in any memory expansion.

Still another way of saving the power consumption of unselected chips is to remove the power source completely. At the moment, this is fully reliable for fast select-deselect only for bipolar PROM or ROM, either fuse linked or mask programmed. The reason for this lies in the recovery characteristic at power-on: Typically, only 20 ns are required for outputs to stabilize at the correct level.

A circuit for this power standby mode is shown in Fig. 4-12. It also serves as the block-disconnect for memory extension; a pass transistor is pulled into conduction by a TTL driver. Since the pass transistor is saturated, its drop is only 0.3 or 0.7 V—tolerable in most systems. (If not, the system buses can be split, one set at 5.75 V and used only for switched memory.)

In principle, this system can be applied to other types of ROM, but the precautions needed may become troublesome. For example, with the 2708 EPROM, the −5 V supply should be applied before the others, and must also be the last off, which complicates deactivation circuitry. Also, the behavior at power application is not given in specifications, and, as far as has been determined, is not controlled.

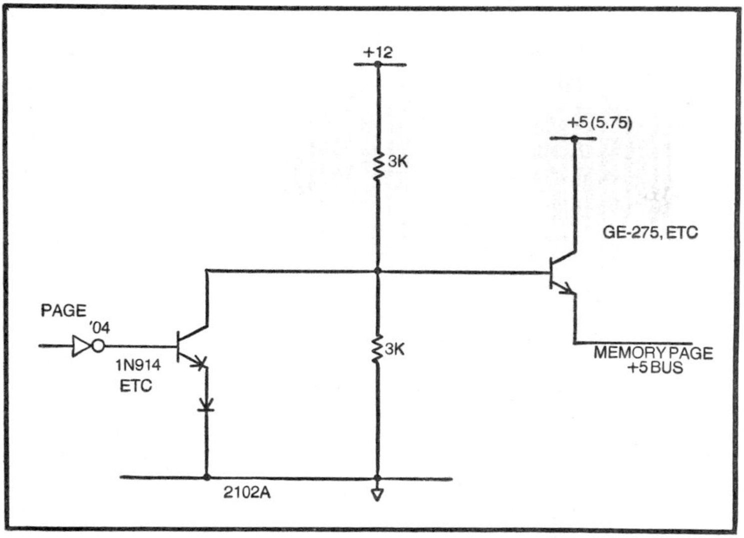

Fig. 4-12. A method of removing power from a bank of ROM when not needed. It is also usable with fuse-linked PROM, but the principle is not recommended for use with multiple supply EPROMs due to requirements for sequencing applied voltages. The circuit is not needed with newer single voltage EPROMs, such as the 2716, which automatically power-down at chip deselect.

Some of these limitations disappear when the memory bank change over (or power savings change only) occurs rarely, with no demand for speed. Many means of combining the principles of Fig. 4-11 and 4-12 are then possible. One is to arrange the power system in two or more sections, each feeding a memory bank. Turning on a power supply activates a memory bank, and can be done manually or under program control.

About the only unusual requirement for this method of memory extension/power saving is that the affected bus elements go to the inactive or hi-Z state when power is removed. The usual precautions relating to leakage, loading, and drive are necessary, of course.

USING SPECIALIZED PREPROGRAMMED ROM

There are a number of specialized preprogrammed ROMs available. The ones most commonly available to the amateur builder are the 2513 character generator family, commonly generating upper case ASCII characters in 7 x 5 dot format. Lowercase is also available. These devices are available in three-voltage and single-voltage supply designs.

Many other types are available. The 2513 is available in lowercase ASCII, in Katakana (Japanese), and in pseudo-random noise coding. Other types are also available for sine-cosine table look up, ASCII-BAUDOT conversion, ASCII-EBOIC conversion, ASCII-Selectric code conversion, "quick brown fox" generation, and for several character sets, including Spanish, German, Scandinavian and different Japanese forms. Some relatively simple conversion devices also use built-in table look up. The 74C915 seven-segment to BCD converter, some inverse converters, and some keyboard encoders are of this type.

Additionally, there are a number of firmware packages available, typically of various forms of BASIC. They are not always easy to use, since they are written for specific machines, with specific I/O configuration. However, if they are available, they might be useful. It is almost certainly easier to construct a board for the special device than to rewrite, say, a 16K BASIC. It might be easier still to copy the code into your own EPROM, but remember to watch the copyright situation.

The first unit to consider is the ASCII upper case character generator. As will be discussed in the next chapter, this is useful with page printout and is necessary with video systems, although it is usually used in these in dedicated rather than general memory form. The math tables are very useful if you are doing much analytical work, and the random number generator if you are doing statistical work. Beyond this, the utility of these preprogrammed ROMs will depend on your particular programming needs and practices.

Table 4-1 lists a few of the available programs.

BUILDING YOUR LIBRARY OF ROM

If you follow the approach of Chapter 3, building your operating system as an extension of NIBL (or some other firmware BASIC), you will find it worthwhile to build up a library of program material in firmware. For one thing, this is a time saver, partly in eliminating loading time, but more importantly, in eliminating loading errors. The main drawback is the incremental cost of the PROMs, but a lot of these can be bought for the price of a good disk system.

There are a few general guides for doing this work that should be kept in mind. Obviously, the program should not be placed in PROM until it has been debugged and runs correctly. Also, unless you aren't concerned with cost, programs in PROM should be

Table 4-1. List of Some Preprogrammed ROM Chips Potentially of Interest in Setting up Operating Systems. At Least One, the LLL or Lawrence Livermore Laboratories Basic is in the Public Domain, and Has Been Fully Described in the Literature (see references).

CHIP TYPE	USED IN COMPUTER	PURPOSE
1903	Exidy Z80	Development Pak
1776	Exidy Z80	Word Processor Pak
1175	8080	Video System Monitor
1314	8080	Teletype System Monitor
INS-8295	SC/MP	NIBL Basic
INS-8298	8080A	LLL Basic
MMS258ET R/N MMS258ET S/N	TRS80	Level 1 Basic
78Ø9 (Basic 1)	TRS80	Level 1 Basic

those commonly used. For most small systems, this includes a good part of the operating system, plus a few common routines.

There are a few program restrictions that must be watched. The program must not set up a table of variables, a stack, or any changeable element within the address range to be placed in PROM. It must not call for program segments to be overwritten during execution, which is common in compilers. Equally, the OS must not call for relocation of the program segment arbitrarily, although some changes, such as page location or alternate contents of a page may be provided for. The code for the range to be placed in PROM must be the same during and following program execution. If it is not, the program must be modified to place the variable elements outside the PROM range and within RAM.

ZIBL-A NIBL DERIVATIVE

While NIBL was originally written with control applications in mind, it was limited by the desire to keep it to one page of the SC/MP memory and by the small number of registers within the CPU. Despite these limitations, its creators came up with some unique and useful features (such as the power-on test for program presence and immediate execution, direct execution of most instructions; LINK and the relative operator).

Dynabite, Inc., has extended the control capabilities of NIBL. They also changed the processor to the powerful Z80. The extended and modified language is called ZIBL, the origin being obvious.

At present the language is available only to purchasers of the single-board control computer marketed by the company. However, a description of language features plus computer characteristics has been released. But some of the concepts can be useful in OS expansion. ZIBL is used as firmware, and it is possible that the chips will be placed on the market in the future.

Part of the ZIBL expansion is concerned with the hardware features of their single-board controller. Major elements are:

- CRT video output, 16 lines of 64 characters
- Keyboard input, eight data lines plus strobe
- RS-232 serial I/O, with 20 mA loop
- RS-232 serial I/O
- Three hundred baud audio cassettes I/O, with listing, machine language and data channels, and with motor control
- Thirty-two memory-mapped flag outputs
- Thirty-two memory-mapped flag inputs
- Eight memory-mapped LEDs
- One eight-bit LED outport
- Four low current relay circuits
- Four high current relay circuits
- One clock

plus the usual expansion capability from data, address and control lines.

Other features are straightforward programming extensions. For example, ZIBL uses three-byte, 24-bit variables instead of the two-byte, 16-bits used by NIBL. Recall that the NIBL interpreter clears all variables on power up, or on a NEW or CLEAR command, and variables are local to a program. ZIBL extends this, allowing up to three sets of local variables to exist simultaneously. This is done by a RUN (program name) command that causes variables to be saved in memory. ZIBL also includes a set of 26 variables available to any program. The 26 string variables are global. Commands are extended to cover the two classes.

As is common in larger interpreters, ZIBL allows named programs. For these, the RUN and END instructions may be followed by a name. LOAD, NAME and RENAME and added, each followed by a name. TOP is extended by adding FREE, which returns the amount of memory unused. TRACE is added, to simplify debugging programs. Operators are the same, plus addition of XOR to the logic group. Some operators reflect the difference in microprocessors: STAT is replaced by SENSE, and DELAY is made an instruction instead of a machine language call.

Several of the extensions arise from the hardware control capability of the specific design. These are:

DTOA	For digital-to-analogue conversion
ATOD	For opposite conversion
TURNON	LITE, RELAY, FLAG
TURNOFF	LITE, RELAY, FLAG
TIME	Sets clock
OPEN	Starts clock
CLOSE	Stops clock
ENTER	From an input channel
IN%	Designates channel

Most of these require a following expression to designate the device or channel. A number of the I/O channels can operate at one of several speeds, as selected by the instruction.

An unusual addition is RESTORE, which sets a pointer to a block of data, defined by a DATA instruction.

Of course, these additions are not accomplished without some cost. Despite the larger instruction set of the Z80, ZIBL requires 8K of ROM for the interpreter, plus expansion of RAM space assigned to stacks, flags and variables.

NSC TINY BASIC

In connection with its single chip microprocessor development, National Semiconductor has developed a new version of NIBL that they call NSC TINY BASIC. Outwardly, it is almost exactly the same as NIBL, retaining emphasis on control.

The COMMAND and STATEMENT set are almost exactly the same. All NIBL statements except END are present. (This has been replaced by STOP). There are four new statements:

- DELAY expression; generates delays from 1 millisecond to 1.04 seconds
- ON number, expression; defines starting address of the two interrupt lines service routines
- INC location
- DEC location; increment or decrement a memory location; intended for multiprocessing

Plus a new command, CONT, that causes program execution to resume after a STOP, a reset or a console interrupt. Automatic execution of a program remains, but its location is changed. The serial I/O rate may be 110, 300, 1200 or 4800 baud, set by the programs or by hardware.

Some of the changes are related to differences between the single-chip series (INS807X) and the SC/MP (INS8060). These are reviewed in a later chapter. Their major effect is to reduce the memory requirement for the interpreter, only 2.5K being required as compared to 4K for the original NIBL.

At present this version of NIBL is available in the form of a preprogrammed microprocessor, the INS8073. This is the programed version of the INS8072 that contains the processor itself, a 64-byte scratch pad memory, and a 2.5K ROM intended for mask programming. (The version without the 2.5K is the INS8070).

NONPORTABILITY PROBLEMS

One of the problems of BASIC and all of its derivatives has been the stream of changes, improvements and simplifications. One result has been *nonportability*, the fact that a program will run only on one machine. The IEEE and others are attempting to rectify this situation, by preparing standards and specs for languages. But there is no review of Tiny BASICs, and equally, none of NIBL as yet.

The following is offered as a suggestion. If you have developed an extension of NIBL which you feel would be useful to others, communicate the fact to the informal users group, SCAMPUS. Include, as a minimum, the purpose of the change, the added instructions, and the machine configuration designed for. If you want to release the details to the public domain, include the source code listing, with full annotation, so the program will be clear to potential users.

The current address is:

Tom Bohon
SCAMPUS NEWSLETTER
P.O. Box 132
Knob Noster, MO 65336

Perhaps, in time, it will be possible to get a standard Tiny BASIC, plus a table of incremental features which provide a bridge to the three levels of BASIC recommended by its originators.

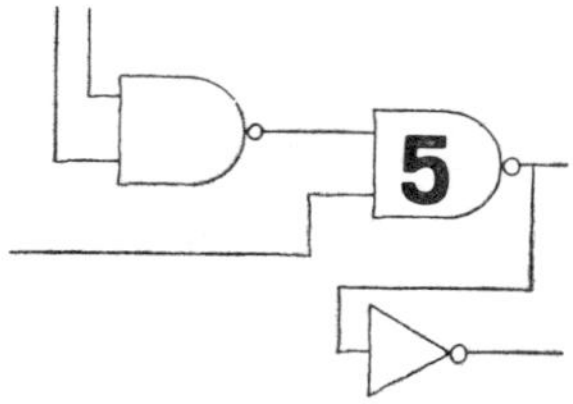

Hard Copy Devices For I/O

I firmly believe that any computer working on real problems needs hard copy output. If a programmed problem is worth running, the results are worth keeping.

Also, any computer doing real work is too valuable to tie up at human speeds. The simple operations of program preparation, input preparation, preliminary debugging, and evaluations should be done *off line*. This frees the computer to do what it does best, and makes program preparation and program execution into operations that can go on simultaneously.

These capabilities require as a minimum a manual input or keyboard that can be independent of computer operation; a permanent copy output device, usually a page printer, that can operate from the keyboard or from the computer; and some temporary or permanent storage fed from the keyboard and capable of computer input. Paper tape is one form and allows editing and checking. Very good equipment is available, for example, in the form of a Teletype assembly with a tape-punch reader.

For the owner of the small computer, the modern ASCII equipment has one major drawback—price. Typically, the combination costs five times as much as the common keyboard-TV screen. Overall, it can double or triple the cost of a home computer. Even used ASCII equipment is quite expensive.

Some relief has developed. The ASCII keyboard is now available in many forms, and is occasionally available, new or used, in surplus at very attractive prices. Page printers for light duty are decreasing in price; this is especially true for *dot matrix* printers. If printouts are acceptable at less than the standard page width, the printers used in adding machine printers still cost less. However, new light duty units still run to hundreds of dollars and continuous duty printers run into the thousands. A further drawback is that the characters formed by the matrix printers are not as easy to read.

Another recent development in the printer area helps. This is a commercial unit, a bank of solenoids that rest over the keyboard of an electric typewriter, actuating the keys under computer control. While this unit is relatively expensive and has not been around long enough to have proven reliability, it should be investigated. The cost is appreciably lower than most dot matrix printers, and it can be used with almost any typewriter.

Another item to investigate is the kit of solenoids sold to convert the IBM Selective to remote operation. The resulting machine is not good for continuous duty, but performs very well in usual home usage and makes very clear copy. Cost is very reasonable.

Neither conversion is permanent, and the typewriter can still be used as typewriters.

The use of the older form of teleprinters or Teletype units, the Baudot five-digit code type designed for wire and radio communication, is also possible. Many computer owners use these for input and output, but they are not popular. This lack of popularity is due to limitations in capability, in particular in the limited character set possible. This drawback to the Baudot devices can be removed entirely by software, partly by some hardware changes, or by a combination. Let's look at what can be done.

HARDWARE CHARACTER EXTENSION

All standardized keyboard Baudot devices, whether Western Union, AT&T, Teletype or military, have blank keys and some nonprinting functions. On many machines it is no great chore to extend the character set, providing up to four additional characters. The simplest approach is to replace existing keys and type bars with those that carry weather symbols, at the same time adding the elements that make the key into a printing character rather than nonprinting. No modifications are needed—just replacement of parts. Excellent illustrated instruction books are available for this. It isn't hard to do, and while it is being done, the machine can be cleaned and lubricated, to be ready for another 20 years of use.

This simple approach gives at least two new characters, the up arrow and the down arrow being standard. The location of the dash (minus) changes. If this is done to a business machine, such as those used for TWX, it may be desirable to change other keys (½, ¾, ⅝, and ⅜, and WRU) to the more common punctuation (;, $, (,), ?, :) that is more useful in computer operation.

On some machines with interchangeable platens, or with platens that are soldered to the keys, further changes are possible. Useful characters are the right, up and down arrows, +, #, ⊕, and ⊗. It is also worthwhile to change from the standard zero to the military Ø, to avoid confusion.

Platens are very hard, and modification involves grinding away unwanted characters. However, interchangeable platens are fastened to type bars by soft solder. In this case the change is easy, except several tries may be needed to get good alignment.

To fully expand the set, it's necessary to make a modification to the function select mechanism, allowing function keys to type and space. This is particularly easy on surplus Kleinschmidt military machines, being done by a ½ × 1½-inch strip of #16 sheet iron fastened inside the function selector guide block. The detail of the modifications will be evident on inspection, or see the instruction book.

Instruction books and parts for these changes and modifications are available from several sources, including the manufacturers. See the ads in the magazines devoted to amateur radio, usually under the heading RTTY.

Figure 5-1 shows an example of the character set made available by these extensions in the form of several lines of

```
ØØ    PRINT "NEW SYMBOLS ARE→, ↓, + AND⊕"
11Ø   REM      ARITHMETICAL USE
12Ø   LET X→3A⊕U 2↓⊕E (B+2)
13Ø   REM      COMMAND USE
14Ø   PRIM  NT
```

Fig. 5-1. Example of a simulated program listing showing use of weather symbols for Baudot-character set extension. A symbol may be assigned a meaning, such as right arrow replacing equals, or another symbol used to designate a new Baudot case (The weather symbol of "completely overcast" is used for this purpose here.)

simulated BASIC. The down arrow is used as the multiplication sign, and the right arrow as the equals. Remaining characters follow customary usage.

SOFTWARE CHARACTER EXTENSION

While these changes and modifications add some needed characters, especially those useful in BASIC or FORTRAN, they don't take care of everything. Further extension of the character set requires software. (This can also be used to avoid hardware change.)

The solution is to arbitrarily establish a new case of symbols, making up to 26 additional symbols or control functions available. This requires reserving one symbol to indicate the new case.

A common symbol for special instructions is the dollar sign. This could be used for the special case, but it seems better to reserve it for BASIC and assembler instructions. Since the only other unique character available on my machine is the lowercase weather symbol ⊕, I used it as the new case indicator. The Baudot code for this is 00000, while in lowercase (000000 in six-digit code, not the same as 100000, upper case null).

There are several ways to construct the new case. The method I use sets the case for the next character only; this character must be alphabetic. The single character rule is practical in actual use, and the restriction to letters simplifies assignment by mnemonic rules.

My complete character assignment is shown in Fig. 5-2. Most of the assignments are mnemonic: ⊕E for exponent, ⊕H for hex and so on. A few are arbitrary, specifically those involving I, J, K, and X, Y, Z. One entry is used as a control for form feed. In practice, there can be duplicated definitions; for example, the back arrow and back space may both be defined as "delete last character".

THE INTERNAL COMPUTER CHARACTER SET

Even though Baudot machines are used for input and output, they do not require internal computer operations be Baudot. There are three good reasons to use ASCII instead:

1) It is easier to order a table of numbers, or to sort alphabetically.

2) It is easier to convert from ASCII numbers to Hex.

3) Many programs include ASCII and would have to be rewritten to use Baudot directly.

In addition, it is not unlikely that you will add ASCII devices in the future.

BAUDOT	ASCII	BAUDOT	ASCII
⊕ A	AT	⊕ N	NULL
⊕ B	BACK ARROW	⊕ O	OPEN BRACKET
⊕ C	CLOSE BRACKET	⊕ P	PERCENT
⊕ D	DELETE	⊕ Q	ENQUIRY
⊕ E	EXPONENT	⊕ R	RUBOUT
⊕ F	FORM FEED	⊕ S	STAR
⊕ G	GREATER	⊕ T	END
⊕ H	HEX NUMBER	⊕ U	UP ARROW
⊕ I	OPEN BRACE	⊕ V	VERTICAL
⊕ J	BACK SLASH	⊕ W	APPROXIMATELY
⊕ K	CLOSE BRACE	⊕ X	DC ONE
⊕ L	LESS	⊕ Y	DC TWO
⊕ M	END MESSAGE UNIT	⊕ Z	DC THREE

Fig. 5-2. Possible assignment for a new case in Baudot. They are easy to remember, since the letter involved is the initial character of the new symbol: B for back arrow, E for exponent, etc.

All of these considerations indicate that internal computer operations should be in ASCII with conversion to or from Baudot as part of the input-output operation. Implementing this requires a pair of conversion tables.

CODE CONVERSION

Since Baudot is a five-level code and ASCII seven-level, there is quite a bit of freedom in table arrangement. One method uses the extra bit space to indicate the case. Using bit D7 high to indicate ASCII takes care of the four ASCII cases denoted by bits D6 and D5 (i.e., control case, symbol/figure case, capital and lower case). Similarly, with D7 low, D6 and D5 can indicate the Baudot case, including the new one denoted by the special symbol.

There is no lowercase in Baudot. However, since it is useful in ASCII work, lowercase ASCII to letter case Baudot should be included: It may not be necessary to provide for special symbols based on punctuation (such as⊕!:), the 26 provided by letters is ample to fill out the ASCII symbol list and to provide a number of control signals.

I decided not to use the special case with punctuation. Instead, another feature was provided—the conversion of Baudot written hex (0, 1, 2---E, F) directly to hex. The alternative to this requires Baudot to ASCII conversion followed by ASCII to hex conversion.

The direct conversion simplifies off-line preparation of hex-coded tape; the tape can be punched directly from the normal keys instead of from relabeled keys, and the tape can be checked easily.

With these decisions made, the Baudot cases become letters, figures, special and hex. However, the hex case normally spreads over 64 characters, since it uses both letters and figures. A little trial and error coding shows that the required memory space can be reduced by using the following algorithms:

- For Baudot hex numbers (0. 1--9), add 45_{16}
- For Baudot hex letters (A, B--F), add 60_{16}

This makes the combination of bits 5 and 6 high indicate the hex conversion code. The two-part algorithm folds the numbers and letters code into a new case that is unique for the 16 symbols needed for hex. The remaining 16 symbols can be set up to return an error code, if desired.

The complete case assignments are:

Case Name	Set D7 D6 D5 to:
Baudot letters	000
Baudot figures	001
Baudot special	010
Baudot hex	011
ASCII control	100
ASCII figures	101
ASCII caps	110
ASCII lowercase	111

Note that some devices using ASCII coding use the eighth bit for parity, while others set it to one and still others set it to zero.(The differences must be handled by software).

With these decisions, all basic elements of the conversion have been tied down. Two hundred fifty-six bits are required for Baudot to ASCII conversion (including hex) and another 256 bits for ASCII to Baudot four-case to three-case conversion. The correspondence is shown in Fig. 5-3 which shows, for each case, the entrance symbol in hex, the symbol itself, and the translated or exit symbol. Unassigned symbols may be filled with rubout, which is the letters shift in Baudot, or with an error indicator.

The use of this table data naturally depends on the look-up algorithm employed. The data can be used to fill a 512-byte PROM or EPROM; the column labeled input being the address, and the column labeled output being the content of that address. More commonly, however, the character to be translated is used as a displacement from the assigned address of the table. Since most

Serial	Baudot - ASCII*									Baudot HEX			ASCII* - Baudot											
	LTRS 000			FIGS 001			SPEC 010			HEX 011			Control 100			Shift 101			Upper 110					
0	00	⊕	FE	20	+	AB	40	⊕⊕	8Ø	60			8Ø	NUL	4C	A	S.P	04	CØ	@	43	EØ		49
1	01	E	C5	21	3	B3	41	⊕E	C6	61	E	ØE	81	SOH		A1	!	2D	C1	A	03	E1	a	03
2	02	L.F.	BA	22	L.F.	8A	42	L.F.	8A	62			82	STX		A2	"	31	C2	B	19	E2	b	19
3	03	A	C1	23	-	AD	43	⊕A	CØ	63	A	ØA	83	ETX	5C	A3	#	54	C3	C	0E	E3	c	0E
4	04	S.P.	AØ	24	S.P.	AØ	44	S.P	AØ	64			84	EOT	5Ø	A4	!	29	C4	D	09	E4	d	09
5	05	S	D3	25	→	BD	45	⊕S	AA	65			85	ENQ	57	A5	%	56	C5	E	01	E5	e	01
6	06	I	C9	26	8	B8	46	⊕I	FB	66	3	Ø3	86	ACK		A6	&	3A	C6	F	0D	E6	F	0D
7	07	U	D5	27	7	B7	47	⊕u	DE	67			87	BEL		A7	'	2B	C7	G	1A	E7	g	1A
8	08	C.R.	8D	28	C.R.	8D	48	C.R.	8D	68			88	B.S.		A8	(	2F	C8	H	14	E8	h	14
9	09	D	C4	29	$	A4	49	⊕D	9Ø	69	D	ØD	89	H.T.		A9	)	32	C9	I	06	E9	i	06
10	OA	R	D2	2A	4	B4	4A	⊕R	FF	6A			8A	L.F.	02	AA	*	34	CA	J	0B	EA	j	0B
11	0B	J	CA	2B	'	A7	4B	⊕J	EØ	6B	8	Ø8	8B	V.T.		AB	+	20	CB	K	0F	EB	k	0F
12	0C	N	CE	2C	,	AC	4C	⊕N	8Ø	6C	7	Ø7	8C	F.F.	4D	AC	.	2C	CC	L	12	EC	l	12
13	0D	F	CG	2D	!	A1	4D	⊕F	8C	6D	F	ØF	8D	C.R.	Ø8	AD	-	23	CD	M	1C	EE	m	1C
14	OE	C	C3	2E	:	BA	4E	⊕C	DD	6E	c	ØC	8F	S.I.		AF	,	3D	CE	N	18	EF	o	18
15	0F	K	CB	2F	(	A8	4F	⊕K	FD	6F	4	Ø4	9Ø	DLE		BØ	Ø	36	DØ	P	16	FØ	p	16
16	10	T	D4	30	5	B5	50	⊕T	84	70			91	DCI	5D	B1	1	37	D1	Q	17	F1	q	17
17	11	Z	DA	31	"	A2	51	⊕Z	93	71			92	DC2	55	B2	2	33	D2	R	0A	F2	r	0A
18	12	L	CC	32	)	A9	52	⊕L	BC	72			93	DC3	93	B3	3	21	D3	S	05	F3	s	05
19	13	W	D7	33	2	B2	53	⊕W	FE	73			94	DC4		B4	4	2A	D4	T	10	F4	t	10
20	14	H	C8	34		AA	54	⊕H	A3	74			95	NAK		B5	5	30	D5	U	07	F5	u	07
21	15	Y	D9	35	6	B6	55	⊕Y	92	75	5	Ø5	96	SYN		B6	6	35	D6	V	1E	F6	v	1E
22	16	P	DØ	36		BØ	56	⊕P	A5	76			97	ETB		B7	7	27	D7	W	13	F7	w	13
23	17	Q	D1	37	1	B1	57	⊕Q	85	77			98	CAN		B8	8	26	D8	X	29	F8	x	29
24	18	O	CF	38	9	B9	58	⊕O	DB	78	2	Ø2	99	E.M.		B9	9	38	D9	Y	15	F9	y	15
25	19	B	C2	39	?	BF	59	⊕B	DF	79	B	ØB	9A	SUB		BA		2E	DA	Z	11	FA	3	11
26	1A	G	C7	3A	&	A6	5A	⊕G	BE	7A	G	Ø6	9B	ESC		BB	.	3E	DB	[	58	FB	{	46
27	1B	Fig.	8Ø	3B	Fig.	8Ø	5B	Fig.	8Ø	7B	Ø	ØØ	9C	F.S.		BC	<	52	DC	\	49	FC			5E
28	1C	M	CD	3C	.	AE	5C	⊕M	83	7C	1	Ø1	9D	G.S.		BD	=	25	DD	]	4E	FD	}	4F
29	1D	X	D8	3D	/	AF	5D	⊕X	91	7D	9	Ø9	9E	R.S.		BE	>	5A	DE	↑	47	FE	~	53
30	1E	V	D6	3E	;	BB	5E	⊕V	FC	7E			9F	U.S.		BF	?	39	DF	←	59	FF	RO	4A
31	1F	LTR.	8Ø	3F	LTR.	8	5F	LTR.	FF	7F														

*Parity bit = 1

Fig. 5-3. Complete symbol set for a four-case Baudot and four-case ASCII system with conversions. This is in the form of a look-up table entered with one case and giving the output in the other case. Columns are labeled in binary, but row data is in hex.

microprocessors consider the displacement to be a signed number, −128 to +127, this method will require a rearrangement of the table with the ASCII elements in reverse order in the lower half and the Baudot elements in the upper half. The table address will be the location assign to the Baudot lowercase null (the special character ⊕ on my machine).

HEADERS AND LEADERS

While this completes the requirements for Baudot-ASCII conversion, there are two other abilities that are very worthwhile and can also be stored in PROM. These are headings for output; one type for the page printer to make a seven-times enlargement of each character; a second for paper tape to make a leader as immediately readable as a type font. The page printer character set limits an expanded printed line to 10 characters, but several such lines can be used. The tape leader can be any length desired.

While these are program niceties, they do have value. The page heading sets off a listing, for example, and makes it easy to leaf through a stack of printouts. The tape saves much time in checking the title or content of a tape, and helps prevent wrong-end starts.

The principle of formation of the page-printer expanded-character is shown in Example 5-2. Two variables enter ROM. One is for the character, the entry being bits 0-5 of the ASCII character. The second is for the line to be printed, determined by the number of times the character is scanned. Scan zero produces only nulls, and sets the minimum space from any previous character line. Line one, the topmost, starts actual character formation.

In the example shown in Fig. 5-4, the letter S, the address for the printed first line is 001 110010, the second line 010 110010, and

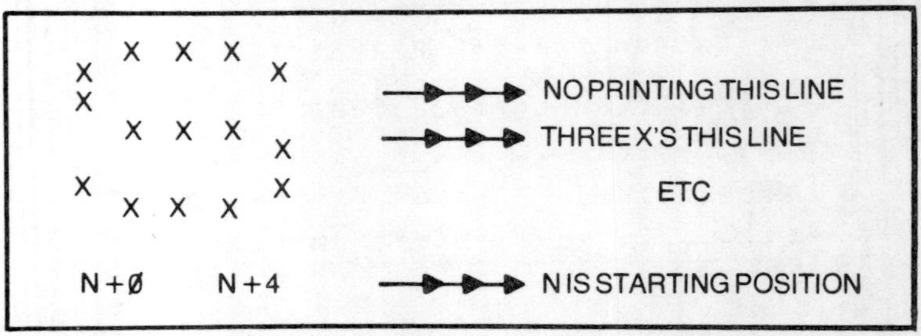

Fig. 5-4. Method of creating expanded size characters by spacing a character, in this case X, in a 5 x 7 matrix. This character uses added spacing along the horizontal axis for emphasis.

```
 XXX      X     X   X   XXXX    X       XXXXX
X   X    X X    XX XX   X   X   X       X
X       X   X   X X X   X   X   X       X
 XXX    X   X   X   X   XXXX    X       XXXX
    X   XXXXX   X   X   X       X       X
X   X   X   X   X   X   X       X       X
 XXX    X   X   X   X   X       XXXXX   XXXXX
```

Fig. 5-5. Sample of the 5 x 7 matrix characters with normal horizontal spacing. This is a good method of heading a printout since it attracts attention. Usually the heading is the name of the program run. Data for these characters are the same as the data in the 2513 character generator.

so on. The output for the first active line is 01110. These five bits are converted to printer characters by a software shift register, or an equivalent masking process. Usually an X is used where inking is desired, and a space for no ink. The first line prints as bXXXb, followed by a space, then the next character. Each line to be printed requires seven scans. Ten characters are the maximum for one line of a 72-column printer.

The process is identical to that used for character formation on TVT screens and can use the same ROM. These are widely available. For example, the 2513 or the 3257, and so on. Note that the addressing may be different. Character ROMs are available for several languages. See the specification sheets for the standard characters, or make your own, remembering that you are working with a 5 x 7 matrix. Figure 5-5 shows a complete header.

The tape leader works on a different principle. Since all bits of a tape character are punched simultaneously, the problem is to choose the character which punches the desired position. For example, as shown in Fig. 5-6, the letter S can be produced from the successive characters L, Y, Y and D, giving a 4 × 5 matrix. Some characters make a better appearance if the matrix is 5 × 5. A font often used is shown in Fig. 5-7, with the resulting tape leader appearing as in Fig. 5-8.

Some variation in the character set is possible, of course, and the set is easily extended to eight-level tape. For this, the 5 x 7 matrix is probably best.

Software to use this character set is also different. The ASCII character is still used for entry. Four or five successive table entries give the punch instructions to form the character on tape. It's best to standardize at five entries, one being blank for the

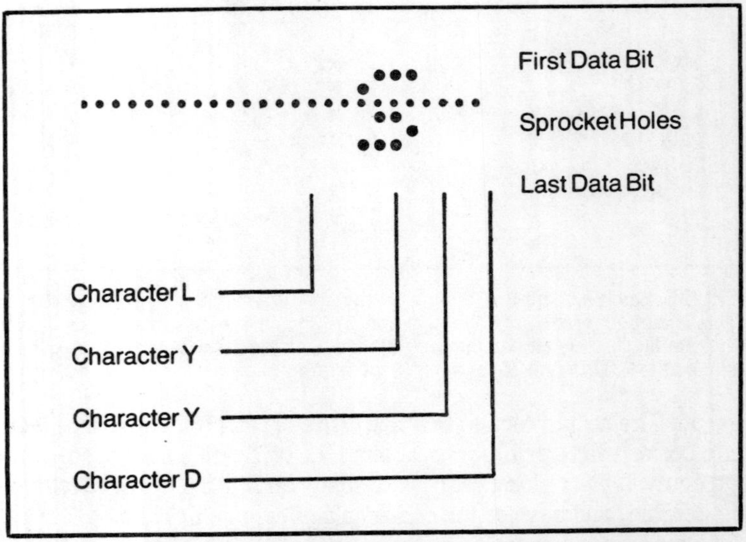

Fig. 5-6. Forming an easily read character in paper tape by punching on a 4 x 5 (or 5 x 5) matrix. Punches are formed by character choice. This is useful in labeling paper tape and is helpful in preventing wrong end or wrong side starts.

four-unit characters. A 50-character set fits into a 256-byte ROM and allows for some punctuation or special symbols. The easiest method of reading is to use the least significant five digits of the ASCII character as the displacement for the first table entry (this value plus 32_{16} for the next entry, and so on). Watch the computer's method of handling displacement, though.

Routines for these conversions and headers should be set up as callable subroutines and made part of the operating system.

ELIMINATING OTHER BAUDOT MACHINE DISADVANTAGES

While these additions vastly improve Baudot machine operations, there are several other disadvantages of these older machines. Specifically, the machines are large and obtrusive, they are noisy, and some are downright smelly. Quite a lot can be done to remedy these things.

Eliminating the Odor

The odor in some older machines was built into the lubricant. It's necessary to completely degrease the machine to get rid of it. This degreasing is a good idea if your machine hasn't been overhauled lately.

Symbol	Punch Baudot 1	2	3	4	5	Symbol	Punch Baudot 1	2	3	4	5
A	V	S	S	V		U	K	T	T	K	
B	LTR	Y	Y	R		V	A	N	T	N	A
C	C	Z	Z	R		W	LTR	CR	SP	CR	LTR
D	LTR	Z	Z	C		X	Z	R	SP	R	Z
E	LTR	Y	Y	Z		Y	E	LF	M	LF	E
F	LTR	S	S	E		Z	Z	B	Y	W	Z
G	C	Z	B	R	CR	Ø	C	B	Y	W	C
H	LTR	SP	SP	LTR		1	L	LTR	T		
I	Z	Z	LTR	Z	Z	2	B	Y	Y	L	
J	CR	T	T	K		3	Z	Y	Y	C	
K	LTR	SP	R	Z		4	U	SP	SP	LTR	
L	LTR	T	T	T		5	J	W	W	F	
M	LTR	LF	SP	LF	LTR	6	C	Y	Y	N	
N	LTR	LF	SP	CR	LTR	7	Z	D	S	A	
O	C	Z	Z	C		8	FIG	Y	Y	FIG	
P	V	S	S	I		9	LF	Y	Y	C	
Q	C	Z	B	C	T	—	SP	SP	SP		
R	V	S	F	P		Period	T				
S	L	Y	Y	D		Space	⊕				
T	E	E	LTR	E	E	⟶	SP	SP	Y	C	SP

Fig. 5-7. Commonly used symbol set for Baudot paper tape. Readability is not impaired if characters are standardized as 5 x 5 elements.

A simple cleaning procedure is to remove the motor, then set the machine mechanism in the open, and spray it thoroughly with an auto engine degreaser, such as Gunk. This is followed by a wash down with hot water, or even with steam. Repeat if necessary, then dry thoroughly and rapidly. Sunshine, plus a fan will do, or try a hair dryer, or an infrared bulb plus a fan.

When thoroughly dry, spray all surfaces with a silicon moisture repellent to prevent rusting and corrosion from fingerprints. Then go through the complete lubrication process.

Cutting Down on Noise

Noise comes from two sources. The operating elements produce a lot as shafts spin, pawls pick up and so on. More comes from the shift mechanism, and from type bars or punch fingers hitting paper. Nothing can be done for many, but there is no reason to put up with the whine of a bad bearing, or the various clanks due

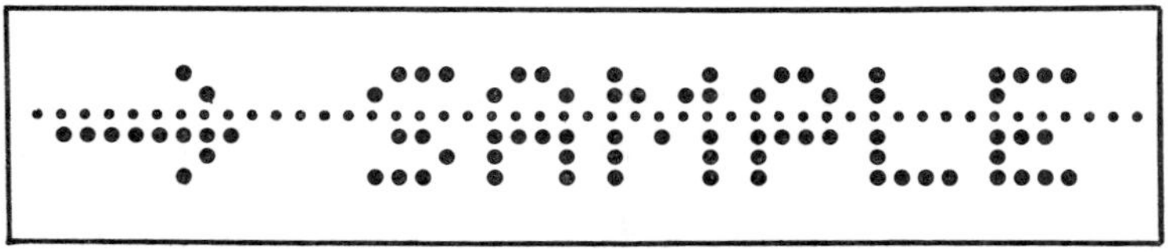

Fig. 5-8. Tape leader using punched characters. The arrow is helpful. Similar character sets for ASCII are possible, and usually use a 5 x 7 matrix.

to improper adjustment. Trace down the sources, and see if noise is inherent or correctable.

The inherent noise will have to be absorbed, or kept from propagating. Be sure that the devices are on isolating pads,such as dense foam rubber. Don't let a printer or punch touch another unit or a wall.

To absorb sound and prevent transmission requires either a cabinet or added material in the existing cases. The absorbing batts used in hi-fi speakers, or under automobile hoods, are good for the inside of cabinets or cases. A combination of absorbing material inside an existing case plus a new outer cabinet, also lined, (it should not touch the device case) will reduce the noise to a very acceptable level.

Pondering Size

Not much can be done about the size of equipment, and the size problem is made worse by acoustic treatment. This is brute-force equipment designed to last—so it is big. However, this doesn't mean that it has to be ugly. There is no reason to stay with wrinkle finish or battleship gray, showing assorted dents and scratches.

If the unit case is in good shape otherwise, a coat of paint will do. Dents, and even tears, can be taken care of with auto-body filler, and there is no reason for not using the wide range of auto colors available. Or you might want to cover the outside of the unit case with the stretchable vinyl material used for auto tops.

A really deluxe installation can resemble home or office furniture, the cabinetwork serving appearance, dust protection and noise isolation. Look at some of the commercial units for ideas.

This approach of using the older Baudot devices has a lot to recommend it. With a little care in buying, you should get a complete setup of page printer, keyboard, tape reader, tape punch and tape preparer for less than one-tenth the price of just one new page printer. With the modification and programming ideas covered here, these will serve all input-output needs—a little slower than newer devices perhaps, but just as well. And, with any care, these older devices should be running long after the newer ones have been sent to the scrap heap.

PAPER TAPE FOR INPUT AND SLOW MEMORY

Paper tape for computer use is one of those things that cause the saying, "Until you try it, you don't know what you're missing."

It's downright handy. One of the main uses is to prepare a problem program for input offline. If you're the only one using the computer, and you're only playing games, this isn't important. But, if the computer is going to be tied up for an hour or so for someone else, you can spend the time getting your program ready for input, free of typing errors, and ready to run at several times your normal typing speed. This is easy with tape.

Another use is preparing data for input. Not only can you get this into input format, you can also check it for accuracy. It helps prevent that stupid feeling that results from transposing two numbers in the first line of input and ruining the whole run. Further, you can set the problem up to start and stop and tape reader, so you don't have to be present—the computer will run happily all night, if need be.

Remote punch and remote read make another step possible, the use of paper tape as slow memory. It's perfectly practical—1K is only 16 feet of Baudot tape, so it's not hard to increase memory capacity appreciably. This mode is especially useful when an early result is needed, or when it must be used as an input at a later step of calculation and available fast memory is filled.

Of course, magnetic tape also has uses as intermediate storage, and should also be used. Its one drawback is that it can't be converted to hard copy without use of the computer, whereas paper tape can. Also, in most systems it is not possible to make a copy of magnetic tape. This can be corrected by adding a second machine and a copy program. I prefer to use the magnetic tape if it is for temporary, intermediate storage only, but use paper tape for other purposes, including the possibility of later reruns of the same problem. Even small systems with magnetic or paper tape have almost unlimited memory available. It may take some effort to use it, but it is far better than wailing that you've only got 8K of ROM.

STANDARD OUTPUT FORMATS

NIBL, like most Tiny BASICS, is very limited in output format capability. There are three modes:

1) A string in quotes, printed exactly with quotes removed.

2) An expression, which is evaluated and printed in decimal format with a leading minus or blank, and one trailing space.

3) A reference, $ (address), printed as ASCII starting at address and continuing until a CR is encountered (CR is not printed).

There is no zone spacing, nor automatic CR after 72 characters.

However, a semicolon at the end of a print statement suppresses the usual end-of-statement CR.

Quite a bit of output formatting can be done with these at the expense of time in writing and executing the program. As usual, PRINT alone is a line feed. Blanks in quotes can be used for creating tables, but if the nulls are used, it will be necessary to use space for this. Depending on the amount of memory available, multiple digit numbers, arrays and tables can be created.

Because of the repetitious work involved, it is better to include a set of one or more standard output routines in the operating system. A useful trick is to set up one of these as a default routine. This is the most commonly used format; you'll have to determine the variation for your system. For scholastic level numerical work a 15-space field is good, and allows easy organization of results into tables. For business, standard forms can be followed.

Standard formats aren't limited to output. In some assembly language systems, tabular location is used to determine the structure of the instruction. Typically, the first column is a line number, the second an address, the third the operator and the fourth the operand, with further entries being remarks. Sophisticated operating systems recognize the end of an item and insert spaces to go to the next tabular point. For example, the Autoline program of some small systems, automatically prints line numbers in increments of 10.

ALTERNATE TELEPRINTER CONNECTIONS

The serial board described in *How to Design, Build and Program Your Own Working Computer System* uses reed relays driven by TTL gates in the Teletype circuit. This allows use of either 60 mA or 20 mA current loops for the printer, simplifying use with some old printers or with equipment used for other purposes.

If the 60 mA loop is not needed, the teleprinter drive can be the circuit used by National Semiconductor in their SC/MP development system. The printer magnets are driven by paralleled TTL gates, the circuit being driven from the 5 V supply. The keyboard circuit uses the –12 V supply to ensure that the space signal goes below the switching threshold of the input gate. This is a Schmidt trigger, used in conjunction with the RC input filter to reduce contact bounce problems. Software test and delay is also used to further protect against false bounce signals.

I prefer to see some added protection in these circuits. It can take the form of 5.1 V Zener diodes connected from the two plus terminals to ground. These give protection against both positive and negative excursions sufficient to damage the gates.

CONVERSION TO RS-232 SIGNALS

RS-232 standard signals swing above and below zero, so the standard 20 mA loop cannot operate with them directly. A simple addition is shown in Fig. 5-9 which handles the necessary level

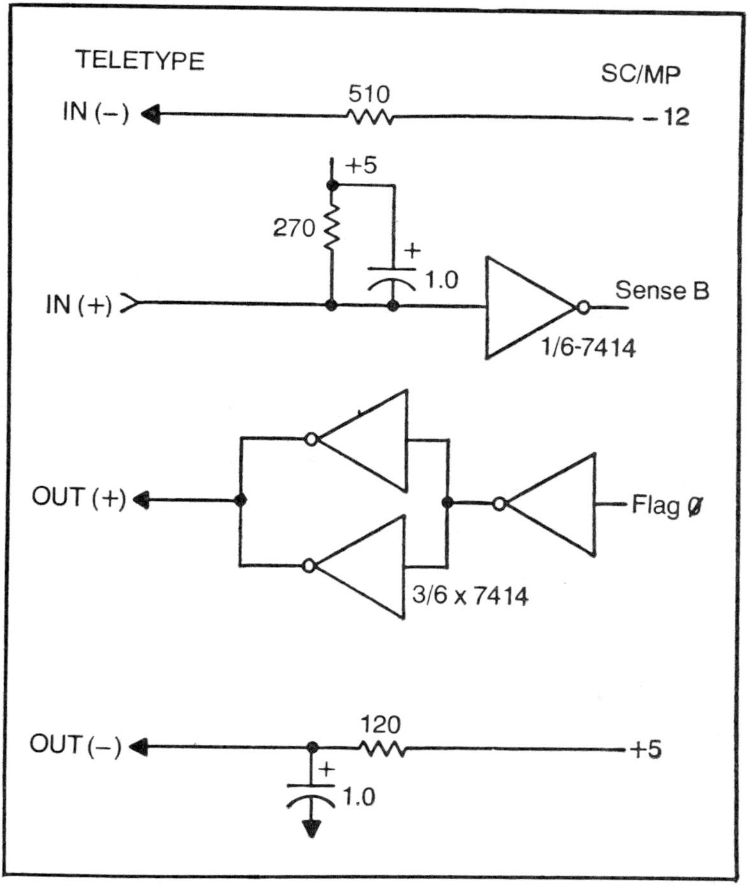

Fig. 5-9. Teleprinter interface elements for an SC/MP based computer, using sense B for input and flag ø for output. A similar output circuit using flag 1 can be set up to control the tape-reader relay. This places tape operation under program control, which is very useful for data input and for establishing a form of batch processing.

shift. As shown in Fig. 5-10, the level shift is done with transistors. The circuit is by C. R. Britten.

USE OF EIGHT-LEVEL TELEPRINTERS

The standard version of NIBL used with the SC/MP chip, and the version in the newer 8073 chip are based on the eight-level ASCII code. Considerable work in reprogramming can be avoided if a suitable machine is used. One that is commonly available in the surplus market is the Teletype 33. It is made in many models, the three main variations being the — RO (a print-only device), the — KSR (a keyboard send-receive model that lacks the tape mechanism), and the — ASR (an automatic send-receive model that includes a tape punch and tape reader).

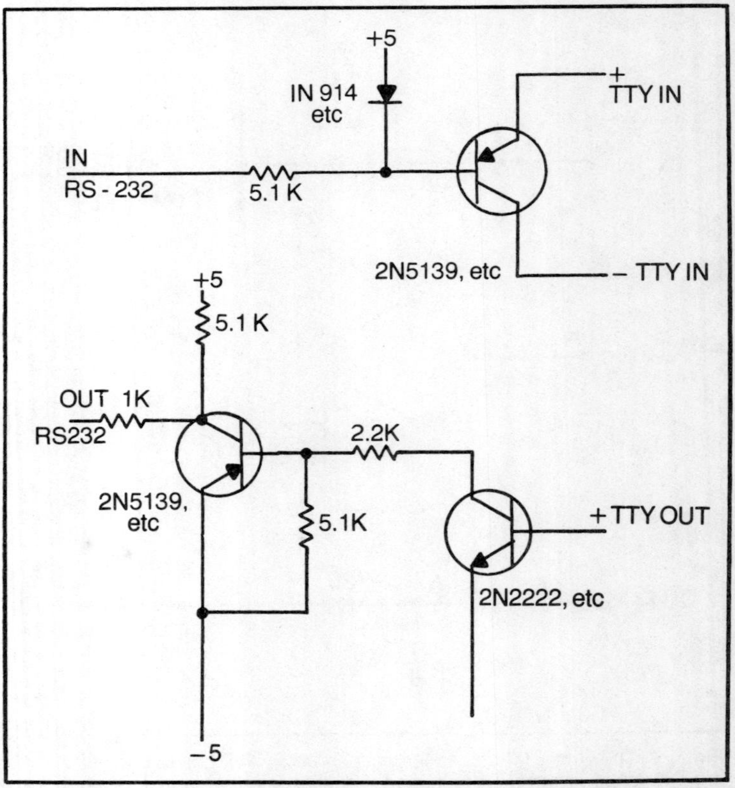

Fig. 5-10. Level changer, for converting the signals of Fig. 5-3 to those of the RS-232 standard. Other flags and sense lines may be treated in the same way, for the various signal conventions of the RS-232.

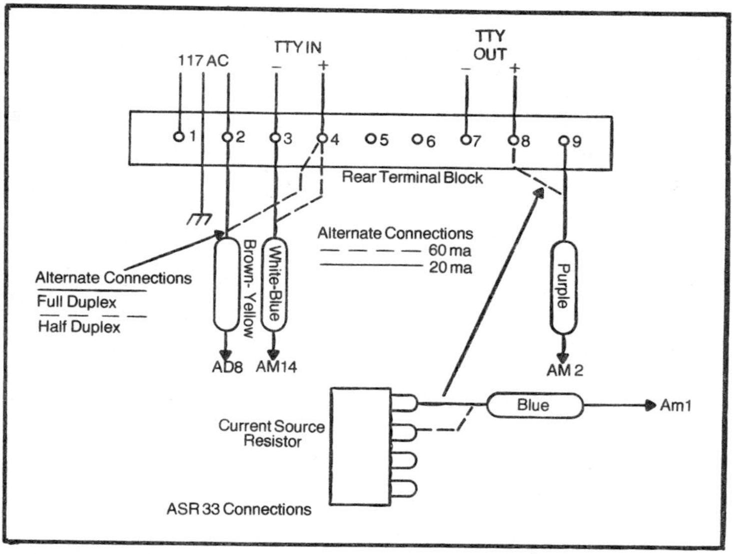

Fig. 5-11. Terminal board connections for using the ASR-32 or ASR-33 Teletype machines with the interface circuits of Fig. 5-3. Connections to other machines follow the same principles.

The most useful, of course, is the 33-ASR. The KSR is usually easier to find at a lower price. Separate tape punches and readers are available, but scarce. The RO model is good as a page printer, and, with a TVT or separate keyboard, perfectly satisfactory.

For the keyboard-page printer models, the external connections are shown in Fig. 5-11. Check the current by measuring the voltage drop across a 100-200 ohm resistor before making the connections.

These machines are considered to be a light-duty design, suitable for about two hours of use a day. Maintenance is recommended each 1500 hours, and full overhaul after 4500 hours of typing. Some units include an hour meter, which indicates motor running time. Typing time will usually be some fraction of this, perhaps as little as 10 percent.

In addition to this design, there are others that use the ASCII code, including some that are as rugged as the older heavy-duty Baudot designs. Since they never wear out they are not easy to find used. And there are a large number of computer grade I/O devices, ranging from light to continuous duty designs. Good buys are scarce, but keep your eyes open at computer fairs, hamfests and even auctions.

ADDING MACHINE PRINTERS

Surplus electrically controlled printers intended for adding machine use are becoming common. Typically these will print 16-30 columns, some with numbers and a few symbols, others in full ASCII. The latter usually use a dot matrix format.

The *SC/MP Application Handbook* includes an interface circuit to drive a Seiko printer. This is quite simple, requiring eight integrated circuits plus a few other parts. Special ICs for driving print control magnets are available. (See the application notes for ideas.)

NOTE ON RECORD PERMANENCE

Permanent is a relative term where record storage is concerned. The thermal print papers are developing a bad reputation, fading completely when exposed to ultraviolet, or even when protected. Also, they run with some glues, and may seal themselves to vinyl. If you use this type, plan on making a copy of everything you want to keep.

Impact type is usually permanent, but brittleness and color change is common in paper exposed to light. Notebooks are usually all right if not kept in full sunlight.

6

A Tape Cartridge Data System and Its Controls

Remember that the decision to forgo the floppy disk route included a decision about the substitute to use. Actually, there are several possible substitutes. One, paper tape, has already been mentioned.

A very good operating system can be built up using paper tape, and at one time this was common. Many microcomputers using, say, the PDP-8, are built around this technique. The main thing that should be added for good paper tape operation is an optical reader, giving a really marked increase in tape input speed. Such units have been on the market for some time and are not difficult to build.

However, the paper tape system includes a drawback you are trying to avoid. This is the need for manual handling of the tapes. Some improvement is possible by providing more than one reader, with tape in endless loops, but this is not too easy with long tapes which tend to tangle. The extra readers, of course, are an expense.

The manual handling needed is reduced markedly with cartridge magnetic tape. And it has the same advantage over cassette magnetic tape. The multiple track capability of cartridges markedly reduces the need for duplicate readers.

There are, of course, disadvantages in cartridges. Magnetic tape, in general, is fragile, and must be handled carefully if dropouts and jammed cartridges are to be avoided. There is also the disadvantage that very little published design data is available, so some pioneering must be done. However, overall, the magnetic tape cartridge system is the best substitute for the disk systems.

CARTRIDGE TAPE SYSTEM DECISIONS

The first system decision that must be made is the type of recording to use. Cartridges are basically designed for audio, usually for the range 50-8000 Hz, at maximum about 20-15,000 Hz. The tape is normally operated in the biased state, with bias essentially equal to the maximum output level. Good tape will have a dynamic range of about 60 dB. Noise in good tape may be 70 dB below maximum output around 100 Hz, decreasing to about 30 dB at 15-20 kHz.

Very good cartridge recorders and recorder-reproducers are available, and an operable system is easily set up using filtered audio from the simple serial (S10) board described in *How to Design, Build and Program Your Own Working Computer System*. In fact, with some care the speed of operation can be increased by a factor of about 2.5, or from 600 to 1500 words per minute, or 25 times teleprinter speed.

However, this approach doesn't satisfy the continued objective of low cost. High quality recorders are expensive, not much less than floppy disk systems, and the middle cost units still seem undesirably expensive. Low cost recorders of the type common for cassette just do not exist.

The only low cost units available are reproduce-only tape decks sold in two types. One family uses a hysteresis motor for power line operation, the other a mechanically speed-regulated DC motor for automobile use. Both types may include a preamplifier.

These play-only decks are in plentiful supply, and competition is keen. One result is that they are often found new in surplus at very low prices. Some caution is needed in buying. The autmobile type with DC motor is not recommended, since the motor speed control is rarely adequate. (The AC type is easily distinguished by the lack of brushes, and by the presence of a *phase-splitting* capacitor.) For either type, there is a good probability that a surplus unit does not meet specifications in some way. It is likely that the trouble will be in the electronics elements, rather than being mechanical in nature.

Since low cost is important, use of this type of material is desirable. This is easy if you abandon the linear reproduction needed for audio and use the method called *Saturation Recording*.

The principle is illustrated in Fig. 6-1. The magnetic field in the tape coating is shown in relation to the input current in the coil that forms part of the tape head. When all particles are magnetized, either N or S polarity, the magnetic field does not increase further as current is increased, and the tape is said to be saturated. Because the magnetic particles retain their magnetism, the relation between intensity and current is different in going from plus to minus than it is in going from minus to plus current.

For audio use, the tape is operated with a steady level of signal plus a varying component. The steady level, or bias, removes the audio distortion caused by the two branches of the magnetism curve, and also reduces the distortion due to curvature of the current-intensity curve. The bias may be DC, or it may be an AC signal well above the normal audio range.

In digital recording you do not need to worry about linear regularity, since only the values 0 and 1 exist, and since they can be set to N_{max}, S_{max}. The hysteresis due to change from one level to another can be ignored, provided the recording speed is not too great. Also, you do not need to be too concerned with the exact value of recording current—as long as it is above the saturation level.

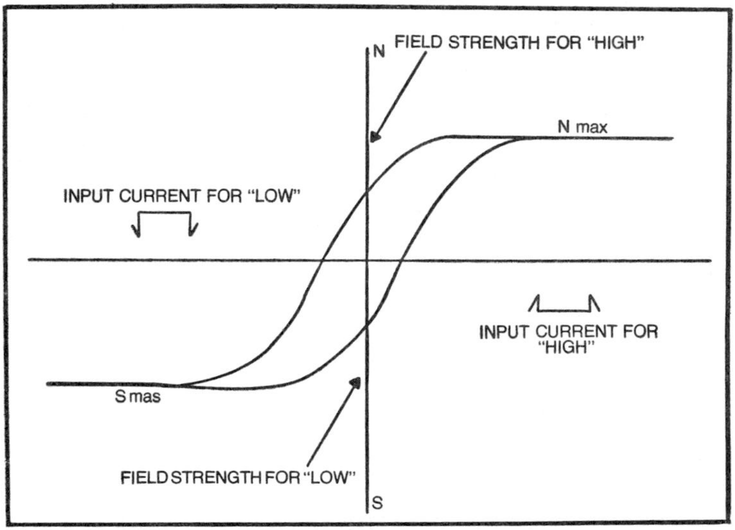

Fig. 6-1. Schematic of the residual magnetic field intensity as a function of magnetizing force for magnetic tape. Areas of interest in saturation recording are shown. The intensity is always at its maximum possible (saturated) value. The polarity does change, depending on whether the input signal is low or high.

The use of saturation recording markedly simplifies the design of the tape decks. Since record and play are separate operations, you can use the same head for both. As you will see, the driving element is also simple, since the current required by most heads is only 1 to 5 mV for full saturation. This is easily secured from a TTL gate. For playback you need an amplifier; while the output with saturation recording is several times greater than normal audio output, it is still only a few mV, which must be brought up to TTL levels.

There are a few control decisions to make. Most decks shift tracks by vertical motion of the head assembly. This motion is established by a cam and pawl arrangement, operated by an electromagnet. Channel selection must be sequential. However, nearly all tape decks have a channel-in-use indication, operated by a switch coupled to the cam. You can use this for software or hardware control.

Nearly all cartridge decks use a pair of mechanical contact fingers, closed by a strip of metal foil at the end-to-beginning tape splice for the drive for the channel change over. (Most also provide a paralleled manual switch for manual selection; this is missing in stripped down decks often found in surplus.) The contact fingers are always in contact with the magnetic side of the tape. This poses a hazard, especially with used decks, in which the electromagnet current may have caused contact pitting with scratch damage to the tape likely.

It is not too difficult to substitute an optical system of an LED and a phototransistor. However, some checks have indicated that smooth contact fingers do not greatly increase the tape wear. Pitting can be avoided by placing the fingers in a low voltage-low current circuit, with a separate drive to the selector magnet. This method is used in the prototype.

The normal deck has a pair of contacts that close when the cartridge is in place. Wiring is customarily arranged to start the drive as soon as the tape is inserted. This arrangement needs to be eliminated to allow remote start. The contacts can be used as a tape-in-place signal.

In all of the decks inspected, the cartridge is held in place by a roller mounted on a spring arm, and fitting into a V section of the cartridge. There is no method of pressure between the drive capstan and the idler roller, except by removal of the cartridge.

It is possible to release this pressure by a solenoid pushing the roller into place, or by a reversible motor operating the spring tension adjustment. But there are complexities, so the prototype retains the simple spring latch. This imposes a user chore—at end the tapes should be partly pulled from the deck or, preferably, returned to dust-tight boxes. (Equally, good practice requires that floppy disks be removed from the machine when not in use, and returned to their protected storage.)

THE BASIC RECORD-REPRODUCE-CIRCUIT

The typical tape deck head contains the two right and left track coils, each having a resistance of a few Ohms to a few tens of Ohms. On normal reproduce, they may give about 1 mV of output, but this increases with saturation recording.

Used as a write head, these units typically require up to a few tens of mV of current to reach saturation. Since the resistance is low, they can be driven directly from TTL gates. One way to do this is to drive one terminal of the coil from a DATA line, and the other by $\overline{\text{DATA}}$. Current through the head will reverse as DATA changes from high to low. A series resistor in each line will set the current to the value that gives saturation, plus a margin.

Since the same head is to be used for read, this write circuit must be removed or disabled in the READ cycle: it should be inactive whenever it is not actually in the WRITE cycle. An easy way to do this is to write with tristate gates and to place them in the high impedance mode by a $\overline{\text{WRITE}}$ signal.

However, since there is ample drive available, you don't really need the push-pull drive this method gives, and can replace one of the TTL gates with a fixed source, the voltage being set to the average of TTL high and low levels. The resistance that set the level can also serve to set the maximum current. The driving TTL should still be tristate.

This circuit, as used in the prototype, is shown in Fig. 6-2. The divider is a 1K potentiometer, which allows fine tuning of the bias setting if necessary. A capacitor is used to remove spikes and ripple that may exist on the 5 V line. Note that this is a bridge circuit, two arms being composed of the totem pole drivers of the TTL gate, the other two of the potentiometer halves. The head coil is in the detector position.

The read circuit must do several things. It must have gain to work at the 1-2 mV input level while giving normal TTL output levels. Further, there must be no mutual interference between the read-write cycles. Finally, while the record on the tape is fully saturated N or S, the head output is zero except at the change from one to the other. Accordingly, the circuit must serve as a zero-crossings detector, with memory.

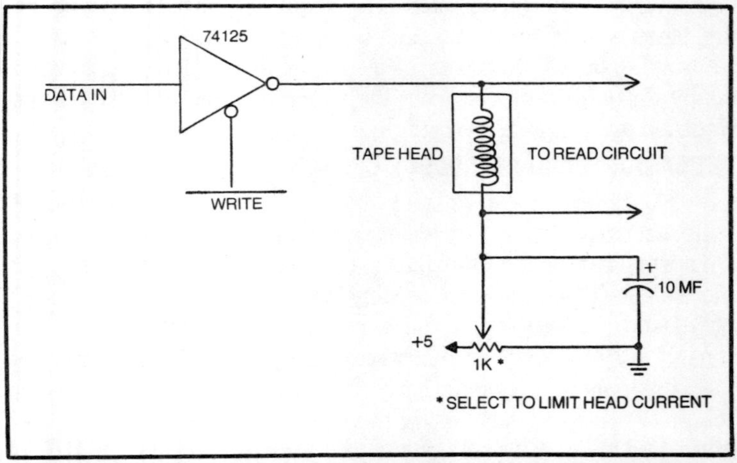

Fig. 6-2. Basic write circuit for saturation recording. Current is provided by a 74125 driver, its totem pole output stage and the 1K resistor forming a bridge circuit with the tape head coil as the bridge detector. The 1K resistor also serves to set the write current, and should be changed in value if head characteristics are different. During the read cycle, the 74125 is placed in the high-Z state.

Since recording speed is not very high (no more than a few thousand cycles of change per second) you can use operational amplifier techniques here. An easy way to accomplish this is shown in Fig. 6-3. One op-amp is used to bring the zero crossing signal to a level of about 1 V; high frequencies can be attenuated by feedback, if noise is a problem.

The second operational amplifier is operated in the regenerative mode, and is sometimes called a threshold detector. It remains in one saturated state until an input exceeds the threshold value in the opposite direction: it serves as the switch and memory element. Threshold is set by the feedback resistors. Output is saturated at about 0.7 V less than supply voltage, or −11 to +11 V. Since this is too great for TTL devices, a limiter formed by a series resistor and a 3.6 V zener diode is used to clamp the output to −0.5 V to +3.6 V approximately. This drives a tristate buffer. Output polarity can be controlled by interchanging coil to input connections.

BASIC CONTROL CIRCUITS

The simplest control circuit is for motor start-stop. Start is initiated by program control, and stop by the end-of-tape contact fingers. The action is secured by an $\overline{R}$-$\overline{S}$ flip-flop driving a relay as

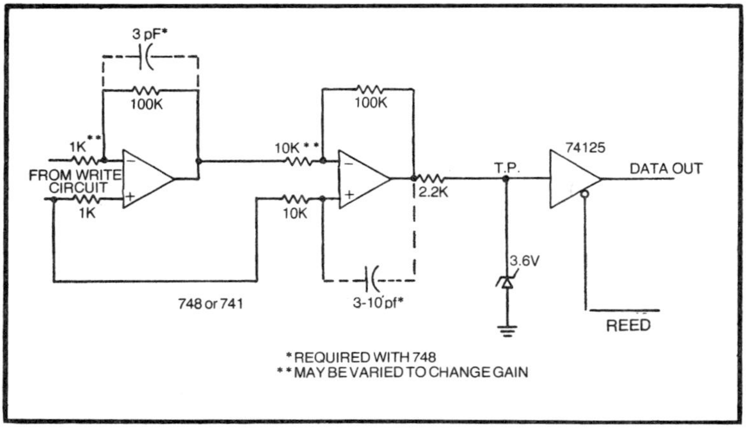

Fig. 6-3. Basic read circuit, using one op-amp as an amplifier, and a second as a regenerative wave-squaring element. During the read cycle, data is fed to the output line by a tristate buffer, its input levels being set by the zener diode, which also functions as a second wave-shaping element. Output will be a series of pulses, changing from low to high or high to low each time the tape field changes direction. Format is controlled by the write program software. Polarity is determined by the polarity of the two connections to the write circuit of Fig. 6-2. A test point can be placed at the zener diode.

shown in Fig. 6-4. The relay should be a low current type, to operate a 3.5 to 4 V with a current of 10-20 mils. A reed relay with 110 V, 1A contact rating is satisfactory. The circuit is shown with a manual switch to allow for test of occasional manual control.

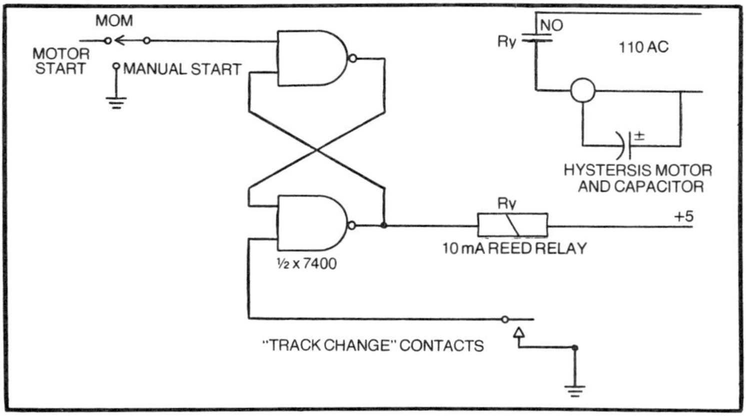

Fig. 6-4. The remote motor-start circuit with automatic stop at the *track change* signal produced by conductive foil on the tape. A reed relay is used to isolate the 5 V and 110 V circuits.

A small decrease in coast time can be secured by shorting the motor when the relay drops out. This saves a few inches of tape, and makes the start-of-recording point a little more uniform from run to run. However, this is not included in the prototype, since the coast time of hysteresis motors under load is reasonably small.

The track select is a little more complex. Make this as automatic as possible. To do this, the control input signal designates the desired track, and the mechanism cycles until this track is reached. Since the motor must be running during this period, there is a further increase in tape use to allow time for the mechanism to cycle through three positions. Use a warning to indicate that track select is not completed. The action should not cycle repeatedly at power application.

Of the many solutions possible, the one adopted for the prototype is shown in Fig. 6-5. A TTL latch is used to receive and hold the track designation, which is indicated by a low. All other track signals are high. The four designation lines are taken to the rotary switch on the track select mechanism. The common arm goes to a sensitive relay; the contacts of this controlling the 12 V electromagnet of the selector. For this circuit to operate, the rotary switch must be a break-before-make type.

When a track line is selected, all unwanted track lines are high. If the selector is in one of these positions, the magnet actuates, giving one step. The break drops the magnet. It picks up again if the following position is also high, but there is no action if this new position is the designated channel.

With two decks, the selectors can be simply paralleled, as shown. This doubles the load on the latch output, but is satisfactory if 10 mA, 2.5 V sensitive relays are used.

While the two-deck relays pick up simultaneously, the track change of the second machine does not occur if its motor is not running. This condition is normal, but cycling occurs if one deck is finishing its cycle as the second deck is started. The overlap is small, so the likelihood is low. In any event, double-track selection does no harm. The track select signal must always be given at the start of tape use.

The circuit associated with each deck relay shows a track-selected tell-tale, the integration time constant being longer than the switch break duration. High indicates that track select is incomplete.

Data line D_0 to D_3 ar used for the track select function. This leaves lines D_4 to D_7 for the start signal, deck 1 or deck 2 selection,

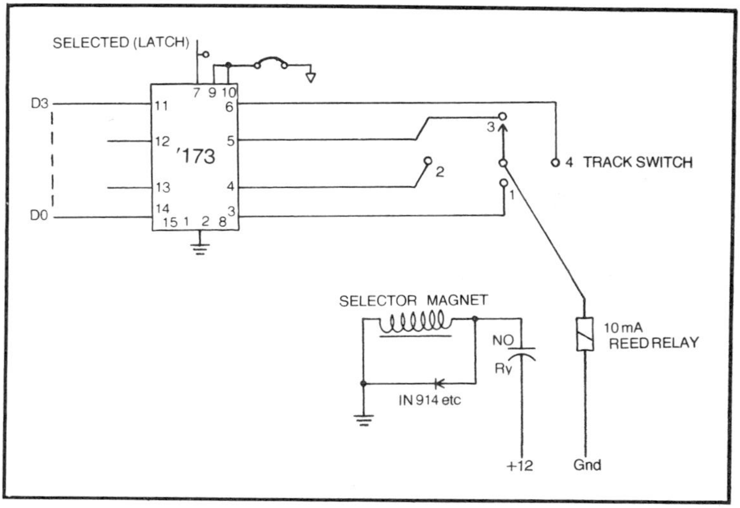

Fig. 6-5. The remote track selector circuit using a 74173 data latch. When a new track is selected, a low is latched at the appropriate track line, the current track line being set high. This picks up the reed relay, actuating the selector magnet that holds until the track mechanism steps and momentarily interrupts the relay current to allow the selector magnet to reset. This action continues until the chosen track and switch position is reached.

and right-left selection. Also, a designation of read or write cycle is needed. This must be independent of the normal read-write strobes, since in this instance read corresponds to *status check* and write corresponds to *accept instructions*.

These operation signals are of the form;

- start/$\overline{\text{auto stop}}$
- deck 1/$\overline{\text{deck 2}}$
- right track/$\overline{\text{left track}}$
- read cycle/$\overline{\text{write cycle}}$

so the necessary instructions can be microcoded into D_4 to D_7. Use a latch to hold them.

The deck, track and mode signals must be decoded to set up the proper control signals for the gates of Figs. 6-2 and 6-3. In the prototype, this operation is performed by a three-line to eight-line decoder, as shown in Fig. 6-6, which also shows the second data line latch. Only one of the eight control gates in the two decks is active, selected by the truth table included. This circuit requires that type 74125 gates as shown in Figs. 6-2 and 6-3.

Motor control circuits are shown in Fig. 6-7. The deck 1/$\overline{\text{deck 2}}$ signal and its inverse are used to steer the start signal to one of

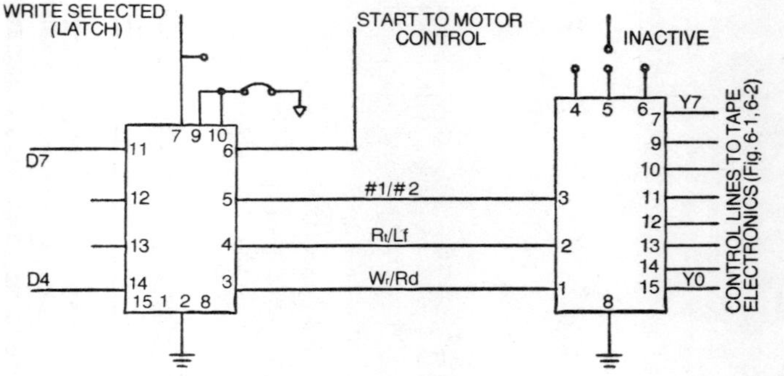

DATA

$D_0 D_1 D_2$	CONTROL LINE - USE	
000	0 RIGHT-WRITE	TAPE BANK 1
100	1 RIGHT-READ	TAPE BANK 1
010	2 LEFT-WRITE	TAPE BANK 1
110	3 LEFT-READ	TAPE BANK 1
001	4 RIGHT-WRITE	TAPE BANK 2
101	5 RIGHT-READ	TAPE BANK 2
011	6 LEFT-WRITE	TAPE BANK 2
111	7 LEFT-READ	TAPE BANK 2

Fig. 6-6. Basic I/O select circuit, using a 74173 latch to hold the select instruction word, and a 74138 three-line to eight-line decoder to generate the specific signal, bank 1 or 2, left or right, and read or write. Bit 7 of the instruction word is the motor-start signal.

the two flip-flops of Fig. 6-4. Provision is made for four status signals: deck 1 start, deck 2 start, deck 1 cycle end, and deck 2 cycle end. Also, the flip-flop status is used to generate a signal showing that neither deck is in use. This inactivates the decoder of Fig. 6-6 to prevent false data being written on the tape.

The address decode circuit is shown in Fig. 6-8. In the prototype the lower four bits are available for tape deck control functions via a buffering gate for each line. These can be available on both the read and the write cycle, or an alternate jumper connection is available to allow these to function on read only or write only. A small change in connections could give two active lines on read and two on write, and so on.

The remainder of the address lines plus the page line are used to generate a *device selected signal*. This is generated by a pair of type 7485 decoders. The comparison input of these is set to the desired 16-bit block of addresses by jumpers. These can be replaced by a DIP switch, if desired.

The device selected signal is combined with the read and write strobes to give the device strobes. The device read strobe gates status signals to the data bus, and the device write strobe latches the select signals present on the data bus, as shown in Fig. 6-6.

There is an added circuit shown in Fig. 6-8 that is not included in the prototype. This is the $\overline{R}$-$\overline{S}$ latch, plus the two spare gates. The two inputs to the latch can be connected as desired to the address buffer gates, say set to A_1 and reset to A_0. For example, the latch could be used for data input, i.e., for serial input of data on the parallel bus. Timing is generated by software. Alternately, the latch can be used for other purposes, for example, to sound a signal

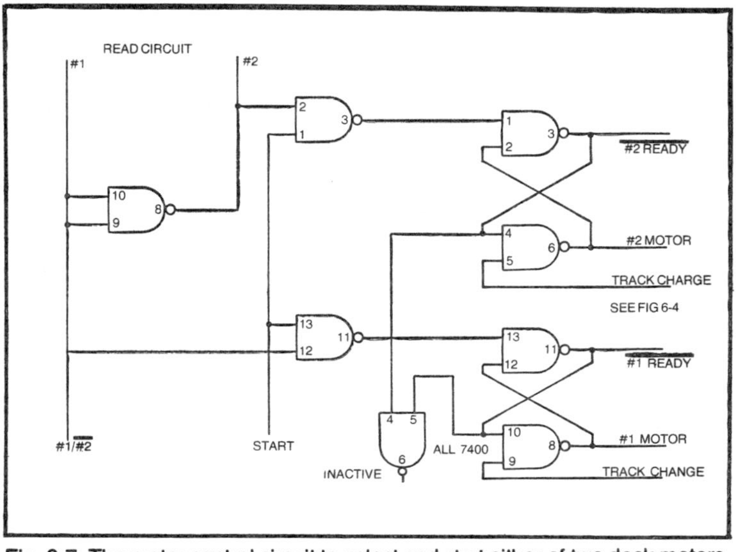

Fig. 6-7. The motor control circuit to select and start either of two deck motors. The track change signal resets the flip-flop, giving auto stop.

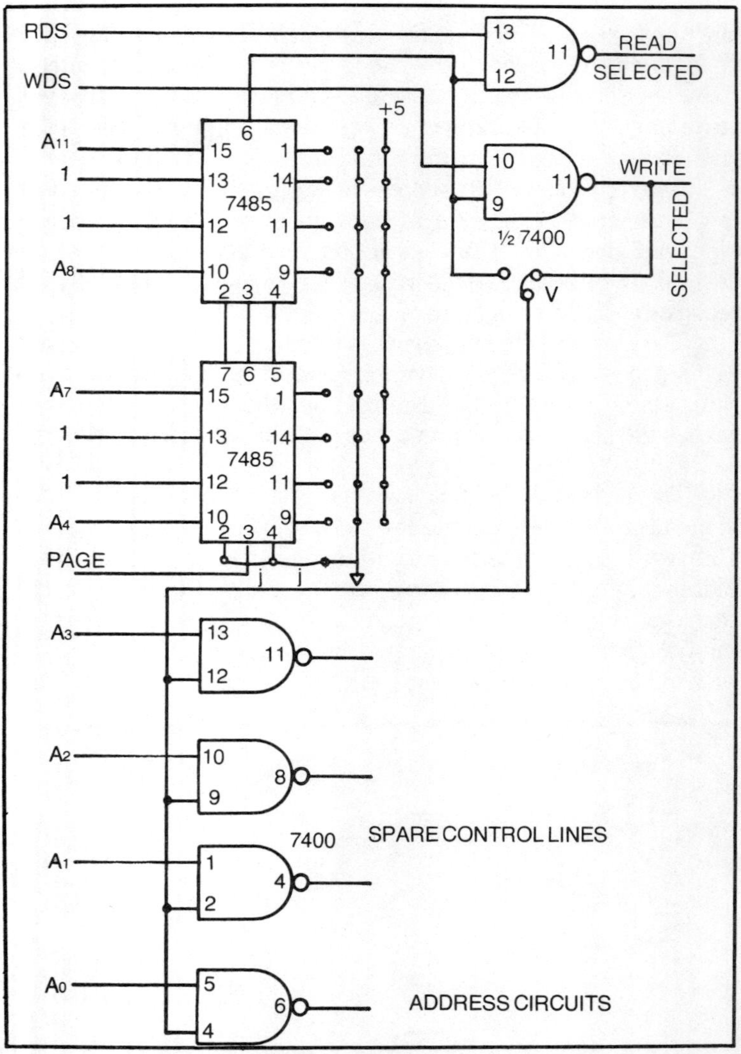

Fig. 6-8. Address circuits for the tape deck and control board. A_0 to A_3 are set up as spare control lines, with A_4 to A_{11}, plus page, being decoded against the preset address. The write strobe places an instruction into the board, the read strobe determines the board and deck status.

as a request to change a cartridge in the tape deck. Various combinations are possible.

The last circuit elements are for data output, as shown in Fig. 6-9. The data is placed on the bus by a pair of tristate buffers, controlled by the read device strobe. Four lines of the prototype are used for status signals, the other four being unassigned. If desired, one of these could be used for data, i.e., for parallel bus input of serial data. (This is no more wasteful of computer time than use of the serial I/O port). The spare latch-buffers can also be used here, for example, to indicate that the tape is to be read as hex rather than Baudot.

The reason these parallel input provisions are not included in the prototype lies in the OS plan. This calls for SIO *a la* NIBL in the

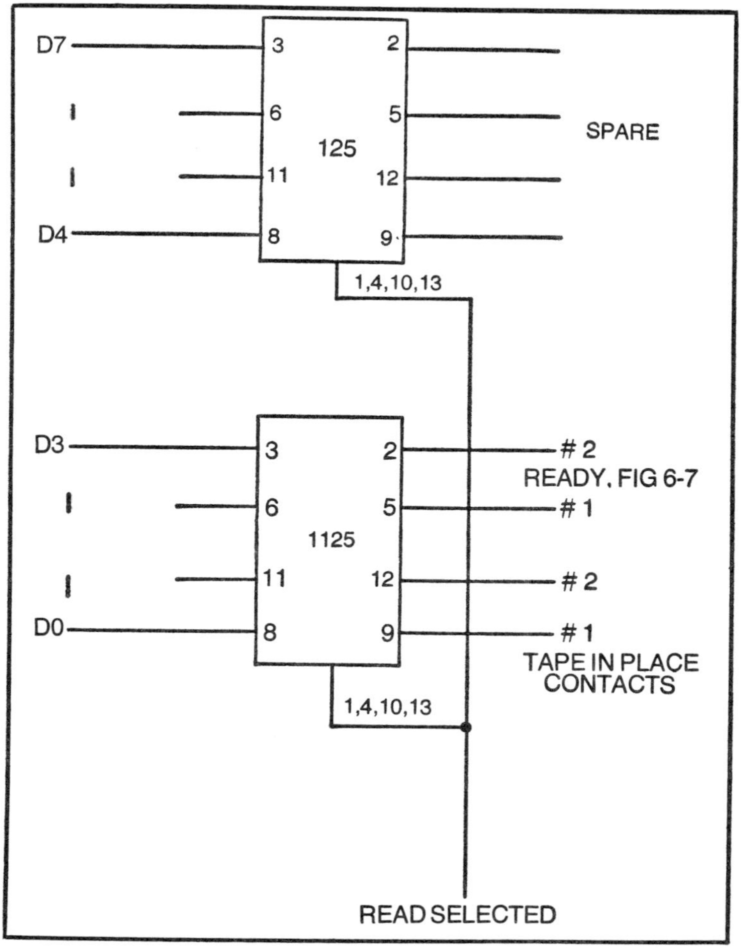

Fig. 6-9. The data read circuit uses a pair of 74125 buffers to place status data on the bus. Four lines are currently spares, and can be used as desired.

early stage, changing eventually to parallel input with parallel-to-serial conversion to save CPU time. A converter board was discussed in *How to Design, Build, and Program Your Own Working Computer*.

BOARD CONSTRUCTION

Because of the large number of interconnections between the tape deck and the control circuitry, the elements shown in Figs. 6-4 through 6-9 are placed on a double-ended board similar to the 4K RAM board. This makes more pins available, and separates the tape signal from the data and address buses.

Board construction is the same as the memory boards, using jogged socket pin connections for the leads on the component side of the board. Board layout is shown in Fig. 6-10, and the completed board in Fig. 6-11.

Because of the low levels involved, the record-play back buffer amplifier elements of Figs. 6-2 and 6-3 are placed on a separate board. The layout used in the prototype is shown in Fig. 6-12. This may not be usable for all decks, since the board size was determined by the mounting space provided for the normally used tape signal preamp. Figure 6-13 shows the board. It is no problem to change the layout to suit another mounting.

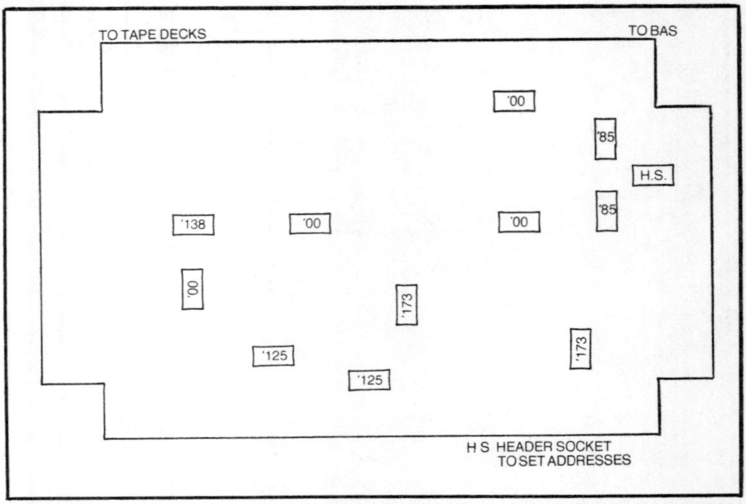

Fig. 6-10. Layout of the tape deck control board. It is double ended to keep the main bus separated from the control lines.

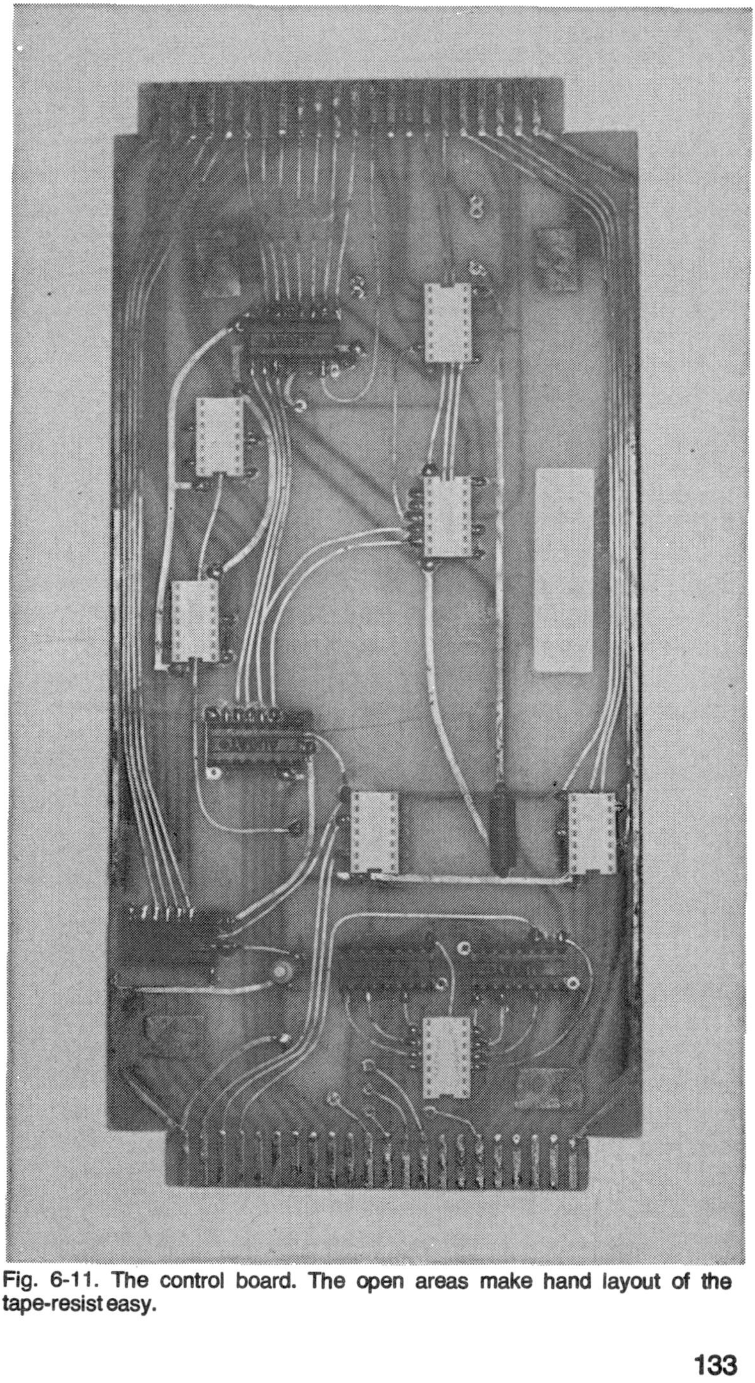

Fig. 6-11. The control board. The open areas make hand layout of the tape-resist easy.

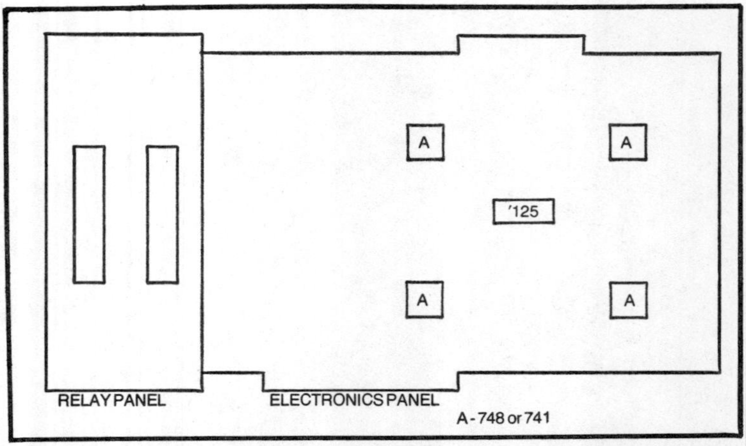

Fig. 6-12. Tape deck relay and electronics panel layout. The boards are fabricated as a unit and cut apart after etching. Board size and shape is determined by available mounting space.

The relays for motor start and track select are mounted on a small piece of PC board, Fig. 6-14. The tape decks used in the prototype has a clear area with mounting holes just the right size for the reed relays. Since these are not critical circuits, they can be mounted anywhere that is convenient.

INTERCONNECT CABLING

With two decks to connect make them separable and plug-in. The leads from each deck are provided with three plug connectors, one a two-wire for ac, the second a 15-pin for the panel of Fig. 6-12, and the last a 12-pin type for all other connections (basically those for motor start and track select). The two multi-pin connectors used are the widely available polarized nylon type made by Molex. The mating receptacles are fed from the second card edge connector of Fig. 6-10.

The pin assignment used in the prototype, and the numbering system, are shown in Fig. 6-15.

TAPE DECK PREPARATION

The tape decks used in the prototype are surplus, advertised as "new, unused, with nonfunctional electronics"—and are just as advertised. All major structural parts are stamped sheet metal with numerous special brackets and lugs formed during stamping. The various stampings give close tolerance fits. The construction is characteristic and differs only in details from other decks examined. See Fig. 6-16.

The first step in preparation is to remove the electronics board, the array of track-in-use lights, and the wiring associated with track change and motor start. Try to avoid altering the adjustments, especially those of the head arm, and the tape guides. It's wise to place a pair of scratch marks on each joining surface as a guide in case disassembly is ever necessary.

There are really only three things to test prior to installation. One is the motor, for which a simple run-no run test is sufficient. Connections are to the junction of one motor lead and the capacitor for one side of the line, and to the isolated motor lead for the other side of the line.

Fig. 6-13. The tape deck electronics panel. The unit is mounted on the deck.

Fig. 6-14. The tape deck relay panel in its mounting location.

The second thing to test is the track change mechanism. In the prototype this requires 12 V DC for actuation. (Some decks use an AC solenoid, indicated by the presence of a heavy conductor across one-half of the pole face.) Track shift should be positive and repeatable in a cycle of four shifts.

If you have a stroboscope, check the motor speed change during track shift, and with a cartridge in place. A small speed decrease during shift is okay, but the tape should not introduce a change. If it does, or causes a marked decrease, first suspect the phase shift capacitor—it probably has lost capacitance. Or, there may be an alignment problem. On some units the motor rotor or fan shows stroboscobic effect under a fluorescent light; a satisfactory stroboscope can be made with a neon bulb and an audio oscillator.

The last thing to check is the head. A resistance check with a low current ohmmeter (1-10 mA on low scale) shows if the winding is open. For a better check connect the head to an audio amplifier and check for good reproduction of a prerecorded cartridge.

If you have a test cartridge, or can borrow one, it is a good idea to check the head alignment at this time. The major things to check are cross-talk caused by improper vertical setting of the head, relative delay between the two channels caused by the two head gaps not being at right angles to the tape, and low volume, which is often caused by insufficient pressure against the tape pad. Follow the instructions with the test cartridge. Lacking a test cartridge, use a good stereo cartridge to test for cross-talk, and a piano recording to test for speed variation.

	Molex 12-pin Power-Control Connector	Molex 15-pin Signal Connector
1	Select 1	Left input
2	Ground	Left output control
3	Select 2	Right input
4	Select 3	Left input control
5	+5V	Left output
6	Select 4	Right input control
7	Select common	Right output control
8	Tape in place	Right output
9	End of tape	NC
10	NC	NC
11	Track change magnet	Ground
12	+12V	+5V
13		NC
14		−12V
15		+12V

Fig. 6-15. Assignment of pins in the 12 and 15-pin Molex connector pairs used in the prototype. The electronics circuit package has its own connector. In addition to these there is a 110 V two-wire connector.

Actually, if you do not intend to exchange tapes between decks, or with another installation , this check can be omitted. If the tape position is the same from run to run, the cross-talk problem does not arise. Head rotation is also relatively unimportant as long as the same deck is used. However, good practice makes sure the head is set to the standard position.

HEAD AND OTHER ADJUSTMENTS

While there may be variation between decks, the following adjustments were found in all the decks I have inspected.

The most obvious adjustment is the motor height relative to the flywheel. It is not highly critical, but too much misalignment introduces unnecessary friction and drive belt wear. The adjustment is easily checked with a steel rule or a depth gauge. It is a good idea to measure and record the belt data (you might wish to keep a spare set on hand).

The second adjustment is the spring tension for the roller that holds the cartridge in place. Insertion and removal are easy, but the important thing is steady pressure against the capstan. Securing this requires the roller to be on the side of the V that is toward the capstan. It cannot touch the opposite side. Position adjustment is unnecessary since the spring has a position lock on all decks I have inspected.

There are several adjustments for the head arm. At some location on the arm, usually near the center, is a screw that adjusts the height of the head with respect to the tape. The arm itself can have a height adjustment at the pivot point, and has some provision for rotating the arm to place the head gaps at right angles to the tape. Finally, the head itself is held in the arm by a clamp. There may be a depth stop, or a depth mark.

The decks I inspected have one or more slots that can be used to check alignment. The head can also be reached from the front. A steel rule with a square end shows if the head gaps are at right angles to the bed. If the end is graduated, it can be used to check the height setting. A depth gauge can be used to check the relative position of the capstan, head, guide and tape end fingers.

Record these values before doing any work. Then if it is necessary to do any disassembly, or any repair later, the deck can be brought into preliminary alignment. For final alignment, use a test tape.

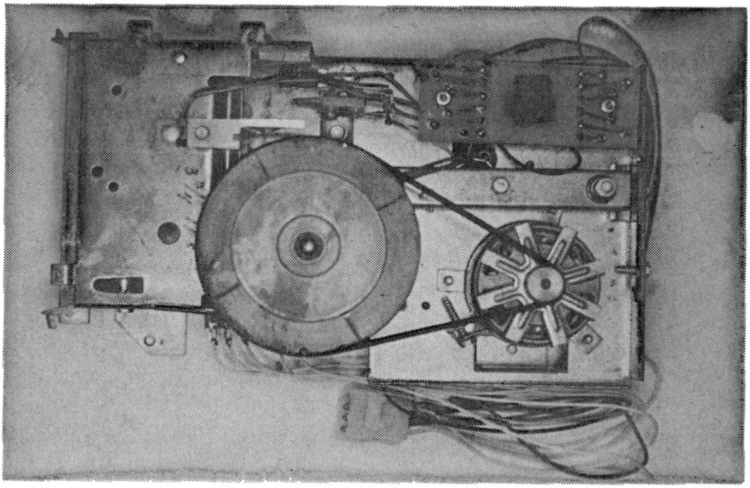

Fig. 6-16. Photo of a completed tape deck.

TAPE

While any old cartridge is good enough for initial tests, only the best tape should be used for operations. It should be prestretched (*tensilized*) Mylar, and not the thin grade: about 1 mil is good. Avoid the thin long-play tapes, partly because they are subject to more stretch, and partly because they tend to jam.

Tape cartridge cases are made in several styles, some being cemented together. These are best avoided, since a jam cannot be cleared. (The only recourse is to tranfer the tape to another cartridge.) Of the types that lock together, those with three lock fingers are easier to open and more rugged.

If you follow suggestions already given, you want cartridges of two or three tape lengths and playing times. Typically, one minute of time, or about 20 feet of tape, will hold 4K per track. Typical use would be:

Short tape	1 minute duration	1 page data
Inter mediate tape	4 minute duration	4 pages data
Long tape	10 minute duration	10 pages data

The average access time for the long tape is five minutes—not unreasonably long.

If you can find tapes of the correct length, it is best to use them. If not, you will have to remove some of the tape, which will require opening the cartridge. Be sure you note the internal arrangement and get the parts the same way. Be careful handling the tape to keep it from picking up dust. A wastebasket with a clean plastic bag inside is good. Measure off the amount of tape you need and rewind it. Make sure the coated face is out. Then make the splice. Special splicing and contact tape are carried by hi-fi stores.

Tape cartridges *must* be released from the deck when it is not in operation to prevent flattening the roller. They should be placed in a dust-proof box, along with the rest of the tape library. Keep this out of direct sunlight, away from other heat, and away from magnetic fields. With reasonable precautions, the cartridges should be good for years of service.

Don't forget—dirt, dust, and even smoke are enemies of error-free records.

FORMATTING THE TAPE RECORDS

The format of the tape record is controlled entirely by software and can be changed at will. However, changes require additional programming, and probably a conversion program plus rewriting of tapes. It is best to select a good format and stay with it.

With one exception there is no real reason to depart from ASCII character format, and many reasons for staying with it. While Baudot is indicated for printer and paper tape use, this is a matter of cost, and it is likely that these older machines will be replaced as opportunity permits. Baudot is immediately converted to ASCII; all internal operations are ASCII or hex. The exception to this is provision of an alternative read/write mode that interprets the eight data positions of ASCII as an eight-bit binary word rather than as a character. However, the set-up for this is easily handled by software, so the format remains unchanged.

The *SC/MP Applications Manual* describes a cassette to SC/MP system interface, and gives an example program for it. The format used with this program is

- Leader of zeros
- Preamble—identification character
- Start address
- Transfer address
- Word count (complement)
- Data block
- Suffix-check sum

In this format the identification character is chosen when the program is written and remains unchanged. Its main function is a precaution against inadvertent use of the work of another programmer.

When the data is loaded, it is checked with respect to word count, then with respect to the check sum. These do not guarantee that there are no errors, but the probability is very low. If correct, program control is jumped to the transfer address. This satisfies most of the needs of the operating system, since the tape can specify return to the keyboard monitor, to an immediate execute, to a machine-language routine, and so on. The desired mode can be specified when the tape is being written by parameter passing.

There is one extension that adapts this format to the planned tape use. It replaces the leader character, zero, by a hex character, with zero being reserved for the single data block or short tapes, and with 1-15 hex available for successive blocks on the intermediate or long tapes. This numbering can be used to catch a programming error, or a specification of the wrong machine, and also allows a simple search for the desired record.

There is a decision that must be made regarding segment. Variable length allows more records, and will give the shortest possible access time. However, records in blocks of 4K fit the SC/MP architecture, and a constant length block allows a timed interrupt to point to the desired block. Such an interrupt system can save valuable CPU time, and is a worthwhile future addition. Thus, a constant length 4K block is best. The block description is summarized by Fig. 6-17.

Automatic shutoff at the end of each does not really change the average access delay, and it prevents use of block 1 on an immediately available basis. The automatic stop just prior to block 1 is preferable.

PROGRAMMING FOR TAPE OPERATIONS

The elements of the OS devoted to tape operations can be as simple or as complex as you want. At one extreme, you can use manual operations entirely, starting and stopping the tape recorder in response to printer prompts. In the simplest form these operations are at teleprinter speeds, and the magnetic tape signals are fed to the serial input circuit. This is the form of a relay in series with the keyboard and tape reader contacts.

With a little more complexity and by rewriting GECO and PUTC routines for shorter timing loops, the operating speed can be increased up to about 1200 baud. These routines could be set up by a LINK command from the keyboard, with a prompt as needed. Extension of other routines can handle hex recordings.

However, the control panel associated with the tape deck is intended for complete control from either the keyboard or a program with several degrees of complexity. The ultimate uses the "name operation-named unit" provided by many high-level languages. READ INPUT TAPE i, n, list is typical form. However, adding these instructions requires rewriting of NIBL, plus extensive additions. It seems better to take a simpler approach through the LINK command with paramater passing.

To handle the control of the tape units, first establish an address block.

Put it on a page reserved for device control. Within the limits established by hardware design (i.e., lowest four bits are usable for microprogramming) the address is arbitrary. It can be changed if necessary.

Second, form a data word; the microprogrammed instructions for the particular operation. Use this form:

D_0)	one high to select track,
)	
D_1)	others low
)	
D_2)	
)	
D_3)	
D_4	high = write, low = read
D_5	high = right track, low = left track
D_6	high = unit1, low = unit 2
D_7	high = set motor relay, 0 = idle

This word must be passed as a parameter (or variable) from the calling program, or the keyboard, to the executing program.

Third, specify the tape segment required. You can simplify this by first recording a chain of null segments (i.e., segments that are correct length and are identified, but have no information in the data section). This permits use of the same segment identification routine for write and read cycles. When the tape segment is located, a simple mark operation on the data word determines jumps to the correct routine.

Finally, pass the start address of the data, the number of bytes if in record mode, and the return address to be used at the

TAPE FORMAT		
Section	**Length**	**Use**
1	128 bytes	Block number 0-15
2	8 bits	Identification number
3	16 bits	Start load address
4	16 bits	Return address
5	16 bits	Length of block
6	Max 4K bytes	Data
7	8 bits	Checksum
8	Min O bytes	Blank to standard block length

Fig. 6-17. Format of a tape record block, up to 15 being allowed per tape. Each block is divided into eight sections, five forming a preamble, and one holding data. The suffix contains a check-sum value to allow error detection, plus a blank block to fill out the block to standard length. This allows timed interrupt, and frees the CPU for other use. Block 8 can be omitted if the operating system is set up like this.

completion of the routine. Normally this is passed by P3 by a LINK command.

The necessary instructions easily fit into one line. The format is:

1000 @ XNNNN = data word; @ XNNNN = low address; @ XNNNN = high address; @ XNNNN = Record length; LINK XNNNN

The routine called by the LINK instruction must first decode the data word, and initiate recorder operation. Then it searches for the record segment, and when found, proceeds with read or write operations, adding or reading the prefix and suffix. On completion, it returns to the calling location. The logic for this is shown in Figs. 6-18 and 6-19. These can be used with the example program in the *SC/MP Applications Handbook*, when writing the OS program.

The special program to prepare a completely blank tape for use can be written into the read/write program, or it can be separate. The latter is probably better. Its major divisions are timing segments for the leader, the program segment identification, and the program block, plus a write routine to place the segment identification on the tape. This program could be written in NIBL to call the read/write routine repeatedly.

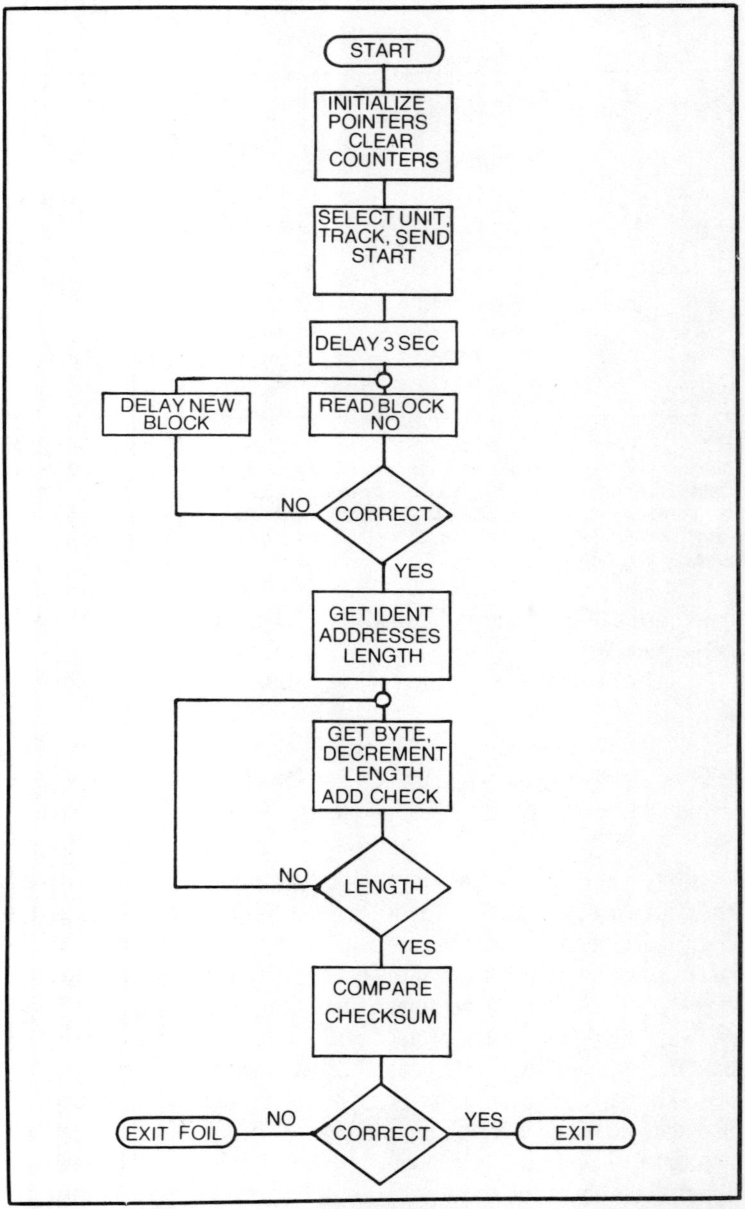

Fig. 6-18. Flow diagram for the operating system tape read section. Three seconds are allowed for track selection. The delay-new-block can be by a loop, or by continuous search, or by a separate time. Good load and bad load exits are indicated.

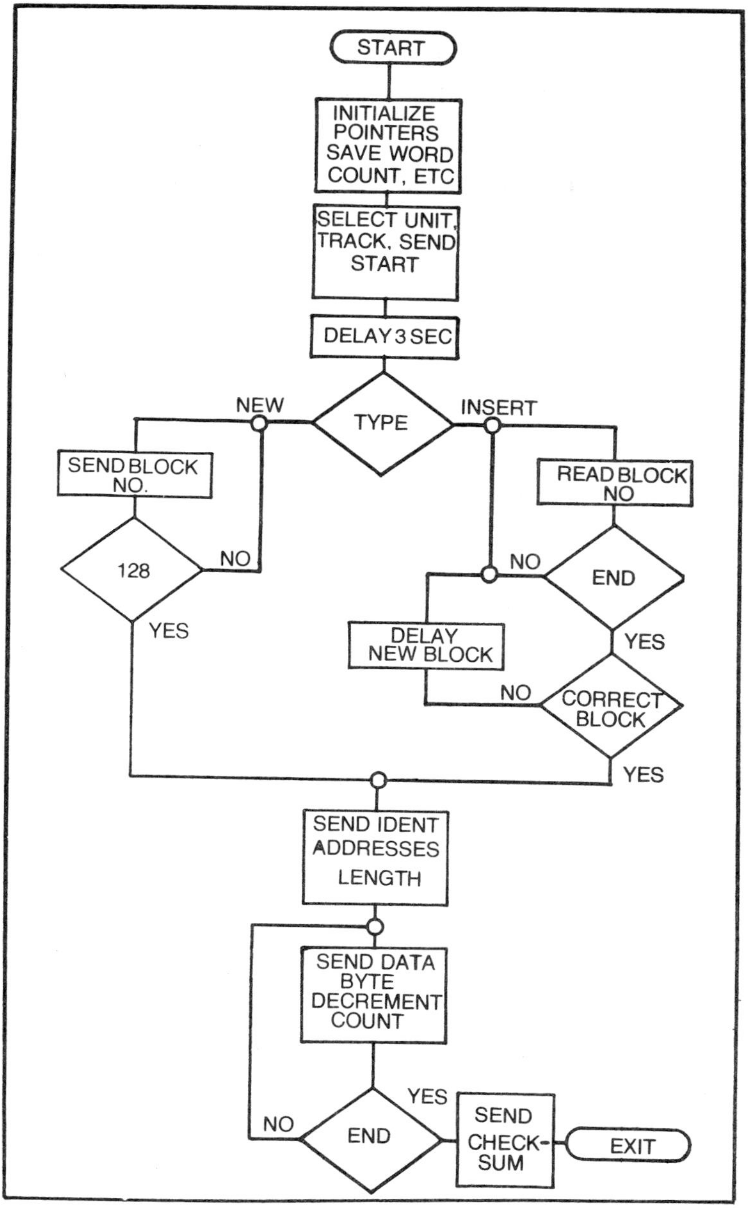

Fig. 6-19. Flow diagram for tape write. Provision is made for new tape generation, and insertion of a block into an existing tape; one reason for maintaining a fixed block length. After writing, the tape can be read for a byte by byte verification, or for a check-sum verification.

MAINTENANCE

Maintenance requirements are simple. After about 10 hours of use, wipe the face of the head, the capstan and the guide with isopropyl alcohol to remove any oxide build-up. A long, cotton-tipped swab is ideal for this. If there is still a stain, the head can be cleaned with one of the special head cleaning cartridges available. However, do this as a last resort, as the cleaner is abrasive.

About once each six months, light lubrication of the motor, bearings and moving parts is a good idea. Sewing machine oil and medium viscosity white grease are suitable.

As the parts of the deck wear, it will become noisier, and errors will start to show up. These decks are not designed for maintenance, and if the usual adjustments do not clear the problem, replacement is necessary. This should not happen for thousands of operations, however.

It is a good idea to change drive belts each year or so. Check the capacity of the phase-splitter capacitor at this time, and replace it if necessary. These operations help hold the error rate down.

ALTERNATIVES TO CARTRIDGE RECORDING

If you want to avoid the high cost of disk, and also keep the many advantages of cassettes, there are two other approaches you

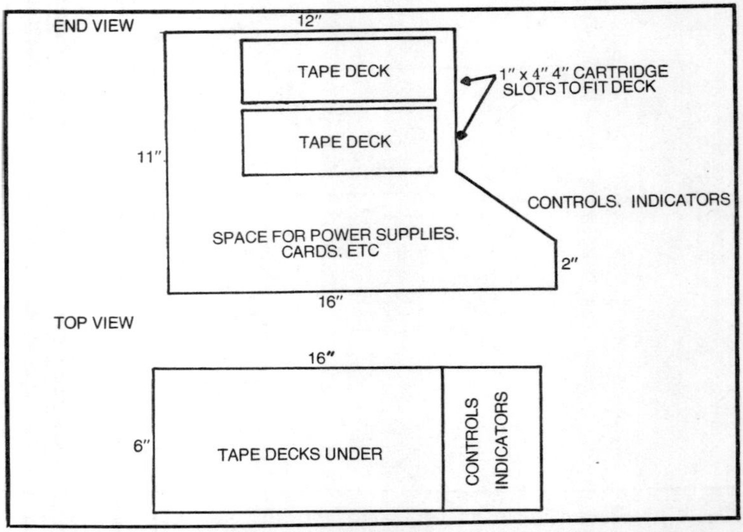

Fig. 6-20. General layout of the tape deck cabinet of the prototype, made to match the cabinet of the earlier book. Manual controls and indicators are not really necessary, but are convenient and worthwhile in the event of trouble.

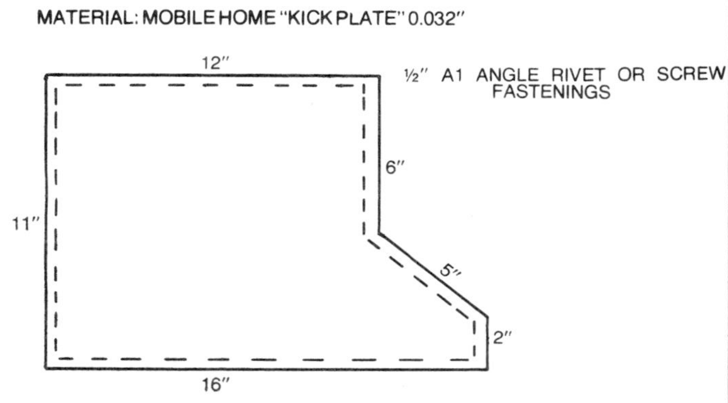

Fig. 6-21. Dimensions of the cabinet end panels. They are identical to the main cabinet ends.

can use. One is to purchase one or two of the special three motor cassette mechanisms made for digital data use. With these and a modest control system you can read tape, rewind it, and have fast forward to skip segments. With a little complexity, you can add read-in-reverse, and scan for a named segment.

As an alternative, you can add a bank of electromagnet actuators to a cassette deck to place operations under remote control. A sophisticated design actuates the latch fingers directly, a simple one operates the finger keys. This approach is not recommended unless you have good machine ship facilities.

MOUNTING THE TAPE DECKS

The cabinet used for the prototype SCMP:Y is hand constructed with no provision for a tape mechanism, since use of a self-contained cartridge recorder was planned. While the tape decks can be left in the usual 4 × 6 × 10-inch cabinets, these offer no shielding, and are not as convenient as they could be.

The approach adopted was to add a small cabinet of the same construction as the original, but cut down in width to suit the tape decks. As shown in Fig. 6-20, two decks are contained in the upper part of the cabinet. The front panel is used for manual control switches and indicators.

This cabinet is permanently fastened to the original cabinet; small spacers are used to give screw head (or rivet) clearance.

Figure 6-21 shows the end panel layout, and Fig. 6-22 the other panels. These are identical to the original design except for length.

The location of mounting brackets for the tape decks are shown in Fig. 6-23, as determined for the original, and may need

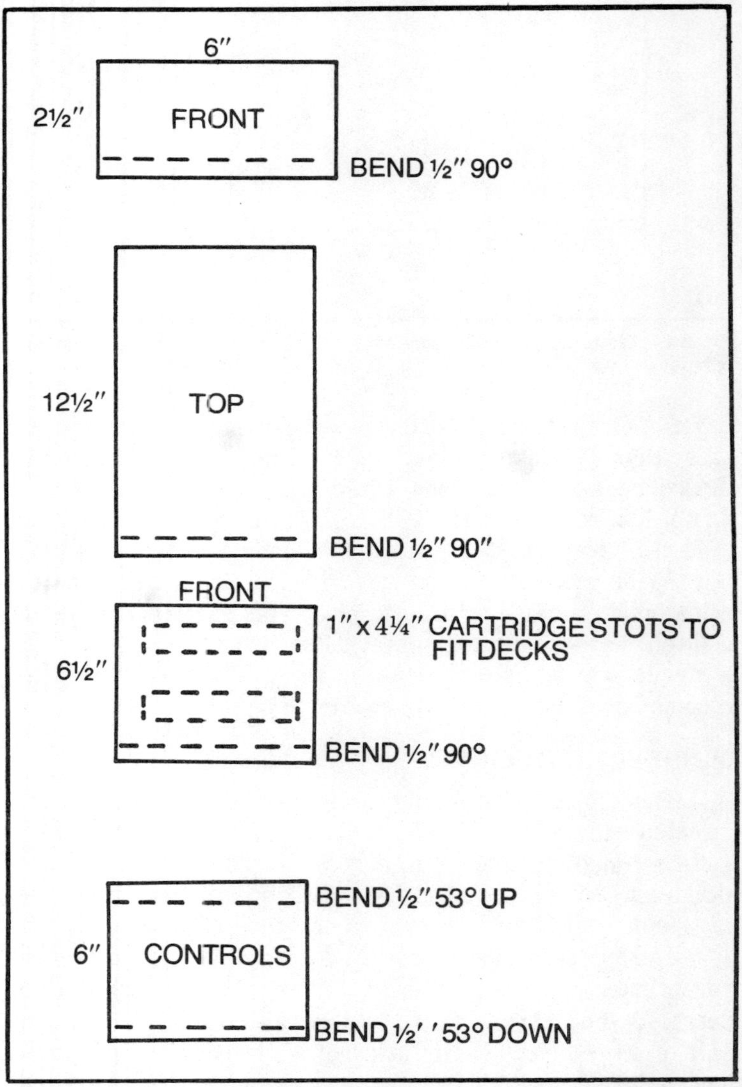

Fig. 6-22. Dimensions of the four pieces that make up the cabinet panels and top. The back is open. They are identical to the panels of the main cabinet in height, but are six inches instead of the 16-inch length of the main cabinet.

Fig. 6-23. The tape deck assembly mounted the main cabinet to make a single unit. Permanent attachment is not necessary. If the units are separated, the electronics panel should be in the main cabinet to avoid excessive bus loading.

modification for other decks. In the prototype, the brackets are 1 × 1-inch aluminum angle, dulled for deck mounting screws. It's easiest to make the decks removable from the rear. The front panel is slotted for cartridge insertion.

Wiring between decks is handled by Molex connectors. The receptacle is passed through a wall slot cut in the original cabinet, rotated 90 degree, and held in place by the wall of the added cabinet. The plug section is wired to the deck. If desired, a spare receptacle set can be set up in a test jig.

The space below the tape decks can be used for a parallel-serial converter card and the control card, or these can be placed in the main section. The space may be used for power supply expansion.

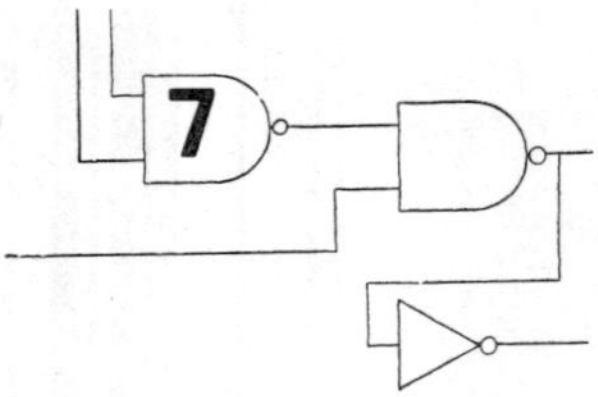

PROM Programming

One of the keys to having a comprehensive operating system in place without using disks is extensive use of ROM. Ideally, this should contain all of the commonly used OS characteristics plus commonly used tables of conversion and look-up. And, if they are commonly used, the ROM should even include programs that are modified during execution; it is faster to transfer these from ROM to RAM than from tape to RAM.

The commercial route to OS development typically starts with an "emulation program" on a large computer. This program simulates the logic processes of the development computer, making possible parallel development of hardware and software. When the OS works correctly on the host computer, it is transferred to ROM, usually a mask programmed ROM. Sometimes there is a preliminary step of PROM.

The home builder must get the same results by another route. Initially, use separately punched paper tape as the feed-in of a new program to your computer, and follow with keyboard command debugging. I assume you are using NIBL as the OS base; if this is not available, control panel manipulation may be necessary.

Since it is wise to allow for changes as the OS grows, it is better to use EPROM rather than PROM. It is also cheaper.

Commercial PROM programming is offered; your local computer dealer may have facilities. However, this service adds to expense and may take a lot of time. Therefore, a good installation needs the ability to program and reprogram EPROM. The ability does not need to be extensive, nor sophisticated, but it should be complete.

REVIEW OF PROM CHARACTERISTICS

At present only the pulse programmed, ultraviolet-erased PROM needs to be considered. The newer electrically alterable PROM are still in relatively short supply, and expensive.

There are several families of the UV from which to choose. Older families require complex program operations that must be repeated many times. Successive designs have simplified the operations and reduced the number of steps, so that the latest design offers programming with a single pulse application.

At the time the most cost effective type of EPROM is the 2708, an intermediate step of development. The older types are becoming scarce and more expensive; the later types are available, but expensive. However, since cost will inevitably drop, it's wise to make some provisions for at least one of the latest designs. The 2716 is representative.

The 2708

The operating voltages and program pulses for the 2708 are shown in Fig. 7-1. The key voltage changes that are required when data is to be written in are: (1) Raise the chip select line to 12 ± 0.5 V from its normal TTL LO/HI operation; and (2) apply a timed pulse of 26 ± 1 V to the normally grounded program pin.

The corresponding timing requirements are shown in Fig. 7-2. Data and programming pulses must be sequenced through all

VOLTAGES AND SIGNALS FOR 2708 PROM								
			Pin 12	Pin 18	Pin 19	Pin 20	Pin 21	Pin 24
Mode	D_0 - D_7	A_0 - A_9	VSS	PGM	VDD	CS/WE	VBB	VCC
Read	D_{out}	A_{in}	GND	GND	+12	LO	−5	+5
Inactive	HI-Z	DC	GND	GND	+12	HI	−5	+5
Program	D_{in}	A_{in}	GND	P*	+12	12 ± 0.6	−5	+5

LO = 0.65 Max P* Program pulse of 26 + 1 volts, See Fig. 7-2

HI = 3.0 Min

DC = Don't Care LO/HI

Fig. 7-1. Voltages and signals for the 2708 EPROM. Voltages for the supply pins change as shown for the read, inactive, and program modes of operation.

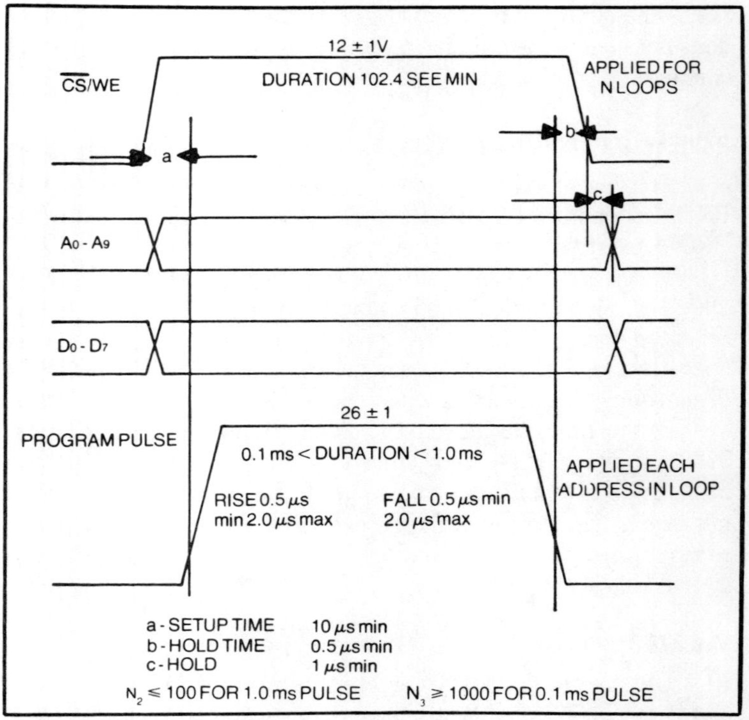

Fig. 7-2. Program pulse specifications for the 2708. This pulse must be repeated between 100 and 1,000 times to ensure correct programming. Locations must be programmed in sequence.

address of the chip and repeated until sufficient program time has been accumulated. For the 2708 this is a total of 100 ms, with the restriction that the program pulse be between 0.1 and 1.0 ms in duration.

As is common in these designs, an unprogrammed device shows all high outputs. Programming brings memory cells to the low state. Once programmed the state cannot be changed electrically. However, any remaining highs can be brought low by going through another programming cycle.

The 2716

The 2716 has a simpler operation. As shown in Fig. 7-3, only 5 V is required for normal operation, and the programming requires only 25 ± 1 V in addition. In timing, as shown by Fig. 7-4, only a single 50 ms period is required for complete programming. This single pulse can be applied to any location, selected randomly instead of in sequence. The reprogramming restriction is the same as for the 2708.

If a unit is completely reprogrammed, it must first be deprogrammed by exposure to UV light. The same is true if a programming error involving a 1 logic change occurs. There are complete specifications for the light source, but remember that *any* UV source will do the erasure if sufficient time is taken. (For room level fluorescent light the time is about three years; and for direct sunlight, about one week. Opaque covering of the only chip may be needed in some applications.) Practical erasure devices, such as one described later, produce higher UV intensity, and give erasure in about 20 minutes.

REVIEW OF PROGRAMMER DESIGN

A large number of PROM programmers have been described in literature. These fall into two basic types. In one, all timing is controlled by software. Typically, they have full address and data buffering, with latches to hold the data and program pulses until the board is addressed again. Board operations are restricted to rise and fall time control, as needed to meet programming specifications.

VOLTAGES AND SIGNALS FOR 2716 PROM

	PIN 18	PIN 18	PIN 21	PIN 24		
Mode	$\overline{CE}$/PGM	$\overline{OE}$	VPP	VCC	D_0 - D_7	A_0 - A_{10}
Read	LO	LO	+5	+5	D_{out}	A_{in}
Inactive	HI	DC	+5	+5	HI-Z	DC
Program	P*	HI	+25 ± 1	+5	D_{in}	A_{in}
P. Verify	LO	LO	+25 ± 1	+5	D_{out}	A_{in}
P. Inhibit	LO	HI	+25 ± 1	+5	HI-Z	DC

LO = 0.65 Max   DC = Dont Care LO/HI
HI = 3.0 Min   P* = Program pulse of 25 ± 1 volts, See Fig. 7-4

Fig. 7-3. Voltages and signals for the 2716 EPROM. Voltages of the supply pins and functions change as shown for the five possible modes of operation.

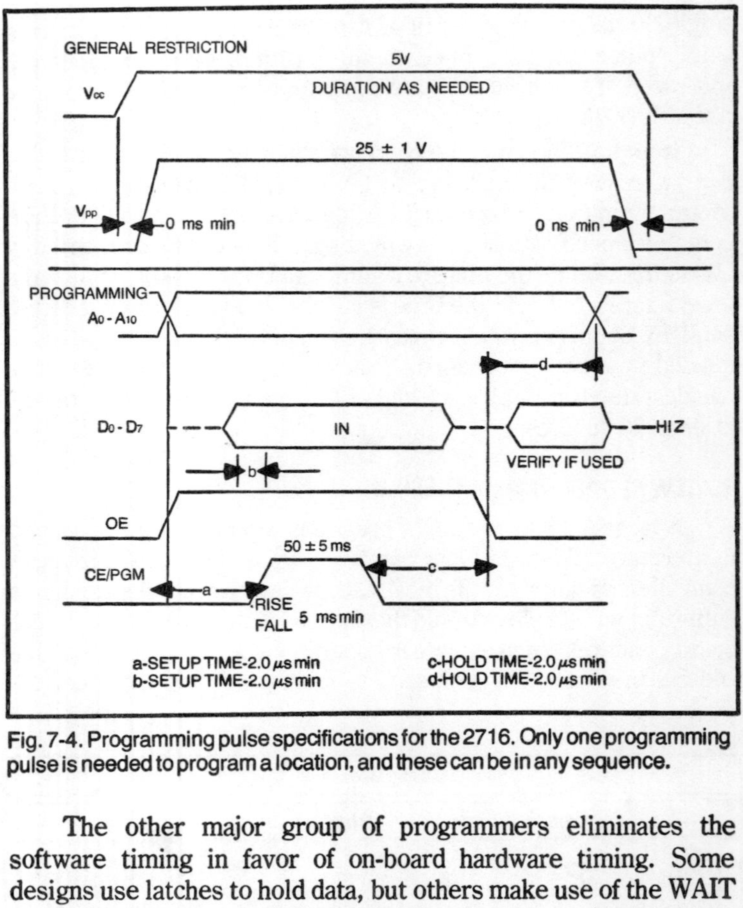

Fig. 7-4. Programming pulse specifications for the 2716. Only one programming pulse is needed to program a location, and these can be in any sequence.

The other major group of programmers eliminates the software timing in favor of on-board hardware timing. Some designs use latches to hold data, but others make use of the WAIT line to force the microprocessor to hold addresses and data for the necessary period. Compatible software design is simple, especially if the CPU has a block transfer instruction.

Since the programmer board is used infrequently, it can be as simple and inexpensive as possible. On-board timing generation can make use of the WAIT circuit. To save further complexity, the programmer board can have its own 4K address block while it is in the computer rack. All buffering can be eliminated. This is no problem in early stages of expansion but may require removal of some other boards when a PROM is burned.

Decide the degree of chip flexibility you want. See Figs. 7-5 and 7-6. The two most likely chips are not plug-interchangeable. The least complex approach may be to provide a second socket mount on the board, leaving connection until it is certain that a particular PROM is to be used.

The following material is based on the prototype, which basically provides for the 2708 type with flexibility for adaption to the 2716 or 2732 types.

EPROM BURNER CIRCUIT DESIGN

The core of a simple EPROM burner is the timing circuit. See Fig. 7-2; there are three parts referenced to the chip address lines:

- A minimum of 10 μs before application of the program pulse
- A pulse between 0.1 and 1.0 ms
- A minimum of 1.0 μs before removal of the address

There is a simple way to generate these. Use a delay multivibrator for the 10 μs lead delay, and a second one for the program pulse itself. The output of the second can be used with the hold circuit of the CPU to keep the address data and write pulse stable during the pulse period. The delay can be held for a short period after the end of the program pulse period by a simple RC integrator.

The timing circuit is shown in Fig. 7-7. At the beginning of the write pulse an $\overline{R}$-$\overline{S}$ flip-flop is set to indicate the hold. At the same time one-half of a 74123 one-shot multivibrator is triggered to give the 10 μs delay. When this times out, it triggers the second

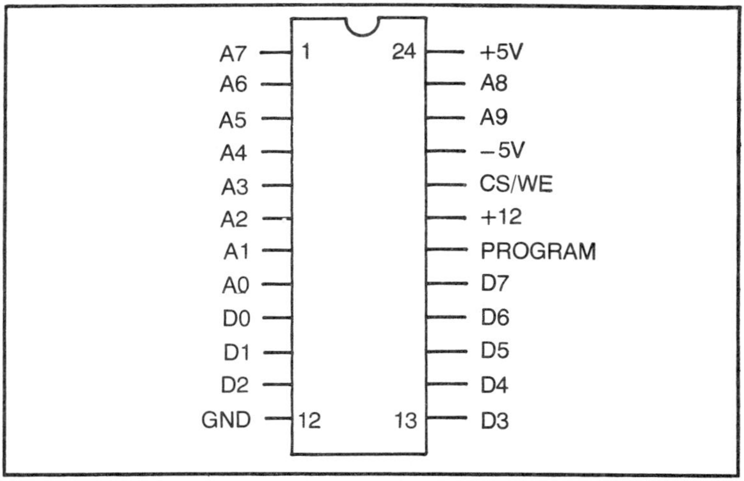

Fig. 7-5. Pin connections for the 2708 EPROM. This is a top view of the package.

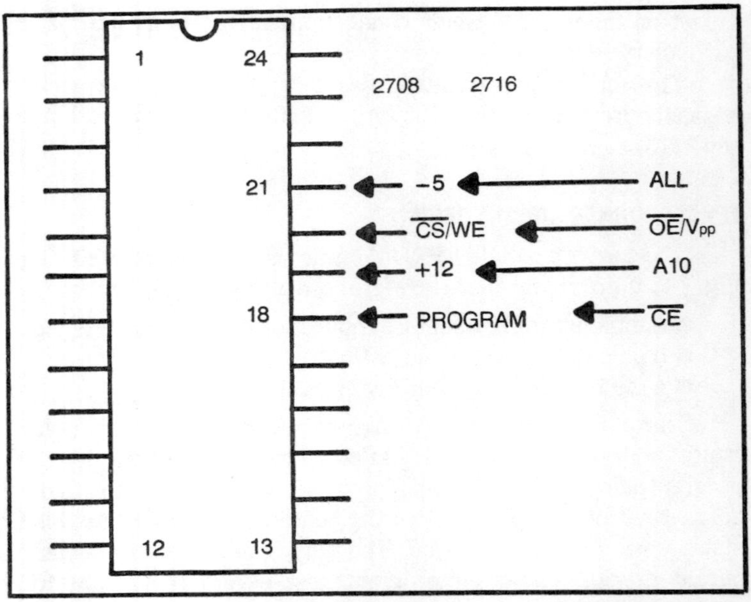

Fig. 7-6. Pin connections that are not the same in the 2708 and 2716 EPROMS. This is a top view.

one-shot, set to a nominal 0.25 ms period. This gives the program time. The end is delayed, and applied to the $\overline{R}$-$\overline{S}$ flip flop to remove the hold.

The times are made variable by potentiometers. Jumpers are provided for various connection possibilities. (This basic circuit is used in a number of PROM programmer designs.)

Before the first programming pulse, the chip select is activated; the line is held at the 12 V program level until all program looping is complete. Then go to a verify mode. Be cautious here, since the $\overline{CS}$/WE program level voltage must be removed before the address change of a read cycle.

The adopted circuit is shown in Fig. 7-8. The $\overline{CS}$/WE line is normally held at the high state. The programming level of 12 V is established by a diode OR circuit, with 12 V applied as the result of a flag 2 signal from the CPU. Or the $\overline{CS}$/WE line can be pulled to the low level by a read generated by the combination of the read strobe and page select. It is not intended that a read will occur during the period that the CS line is at the program level.

The write circuit also combines the write strobe and page select. This provides the initiation of the $\overline{R}$-$\overline{S}$ flip flop and the

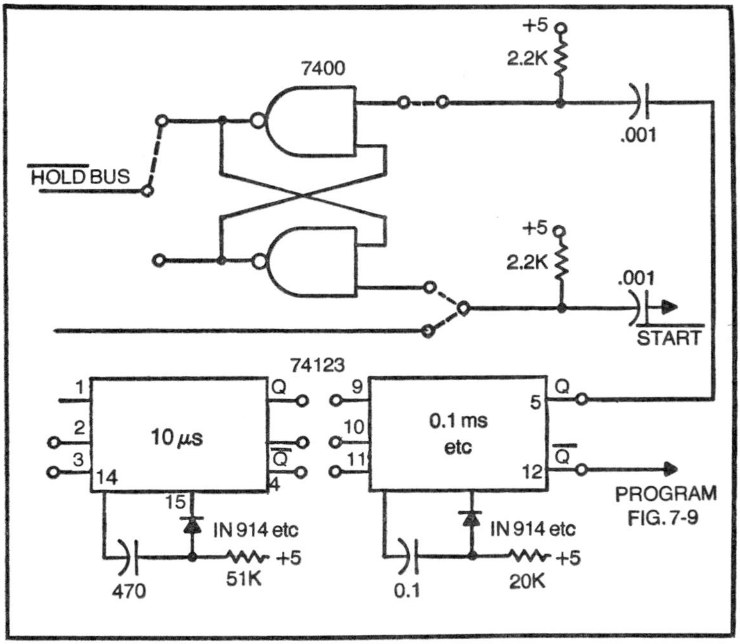

Fig. 7-7. Timing circuit to generate the programming pulse, after delaying it for its minimum required time.

timing MVs. Note the Zener regulated – 5 volt supply, also shown here.

Because the computer main frame does not have a +25 V source, the board must generate it. This is done by the circuit in Fig. 7-9A, which also provides voltage regulation and programming pulse generation. The 12.6 V AC is fed to a voltage doubler, whose output provides 35 to 37 V of well filtered DC. This feeds a zener-referenced pass transistor that provides further filtering, voltage regulation, and pulse generation. The last operation is controlled by two switching transistors; one pulling down the voltage across the reference zener, the second pulling down the output voltage. The two give a measure of independent control to the program pulse rise and fall times, which can be varied by changing the output and zener shunting capacitors. (These times are specified between 0.5 and 2.0 μs.) The pull-down transistors are dirven from the second one-shot MV.

The drawing also indicates the position of the data and address lines. They are not buffered, since it is anticipated that PROM

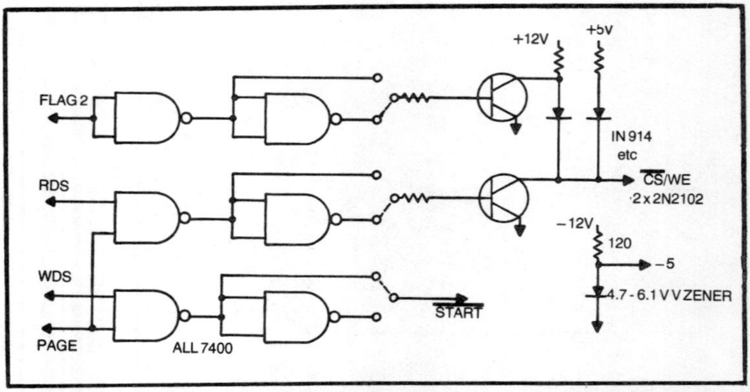

Fig. 7-8. Method of generating the chip select, and write enable signals for the 2708 in the normal and program modes. A – 5 V regulator is also shown.

programming is done with all (except necessary memory and I/O) devices removed.

BOARD CONSTRUCTION

Zero-insertion-force sockets are readily available, and their use with EPROMs is recommended. The commonly available design is relatively large, and requires at least 7/8 x 2 inches of space; 1 x 2¼ inches is better. Because of their large size, jogged construction is not feasible.

A few design trials show that chip flexibility can be provided with a little complexity. The connections for alternate chips are omitted from the prototype. But pads for a second EPROM socket are provided with solder pads for each lead. Further, the four leads that have different 2708 and 2716 designs are not connected directly, but rather go to a 16-pin socket. Plug-in headers are used; one for each type of EPROM to be programmed. See Fig. 7-9B.

The other thing that requires some special attention is the voltage-doubler regulator of Fig. 7-9A. It is placed at the end of the board, as far away from the contact fingers as possible. This guards against accidental shorting to ground or a bus line.

The remainder of the circuit is placed on the board where convenient. Figure 7-10 shows the layout of the prototype, and Fig. 7-11 is a photo of the component side. The layout could undoubtedly be improved.

In constructing this board, pay particular attention to the spacing between the conductor strips carrying the power leads. Avoid shorting the circuits.

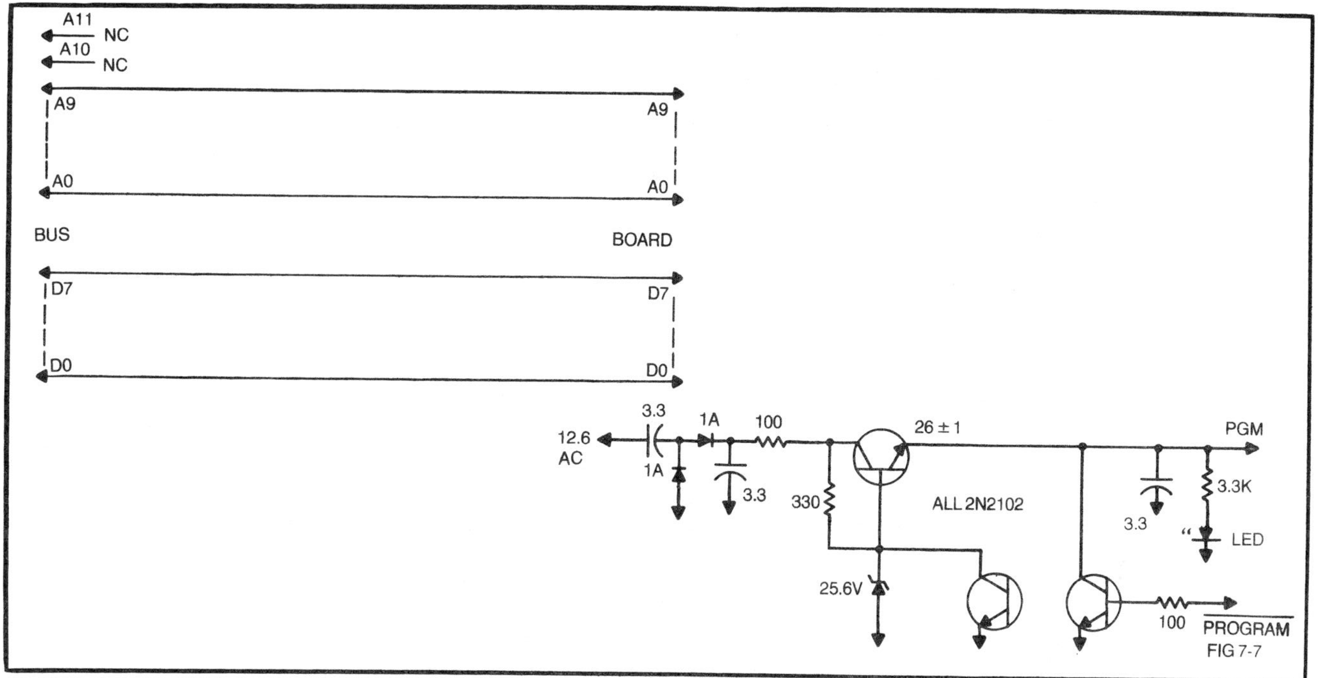

Fig. 7-9A. A method of deriving and regulating a 26 ± 1 V supply from 12.6 V AC, a controlled program pulse is generated from this. The capacitor is used to control the rise/fall time. Address and data lines go directly from bus to chip.

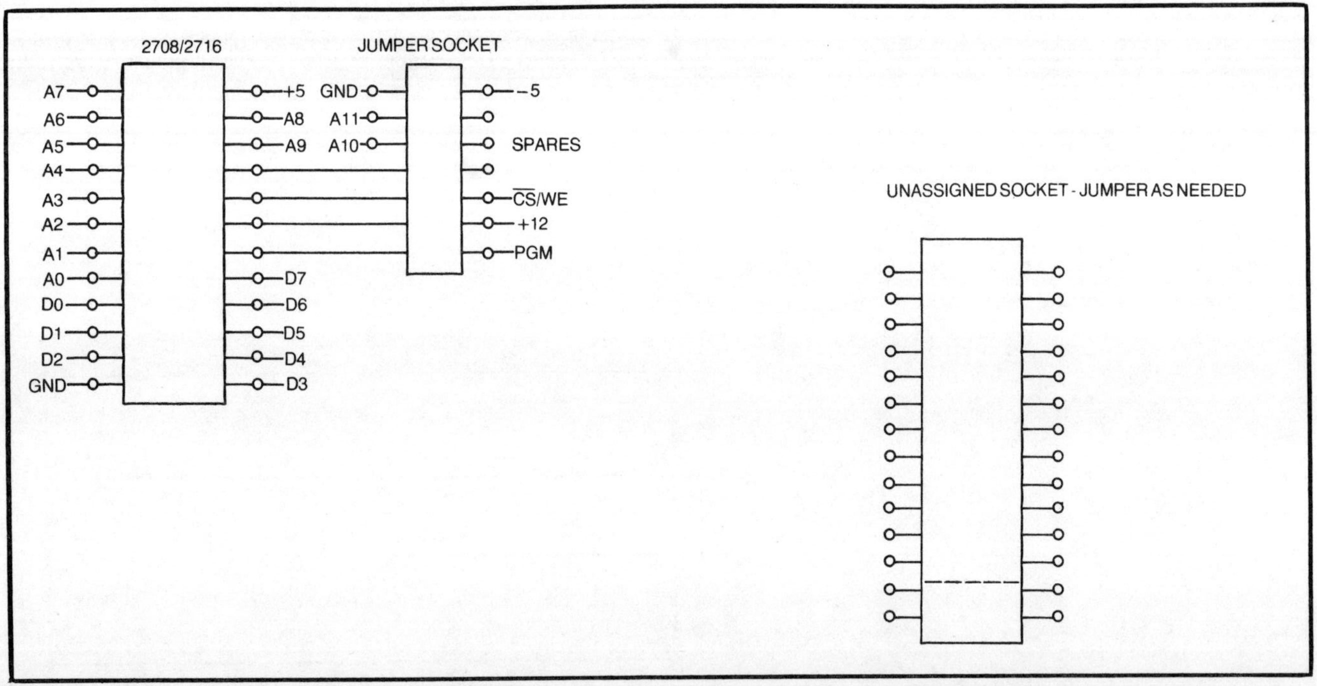

Fig. 7-9B. Method of using a jumper socket to change connections between 2708 and 2716 programming. A spare socket is provided, and can be jumpered to the circuits as needed.

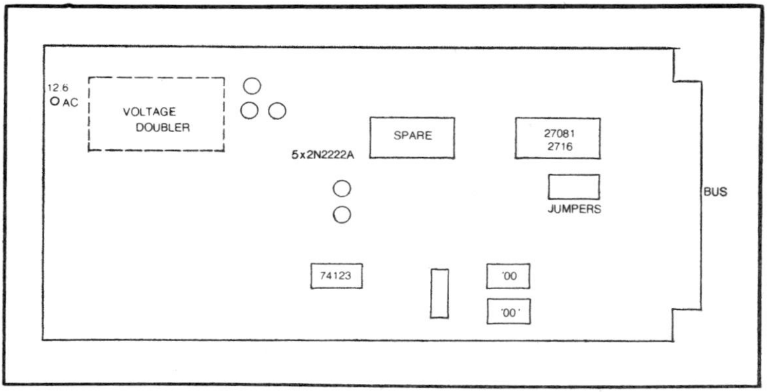

Fig. 7-10. Layout of the prototype PROM-burner board. The 10.6 V lead is connected to this board by a jumper from a jack installed on the control panel face.

MODIFICATION TO COMPUTER CONTROL PANEL

The original design of the computer control board uses the hold line to generate single-step capability. The circuit is shown in Fig. 7-12. When the RUN switch is closed, the next ADS strobe is gated to the clear input of a type D flip-flop. Its $\overline{Q}$ output goes high and is fed to the HOLD buffer of the CPU, placing the unit in a hold that lasts until the FF is set by the step switch debounce circuit.

Since the 7474 has a totem-pole output, you cannot tie another gate output in parallel to the wired OR connection. A circuit change is necessary. The easiest way to do this is to move the hold line to the Q output of the FF, feeding this line to the CPU via an inverting gate. This maintains signal polarity and allows additional HOLD input.

The additional gate can be one of the spare two-gates already on the control board, for example, in packages R and U. It can be connected as shown in Fig. 7-13, to produce a hold bus in addition to the single-step function. The bus requires a pull-up resistor and open collector gates if more than one hold input is expected.

Figure 7-13 also shows an alternate hold buffer, the 7430. Each gate input can be connected to a hold source, which can be conventional gates. This might be convenient if a number of slow memory chips are to be used for some reason.

BOARD CHECK

The usual board check needs to be supplemented with an oscilloscope. After the usual inspection and check for opens and

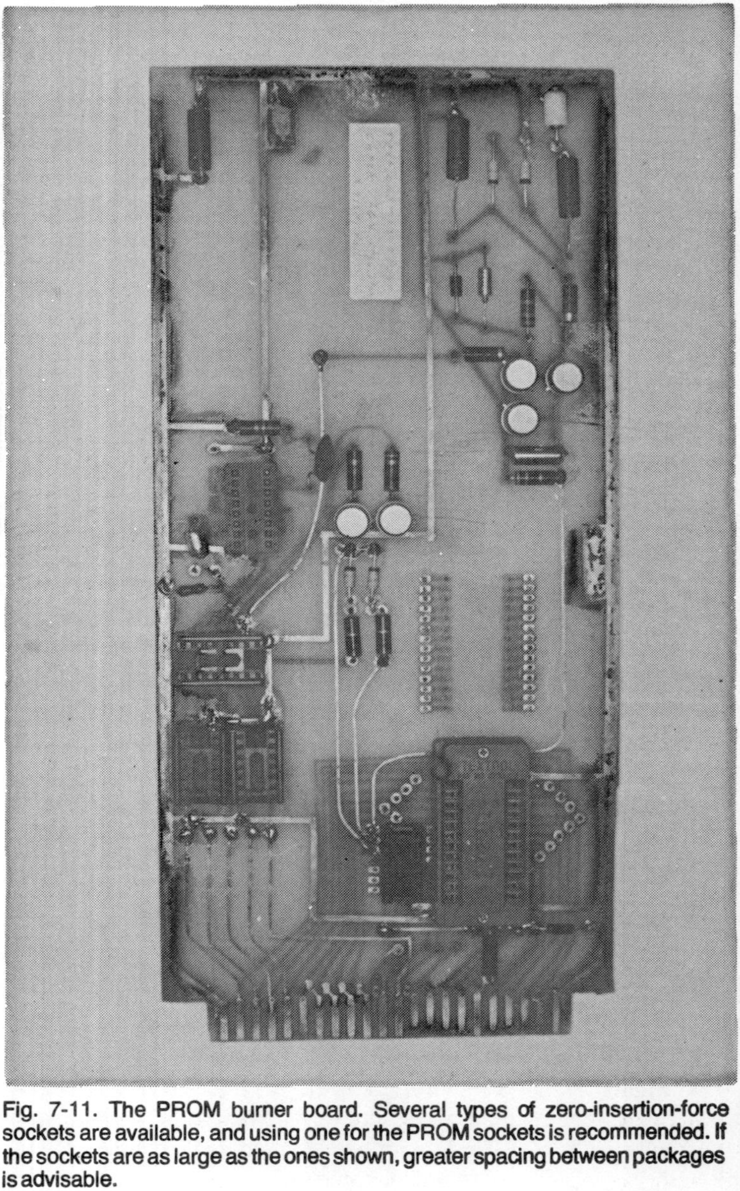

Fig. 7-11. The PROM burner board. Several types of zero-insertion-force sockets are available, and using one for the PROM sockets is recommended. If the sockets are as large as the ones shown, greater spacing between packages is advisable.

shorts, apply the 12.6 V AC and check the 25 V supply. Then the two switching transistors can be installed, and the transient performance of the programming pulse checked.

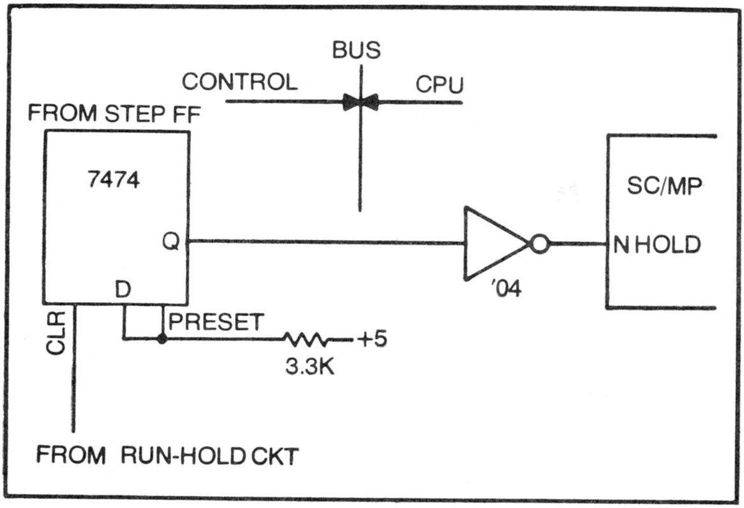

Fig. 7-12. The hold circuit originally used in the prototype. Because the 7474 has a totem-pole output, this circuit does not allow another source of hold signals.

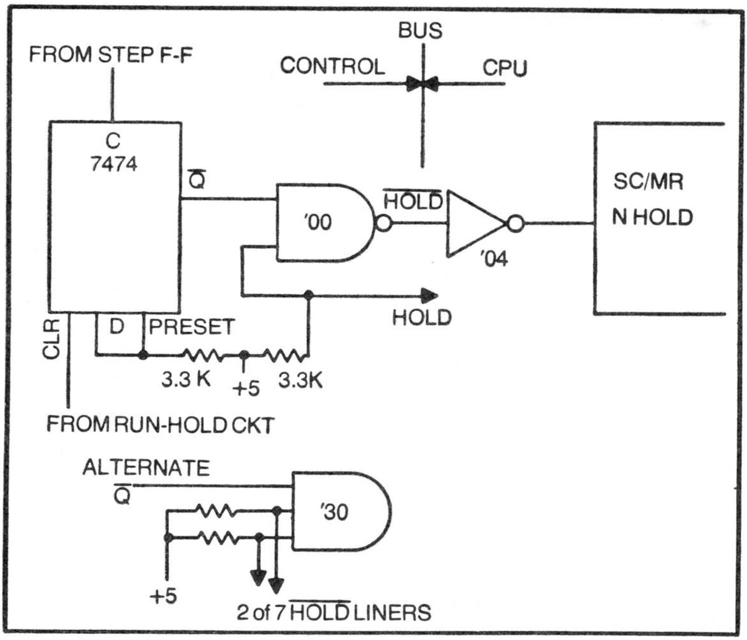

Fig. 7-13. Modification to original hold circuit to allow external hold inputs (by pulling down the hold line).

Follow this by checking the chip-select supply circuits. Once you are sure there is no cross-feed to other circuits, the TTL ICs can be installed. Check the page, read and write circuits in the usual way, then check and adjust the MVs. This is easier if a repetitive trigger is applied to the write line. Finally, check the rise and fall times of the program pulse.

A further check with the board in the main frame is worthwhile. This is also a good way to check the program. Since the cycle time is set, and since the data locations are to be pulsed in sequence, the address lines should show an active-inactive cycle with durations in the 1, 2, 4 - - - sequence. A further check can be made by writing a number sequence into a block of memory and using the program to transfer it to the EPROM programmer. The data lines should reflect the sequence.

THE PROM-BURNER PROGRAM

Data to be placed in firmware must be put into memory by a load routine. The most accurate method is to punch a paper tape, check it for errors, then load it into a convenient location.

Because of the hardware timing, the transfer program is a simple block data transfer, plus a short preamble and closeout sequences. The preample sets up the addresses and the loop counter, then prepares for programming by setting flag 2. Then block transfer takes place. The closeout removes the programming level, preparing the EPROM for read, and returns to the specified address (normally the keyboard monitor).

Good practice follows the write routine with an error-detect routine. The corresponding locations of original and transferred data are compared. If they are not the same, the two values and the address are printed out and further addressed are checked. A printout of the firmware values, to be checked against the source data, concludes the routine. Errors may occur. If the computer normally runs well, it is likely that errors are due to faulty input data. In any event the source of the error should be tracked down. If there is only a one bit error, there is a fifty-fifty chance that it can be corrected by rerunning the sequence with corrected input. This can be done if a high bit must be changed to low. If there are several errors, the package will have to be erased and reprogrammed.

PROGRAMMING THE EPROM

If you have gone through the test sequence, transferring a program or data to a package should be easy.

1.) Shut down the computer.

2.) Remove unnecessary memory.

3.) Install the PROM in the programming board, taking the antistatic precautions needed with MOS devices.

4.) Install the board in the card cage.

5.) Bring up the computer and load the memory with the data to be transferred.

6.) Go through the transfer process.

7.) Go through the check process.

8.) If successful, shut down the computer, remove the PROM board, and transfer the now programmed EPROM to its designated ROM board. Replace any boards removed and resume operations.

9.) If an error occurs, track down the problem, correct it.

Because of the shut-down and start-up process, schedule operations to include EPROMS in a batch.

EPROM ERASING

When it's necessary to change a program in EPROM, the unit must be erased, then reprogramed. The technical specifications for this are rigid. Exposure at a distance of one inch to a source having a wavelength of 2537 angstroms until an exposure of $15W\text{-}sec/cm^2$ has been accumulated, the source having a rating of 12,000 $\mu W/cm^2$.

Actually, the requirements are not nearly as restrictive as this seems to indicate. Erasing can be accomplished by any source having radiation less than about 4000 angstroms. This is just outside the range of human vision. On a clear day, the intensity of the sun is about 50 percent as great over the range 3000-4000 angstroms as it is in the visible range, so the sun can be used.

In this range a tungsten lamp in a quartz bulb has only about two percent intensity, but a number of common sources have greater intensity. They generate light by a discharge of a gas, mercury vapor being best. Sources based on such discharge are found in fluorescent lamps, high intensity street lamps, sun-tan bulbs, lights for special effects in night clubs and theatres, and germicidal bulbs. However, only the last three have appreciable UV radiation, since the bulb glass of the others is chosen to keep the UV level low.

The specification states a minimum exposure, but longer exposures, at least within reason, do no harm. ("Within reason" may be up to ten times the minimum, or so.) Also, the minimum is easy to determine; place the EPROM in the computer and reread the output. If none of the data has changed, the exposure is below the minimum. If some of the zeros have become ones, the exposure is approaching the minimum, and if all locations read ones, its time to stop.

Some of the commercial UV bulbs available include the R-52 that requires 20 minutes, the S52 and S58 that require 21 minutes, and the UVS-54 and UVS-11 that require 44 and 46 minutes respectively. In comparison, sunlight requires about seven clear

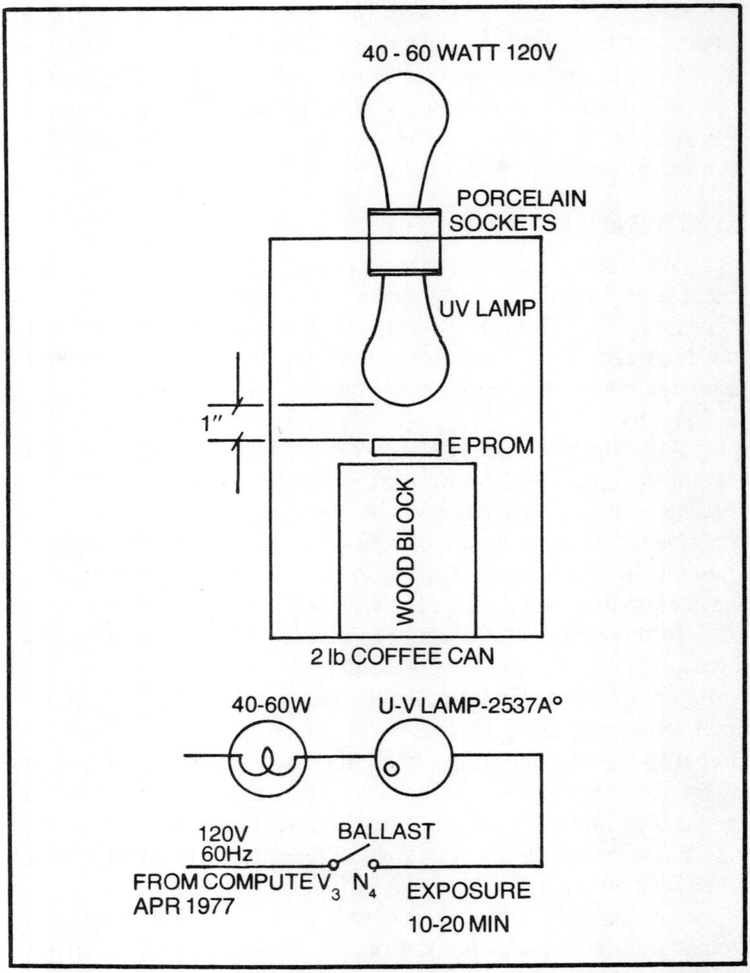

Fig. 7-14. General layout of a UV source for EPROM erasure. The UV lamp should produce 2537 angstrom UV, often called short UV. See test.

Fig. 7-15A. A low cost EPROM eraser using solar UV. Front surface reflection is used to prevent absorbing the UV. The intensity at the EPROM is about three times normal sunlight. If pointed to the noon sun and left fixed in position, several days exposure are required for erasure. Moving the unit each hour, to follow the sun erases in about two days.

days of exposure. This can be reduced to about two days by using *first surface* reflectors, such as four aluminum sheets, and is certainly the cheapest eraser.

There have been several descriptions of home-built EPROM erasers in computer magazines. One is shown in Fig. 7-14, which also shows suggested wiring. Figure 7-15A shows a photo of the simple unit I used. Also, see Fig. 7-15B.

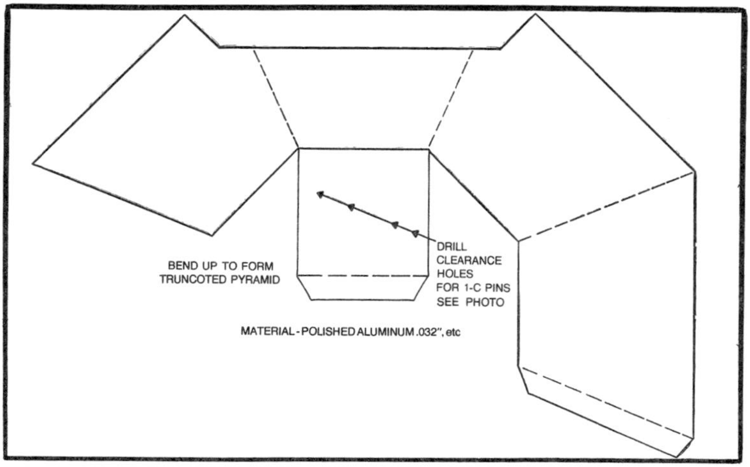

Fig. 7-15B. A diagram of Fig. 7-15A, the solar EPROM eraser.

Germicidal bulbs are becoming scarce, and it may be necessary to place a special order to set one. They can sometimes be found in repair shops.

If the germicidal bulb is not available, some black light can be used instead. To be usable the bulb must have a tubular plastic filter that slides off.

Another alternative is the sun lamps bulb, either tubular and reflector. These take more power than the germidical and black-light bulbs Experiment to determine the required exposure. The only precaution is to keep the PROM from becoming too hot. The maximum erasure temperature is not specified as 125 degrees. The maximum erasure temperature is not specified.

Erasure and reprogramming is limited to about 50 cycles, unless a conditioning process is included (see reference list). In principle, the conditioning process is simple—bake the PROM in an oven at 400 degree Fahrenheit (204°C) for 45 minutes. Practically, the process must be done with care. The main problem is that the oven temperature can not exceed 450 degrees. The usual kitchen oven is unreliable due to cycling of the thermostat and to calibration errors. Temperature averaging is needed. It can be provided by a good-sized block of brass, just wide enough to allow the package pins to slide over it. Silicone grease can be used to improve heat transfer. A separate thermometer should be used to set the temperature correctly.

ALTERNATE DESIGNS

It is convenient to have a programming unit that is independent of the computer. Because of the repeated cycling required, this is not easy with the 2708. With the 2716/2732 it is not all difficult, since a single operation is sufficient for programming, and since addressing can be in any order.

Package units and kits for manual control programming are available. It is not difficult to arrive at a design. A simple one uses two switch-banks, one for data, and another for programs. For checking, a pair of nine-digit LED sticks can be used, like the prototype control panel. A single-step switch can initiate the programming pulse.

It saves time to separate the process of data preparation and actual programming. This can easily be done by using paper tape. Addresses can be incremented by one for each two tape characters that provide the data word (Baudot machines). It is a good idea to maintain a check sum.

NOTES ON OS PROGRAMMING: STRUCTURED PROGRAMMING

Programming the operating system differs markedly from general computing. A major difference is that mathematical operations are relatively rare and simple—usually calculation of jumps, and equivalent operations. Logical decisions, such as matching an entry against a list, or selecting a subroutine from a number, are much more typical. This is not to imply that OS programming is simple. One segment can be complex, and the ensemble very complex.

This emphasis on logical processes and the associated complexity have led to recommendations that OS programs be written as *structured programs*. The recommendation is good, but it must not be thought that this requires a special programming language. The term refers to the process of organizing the program so that it is logically correct, logically clear, and simple to execute. Such programs are well structured. However, the term *structured program* has come to mean one in which the program instructions are reduced to the minimum possible.

An advanced paper shows that only two types of statements and a total of three requirements are sufficient for writing correct programs. The requirements are: (1) simple statement sequencing, (2) IF-THEN-ELSE conditional branching, and (3) DO-WHILE conditional iteration. All languages known to me have a form of the two statements. With the first requirement, all of these languages can be used to write a well-structured program.

The form of the statement may vary with the language. In NIBL, for example, IF-THEN is used, but ELSE is implied; it must be the next statement. DO-UNTIL is used instead of the DO-WHILE, a simple logic inversion.

The original BASIC and other Tiny BASICs do not specifically include the DO-WHILE construct. Instead, the FOR, NEXT and GOTO instructions must be combined for the same result. This increases complexity, (the source of criticisms of the original BASIC); complexity makes errors easier.

One little trick that is often used in writing structured programs is leading edge indentation, by one space for each successive nesting of a loop-within-a-loop. An elementary example of this is shown in Fig. 7-16. This calls attention to the structure and is a good idea. Of course, attention must be paid to the capacity

```
100    LET A = X
110 LET B = Y
120 IF A = B THEN 180 ; REM GCD (X,Y0) = GCD (A,B)
130 IF A   B THEN 160
140 LED B = B – A
150 GOTO 120
160 LET A = A – B
170 GOTO 120
180 REM GCD (X,Y) = A=B
190 RETURN
200 END
```

Fig. 7-16. Example program using indentation of lines to emphasize program structure. While this increases program storage space somewhat, it improves readability in a list.

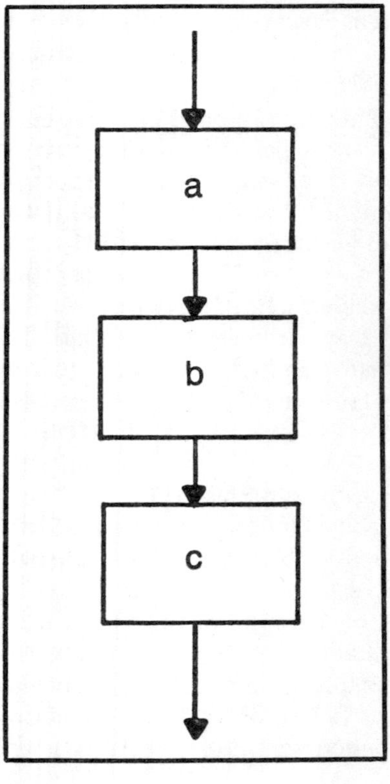

Fig. 7-17. The first and simplest basic flow pattern in a program, direct sequencing from item a to b to c, etc.

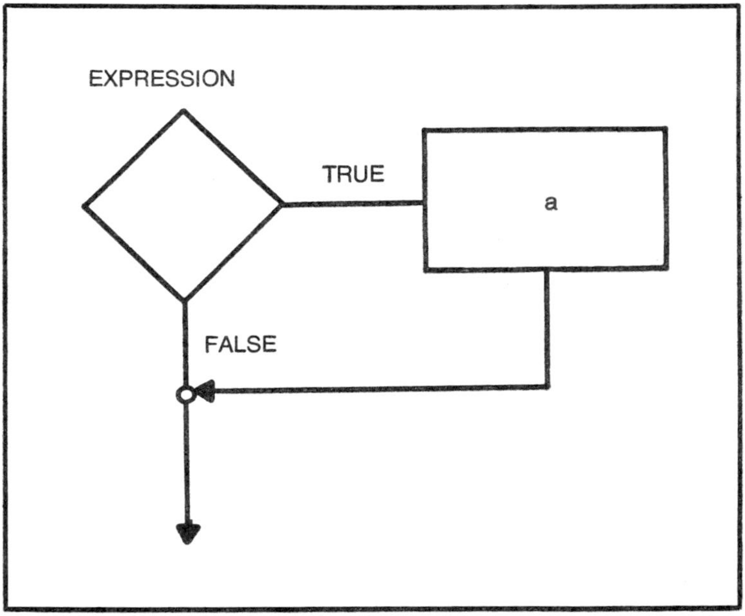

Fig. 7-18. Another basic flow pattern, the IF-THEN construct. A is executed if the expression is evaluated as true, otherwise it is skipped.

of the computer and its language. NIBL allows four entries to the DO stack before returning an error code. (In long programs it may return a "no memory" error before the stack limit is reached.)

The power of the two statements may be easier to see if they are shown in the form of the six basic operations they include. The first, and simplest, is the straight sequence of Fig. 7-17. In BASIC this sequence is enforced by the execution of lines in order by number. NIBL can have several statements on a line. These are executed in order if they are simple statements, but a conditional or absolute branch may (or will) cause a later statement on the same line to be skipped.

The second construction is the IF-THEN conditional of Fig. 7-18. It causes an execution to occur if a condition is true. An extension forms the third class, the IF-THEN-ELSE form of Fig. 7-19. It is sometimes called the OR structure.

DO statements also have two forms, the first being the DO-WHILE flow of Fig. 7-20. The associate structure is the DO-UNTIL of Fig. 7-21. It is sometimes termed a repeat UNTIL form.

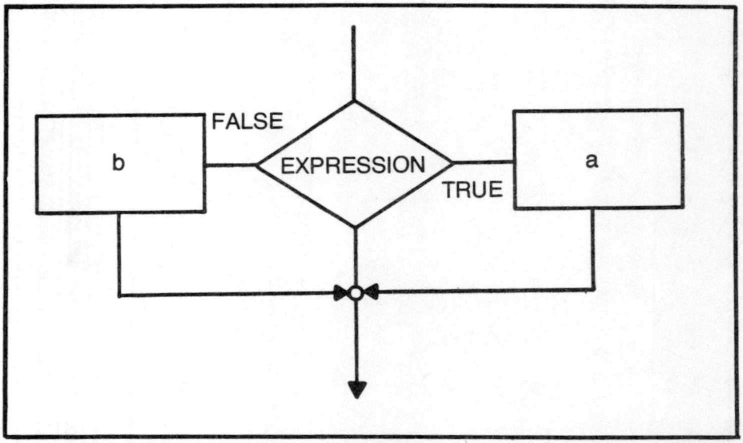

Fig. 7-19. The more complex IF-THEN, ELSE construct, where A is executed if the expression is true, and B if the expression is not true.

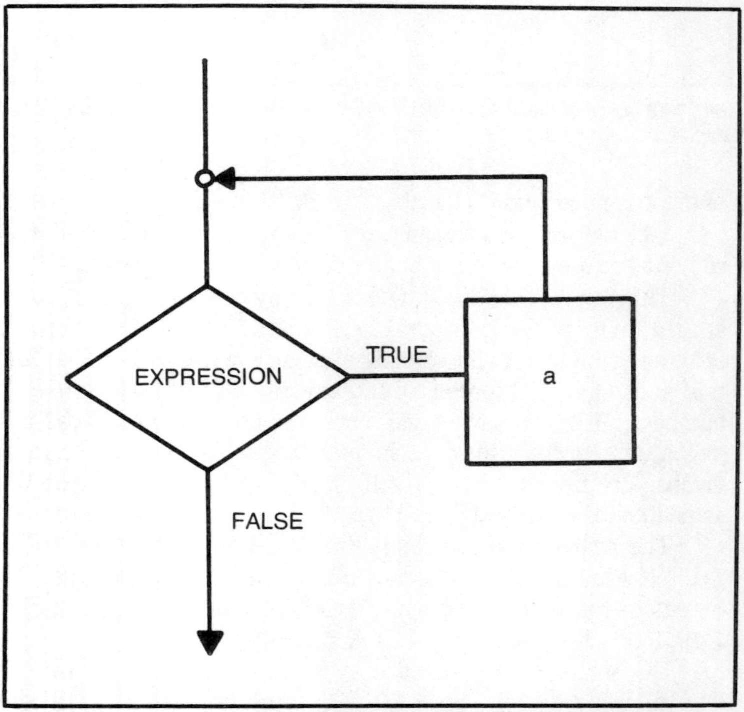

Fig. 7-20. By looping back the DO-WHILE construct is obtained. A is executed as long as the expression is true, often by an expression involving the LESS-THAN or GREATER-THAN equivalence.

The final form is the SELECT or multiple branch structure of Fig. 7-22. Many times this is shown as a chain of successive IF-THEN constructs, with the end of the THEN operations returning to a common program point. This is found in many programs. Many languages, including NIBL, provide for direct use by the GOTO expression, where the calculated value of the expression sets the line number of the routines a_x of the figure.

Machine language does not allow the simple and easily visible form of the IF and DO statements. Instead, a series of operations must be specified. However, these structures can be recognized with them. An example is included in the Appendix.

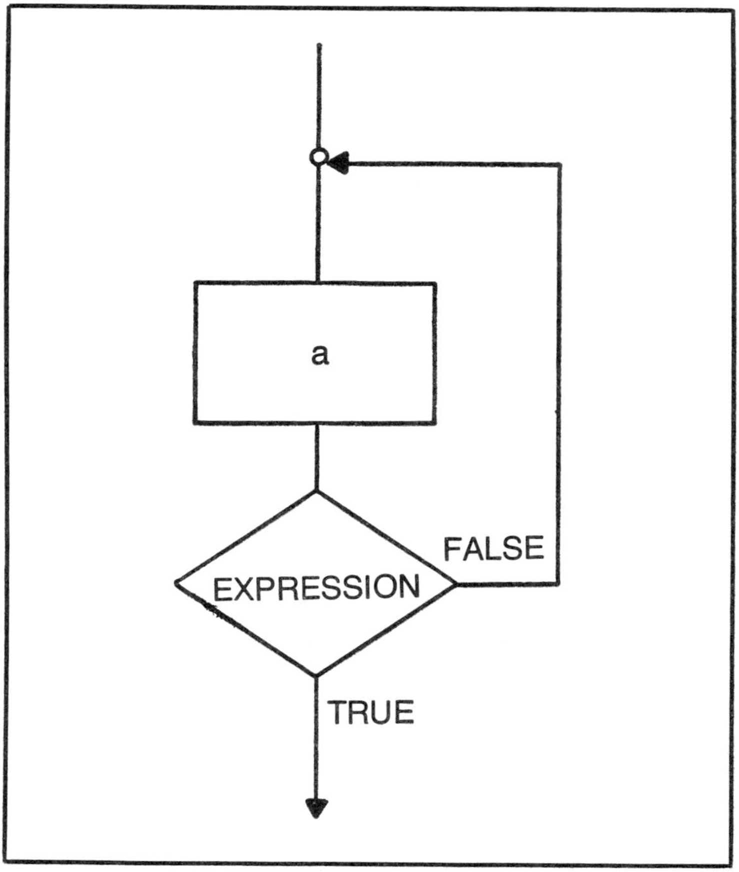

Fig. 7-21. Looping back is also used in the REPEAT-UNTIL construct, which is an alternative to DO-WHILE. However, in this construct, A is executed once for the test.

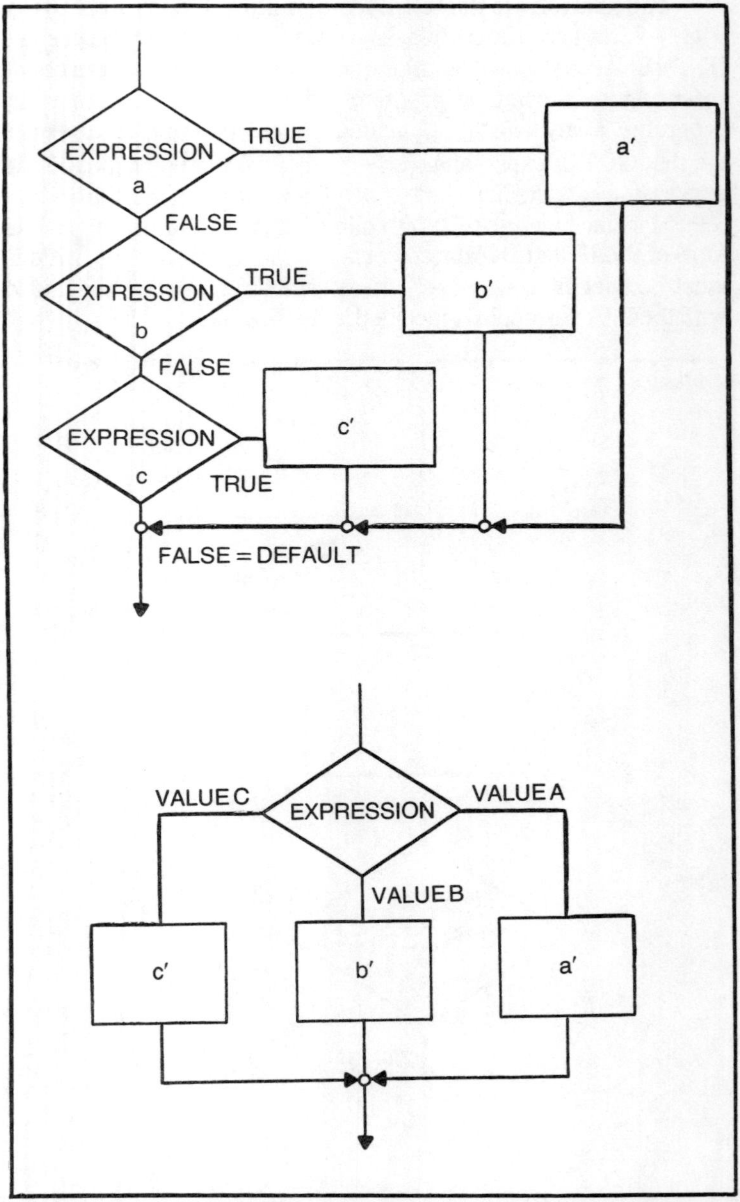

Fig. 7-22. Two forms of select constructs to choose between one of three (or more) paths. Literature on search theory should be consulted on these forms. Often the upper one results from a sequential search, the lower from a hash function.

Simple problems can be handled by simple programs, generated by simple procedures. As the problems become more complex, the programs do too, and there is greater and greater need for rigor. A full OS becomes very complex, so great care is needed to plan it. Visualize the full extent of the system you want to undertake, based on your own version of Figs. 3-1 and 3-2, and your software plan as summarized by your version of Fig. 3-5. With the basics set, the particular programs and program segments that you need can be identified and planned.

As always, remember that time and trouble is saved if a good record of the steps is kept. It's helpful to have this in two forms—the file record, and a condensed version kept right at the keyboard.

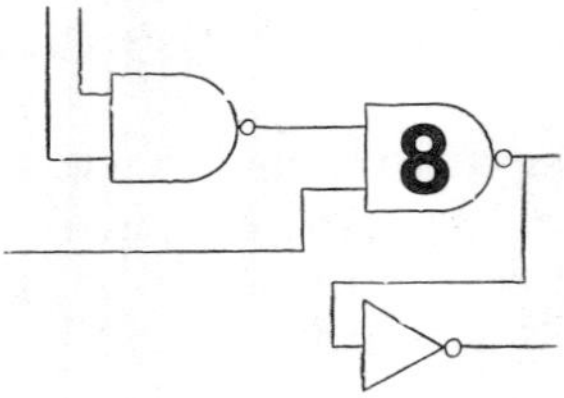

An Auxiliary Processor Panel And Its Programs

Any microprocessor chip on the market has a lot of overall capability, but the arithmetic problems it can handle directly is strictly limited. With some exceptions, microprocessor chips can add and subtract; subtraction being a complement and add operation. Some newer and more complex chips can multiply and divide, but even these perform only the basic operations.

If a computer built is to have extended mathematical capability, it must either have special hardware, or it must use software. Software works because multiplication and division are short forms of repeated addition and subtraction, and because all higher order mathematics can be built up from these basics. The disadvantage of software is that it requires considerable central processor time and a large section of memory.

The *SC/MP Applications Handbook* gives routines for a number of operations. Program memory requirements for some of these are:

16 bit add	16 bytes
16 bit subtract	17 bytes
unsigned 8 bit multiply	40 bytes
16 bit by 8 bit unsigned divide	99 bytes

The unsigned multiply requires two to three milliseconds, and the divide from five to 20 milliseconds.

These are just basic operations: The amount of program memory increases rapidly as higher functions are attempted, and it is not uncommon for the mathematical elements alone to require 1000-2000 bytes. Storage of variables, keeping track of exponents and the decimal point, etc., add to the memory needed. Typically, getting good math capability by software requires 12K for the program plus 8K for RAM. However, 32K can be easily used.

It is natural for the home builder to think about the hardware route, in particular, the chips of the ubiquitous hand-held calculator. This can yield a vast improvement in math capability, but is a mixed blessing. The processor chips used in calculators are optimized for keyboard input and visual output and, therefore, require considerable interface circuitry to adapt to computer use. These chips are still relatively slow, and may require as much as a second to handle some operations. And, of course, there still must be appropriate software. Yet, adding a calculator chip can double the power and speed of advanced mathematics problems.

Some of the drawbacks to use of the calculator chip have been removed recent developments. Several manufacturers offer special chips intended for math processing; these may be adaptations of calculator chips, or may be a special design. While no more powerful, they are much easier to interface, and are faster. The simpler interface means that programming is also simpler.

However, before looking at design specifics, ask yourself if you really need to double the mathematics capability. If your anser is "maybe," perhaps you should consider an alternative. Let's look at the possibilities.

ALTERNATIVES TO COMPUTER MATH UNITS

By far the simplest and cheapest alternative is a standard hand-held calculator. Even a $7 to $15, four-function hand-held, kept at the program work table and at the console, will accomplish a lot. The procedure with these is to let the computer keep track of the variable, and of the program steps. It can present one or more data items, a set of steps, and request input. True, the manual operation will be slower than it would be on the machine, but not as slow as you might think.

The next approach is to go to one of the scientific calculators currently offered in the $15-$35 class. These have far greater capability: in fact, their internal stored program memory is typically 1 to 2K of instruction words. They are operated in the same fashion as the simpler calculators, but can handle more complex problems.

You could abandon the computer for math and use a programmable calculator instead. They are available for $40-$80, but the most advanced models cost about $200 for nonprinting types, and $400 to $600 with printed tape output. These units are almost full computers. In the low cost varieties programs must be hand loaded often several times during execution. The expensive units load programs from cards or plug-in modules. Their powerful attack of math problems can make a computer a luxury, not a necessity.

Another alternative to a computer math board is the *compulator* (see references). In its original form this used a paper-tape reader with a relay tree to simulate keyboard input and the usual multidigit display for output. The design and the relay tree are shown in Fig. 8-1. The paper tape reads the program (or data) as key strokes, programming the calculator just as the human operator does. Looping is handled by forming the program tape into an endless loop.

The compulator is a powerful math problem-solving tool. Programs can be up to thousands of steps in length. Execution speed is much greater than for manual input; but manual input is always possible through programmed halts. As with all calculators, program tricks can be developed. An example is the use of several NO-OP characters to introduce a wait while the calculator runs through a complex sequence, such as a trig function calculation.

The build-or-buy decision reviewed in Chapter 1 is a little different here. It is "Should I build, or buy, or use a calculator?" If your decision is to build, let's look at ways to do it.

BASIC CHARACTERISTICS OF CALCULATOR CHIPS

For the purpose of computer processing, the available calculator chips can be divided into families:

- Early chips, relatively slow, with 12-digit capability and elementary arithmetic
- Basic chips, six to eight digits, with elementary arithmetic
- Simple chips, usually eight-digit, with elementary arithmetic plus a few functions (many have a simple form of memory)
- Slide rule chips, usually eight-digit, with a number of functions (most have a single level of memory, but allow a number of pending operations)
- Scientific chips, eight to 12 digits, many functions, exponential notation, several to many levels of memory

In addition there are variations on these: some for teaching: and some for special problems, such as business, statistics or units conversion.

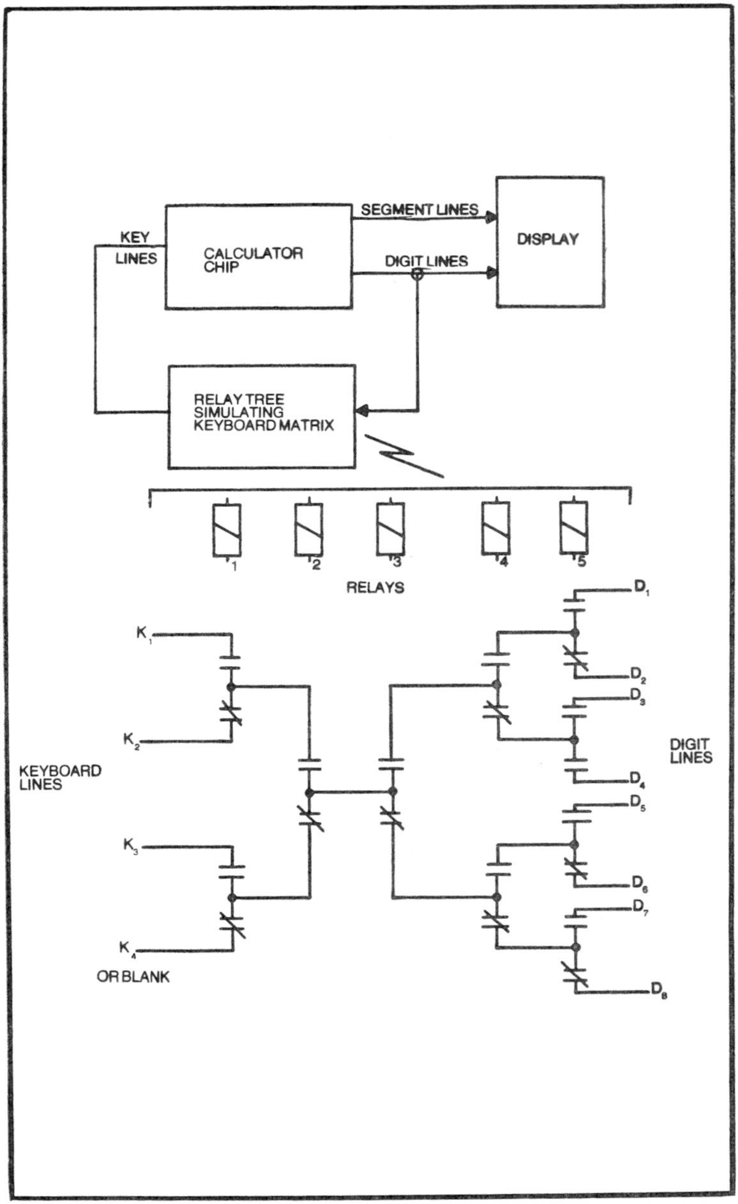

Fig. 8-1. Design of the compulator. A relay tree with contacts connecting to key and digit lines is used to provide input, a five-digit word being eqivalent to a key stroke. Relay action is controlled by paper tape, and can be as long as desired. This is a form of programmable calculator.

The block diagrams of these chips, as far as externals are concerned, are essentially identical. Figure 8-2 shows a typical schematic. Key lines may range from two to five. Early chips used direct BCD input; a few modern chips do too. The rest use multiplex from the output, to reduce the number of chip pins needed. The feedback is from the digit lines, except for a TI family, which uses the seven-segment and the positive bus. Seven-segment plus decimal point output is nearly universal, but a few chips use BCD output, or both. There is a difference in character shape for some. There may be two families, one for LED displays, one for fluorescent.

Digit output is universally one pin per display digit. One digit can be reserved for sign and error.

Power, for most chips, is +9 V. Some use a voltage doubler circuit. A few will operate on +5 V. (All chips are MOS devices.)

Control functions are not provided on many chips. There may be a ready signal. Some provide for external oscillator input, for speed change, or for change in notation or function, as from degree to radian measure, or fixed decimal location selection. A single manufacturer may have dozens of design variations.

One feature of all keyboard chips is dictated by the poor bounce characteristics of low-cost switches. There is a delay before input is accepted, perhaps by counting a number of cycles before accepting input. Similarly, there is some form of delay before a second input is accepted. These delays may total 20 ms, severely limiting the rate at which data can be fed to the input. A few chips have an input pin to reduce the delay by a factor of 10 for test purposes.

In addition calculators work by repeated addition, so there may be appreciable delay before an answer to an operation is available. Most blank the display during this period, but in some the display or a single digit may flash. A few provide a ready signal, which indicates when output is valid and when new data can be accepted.

Other variations exist. In some calculators, the display starts at the left, but in most, it starts at the right. In some, the sign and error signals are at the left, in others, just preceding the left most digit. Some calculators have a *battery save* feature, presenting only decimal points or shutting off 30 seconds after a key is pressed.

For additional data see the reference list.

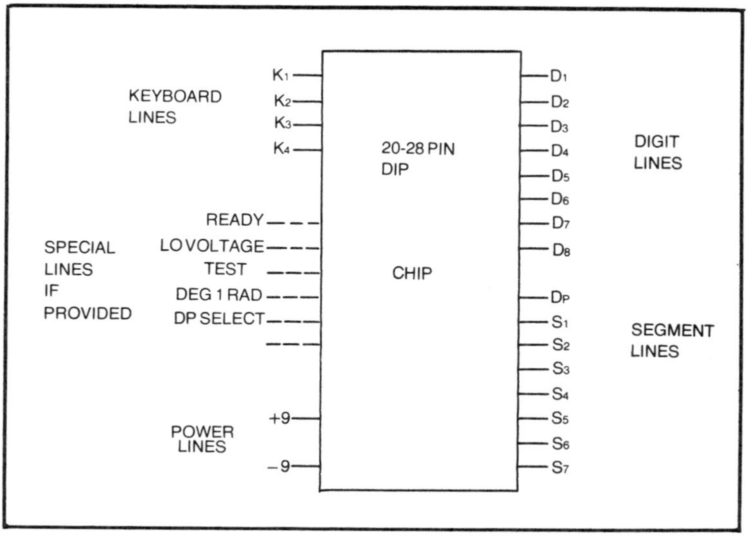

Fig. 8-2. Schematic of the five buses of a generalized calculator chip, with segment line, digit line, key line, control and power buses. The number of digit lines, key lines and control lines varies with chip design.

COMPUTER ARITHMETIC CHIPS

In addition to the calculator chips designed for keyboard input, there is a much smaller family of arithmetic processors designed specifically for computer operations. These are related to a calculator family, or to a CPU with ROM. They are not optimized for keyboard operation, but, at least to a degree, to computer use. They do not include the long delay on input, or the multiplexing of input to reduce pin count. Also, they have control signal pins for at least basic control.

One of these processor chips is the AM9511; a 16 and 32 point arithmetic unit with a set of derived floating point functions built in. The capability can be seen from the command list of Table 8-1. The stack is an eight-level 16-bit wide memory, which allows the usual X, Y, Z, T arrangement of reverse polish calculators. The chip is fast, and can operate with a clock of 4 MHz. Cosine generation requires less than one ms, and 16-bit add only four μs.

Table 8-1 shows the command code, and the combination of the read, write and command strobes that are used to control operations. Additional control functions are provided, as shown on the pin-out chart of Fig. 8-3. These additional functions can be used in a "not busy" poll mode, and interrupt mode, or in some

Table 8-1. Command List of the AM9511 Arithmetic Function Chip, Designed for Number Processing in Computers.

16 Bit Fixed Point Basic Operations
- Add
- Subtract
- Multiply (Upper, Lower Halves)
- Divide

32 Bit Fixed Point Basic Operations
- Add
- Subtract
- Multiply (Upper, Lower Halves)
- Divide

32 Bit Floating Point Operations
- Add
- Subtract
- Multiply
- Divide

32 Bit Derived Operations

Square Root	Arc Sine	Log (Natural)
Sine	Arc Cosine	Exponent
Cosine	Arc Tangent	Power
Tangent	Log (10)	Pi

Data Manipulation

No Operation		Push Stack	(3)
Fixed-Float Conversion	(2)	Pop Stack	(3)
Float-Fixed Conversion	(2)	Exchange	(3)
Change Sign	(3)		

multiprocessor modes. Details are available from the manufacturers (one of the most interesting items being a discussion of the formation of the computing algorithms used, and of their accuracy).

Another arithmetic unit is the MM57109. Unlike the preceding chip, it is decimal oriented and can handle from one to eight BCD digits. The instruction set is shown in Table 8-2. The arithmetic function list is slightly more extensive, but it should be noted that the execution speed is lower. This is established by a 400 kHz oscillator, divided internally to the actual clock of 100 kHz. Cosine generation requires over one-half second on the average, and floating point add requires 22 ms, average.

The chip is a mask programmed version of a simple processor, the MM5799, a design oriented to control operations. Many of the instructions reflect the original use. They can be useful in system extension, for example, by adding a special memory package that allows the arithmetic unit to conduct an entire routine while the main CPU is on other tasks.

This flexibility is also evident in the pin-out connections shown in Fig. 8-4. A minimum interconnection is done with only 18 of the pins. The others are intended for stand-alone operations, but may be used in other ways.

Contrasting these two units, the major difference is that the AM9511 is designed for computer clock speeds, while the MM57109 is designed to operate at calculator speeds. The cost differences is related to speed.

There is another approach to arithmetic unit development that may appeal to experimenters. This is use of the ALU, the arithmetic-logic units of the standard TTL line.

The 74181 series of ALUs can perform a total of 32 arithmetic or logic operations involving one or two four-bit words. Some of the operations are trivial, but all of the basic logical and arithmetic operations are available. The units can be cascaded. The 74S381 is a simpler version, providing only eight basic arithmetic and logic operations.

I have not used these units in any hobby or home microprocessor work. The only reference I've found is for a logic demonstrator for training. They are mentioned to show that there are untouched fields for the experimental minded.

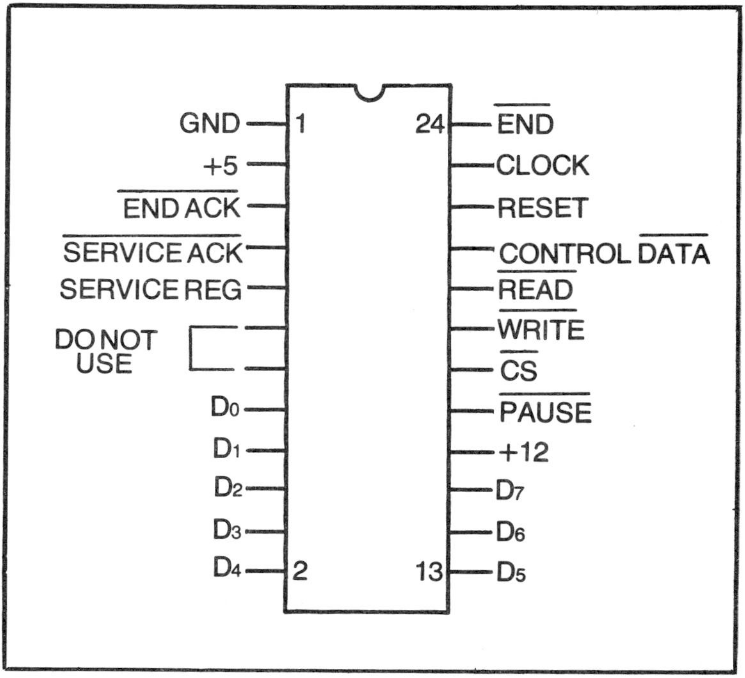

Fig. 8-3. Pin connections for the AM9511 chip.

Table 8-2. Command List for the MM57109 Number Cruncher, Another Arithmetic Function Chip Designed for Computer Use.

Monadic Operations

Square Root	Log (Natural)	Arc Sine
Square	Log (10)	Arc Cosine
Ten to X	Sine	Arc Tangent
E to X	Cosine	Degrees-Radians
One over X	Tangent	Radians-Degrees

Dyadic Operations

Add	Power	Memory Divide
Subtract	Add to Memory	
Multiply	Subtract from Memory	
Divide	Memory Times	

Control Operations

Decimal Point	Change Sign	No Operation
Enter Exponent	Enter	Halt

Move Operations

Roll Stack	Memory Store	Left Shift
Pop Stack	Memory Recall	Right Shift
X-Y Exchange	X-Memory Exchange	

Branch Operations

Jump To	X=0?
Test Jump	X<1?
Test Error	X<0?

Miscellaneous Operations

Master Clear	Pi
Clear Error	

OBTAINING CHIPS AND DATA

There are several ways of getting a suitable calculator chip for this board. However, it is not always possible to get a particular chip at a particular time. Decide on the general type of chip desired, then to get one of the family, and only then complete board design.

If you are going to base your work on calculator techniques, I suggested a scientific chip. The cost and complexity are not much more than for simple types, and the power and flexibility are far greater. You can get even more flexibility if the chip includes statistical, business and other types of internal programs.

The simpler chips are available from the supply houses that advertise in electronics and in computer magazines. Slide rule and scientific chips are also available; probably end-of-production surplus. If you watch the ads, you may find a suitable chip by the time you are ready to build. Be sure to request a copy of the spec sheet when you order.

The most reliable way to secure a chip is to buy a suitable calculator and use its chip or chip-board assembly. The cost is not too great, the chip is obviously working. Sometimes a friendly storekeeper will let you snap the back cover off, to check the type of chip, and the mechanical suitability.

A few calculators do not work well for this use. For example, the TI-30 uses a chip with nonstandard pin spacing and with segment multiplex. Considerable software manipulation is required to make the data usable. Also, the chip is mounted on the LED board. leads will have to be cut to get the output signals. Avoid some very low cost calculators, such as the NOVUS, that glue an unpackaged chip die to the display board.

The third way to get a suitable chip is to order it through distributor channels. Cost will be relatively high, since OEM pricing applies. However, this may be the only way to get a specific chip.

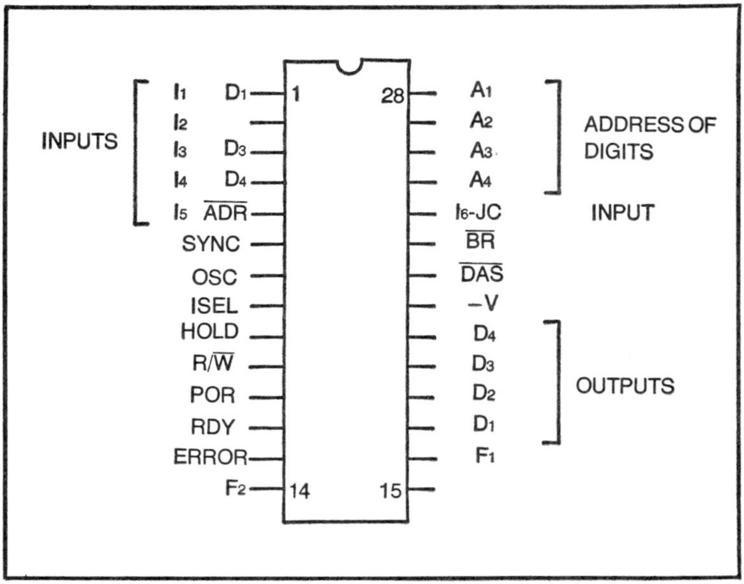

Fig. 8-4. Pin connections for the MM57109.

Several chip manufacturers publish full data in their specification books, and others provide spec sheets on request. The practice is not universal, however. Calculator manufacturers are often unresponsive to requests for data. However, it is often possible to identify the chip manufacturer, perhaps by a logo, or by the chip-type number. Here, presence of a dash and a suffix probably indicates a special design. Information can be hard to get, but the changes probably don't affect the pin-outs or signals. The key functions and the internal programming can be determined by experiment once the chip is in operation.

If a calculator is used as the chip source, the pin use should be evident from the connections, and the signal levels and timing can be measured with an oscilloscope. Blank pins may represent unused features. Check first with an oscilloscope, then by connection to the minus and plus buses.

All chips I am familiar with are MOS designs with gate protection. It should be possible to identify the pins for V_{ss} and V_{gg} or V_{dd} by measuring low voltage resistance. For most chips identification of other pins is easy, since the internal oscillator starts when the power is turned on.

At the present time using a calculator chip appears inferior to the simpler route of using a specially designed arithmetic processor. Purchasing a chip separately from a calculator makes the board circuits simpler, and the software easier to write. Cost may be about equal, although a high performance processor chip will not be cheap. Specs are probably much easier to get.

A universal design for an arithmetic board is given that can use any known calculator or special chip, and probably any calculator-type chip for which the pin connections can be determined. While the design may be more complex than needed with some processors, it's easy to simplify. An example, for the MM57109, is discussed later.

DESIGN OF A UNIVERSAL CALCULATOR-PROCESSOR BOARD

Since there are so many types of calculator and special chips available, let's use a two-step approach to the processor board design. The first is the design and construction of a universal board with basic signals, space for mounting parts for additional signals, plus a calculator chip mounting area. The second step is an adapter for using the board with a selected calculator chip. Only one example of adapter design is covered. Design others from the chip specifications and data.

To use the calculator chip you must be able to insert a digit byte or half-byte on command, initiate an operation on command, and readout of a digit or byte on command. Since there are delays involved, you need signals for "ready to accept data," and for "data available." The first two are identical, except for a few operations in some calculator chips, such as changing from degree to radian measures. Since most scan systems work on the basis of digit signal selection, remember that there can be close coupling among all three function circuits.

The data bus is used to enter data and operations, to select the output digit and to transmit output. Your first two decisions are dictated by the number of entries possible, from 20 to 40 (or more). The next two follow from the close connection of the scan system.

For calculator chips input data must be in an arbitrary code dictated by the scan arrangement of the particular chip; the same is true for operation selection. In principle you can provide a hardware code converter, say from ASCII to the calculator code, but this is easier to do by software in the CPU. Similarly, for calculator chips, output is in seven-segment code. This includes sign and error information. Decoding, again, can be by hardware, but software is easier.

Since the data bus is used for input, it would be possible to use a single address location. However, for flexibility allow for some control by address change. Eight arbitrary locations are used in the prototype, specifically lines A_0 to A_2. They can be used for such functions as lock-in, lock-out, and degree-radian selection.

The RDS and WDS strobes are needed. The write strobe corresponds to entry, the read strobe to readout. However, beware of the input delays, and the possible delay on output, digit read-out, hold and ready circuits must be provided. Figure 8-5 shows the block diagram.

CIRCUIT DESIGN

The address select circuit need not be much different from those already used. You can have leeway locating parts by using 7485 comparators. A 74138 can be used to decode the lower addresses, generating the individual control signs. Or, they can be microprogrammed.

The circuit used in the prototype is shown in Fig. 8-6. It requires only two comparators and two gate packages. It gives full freedom of address location, but only three control signals. It also

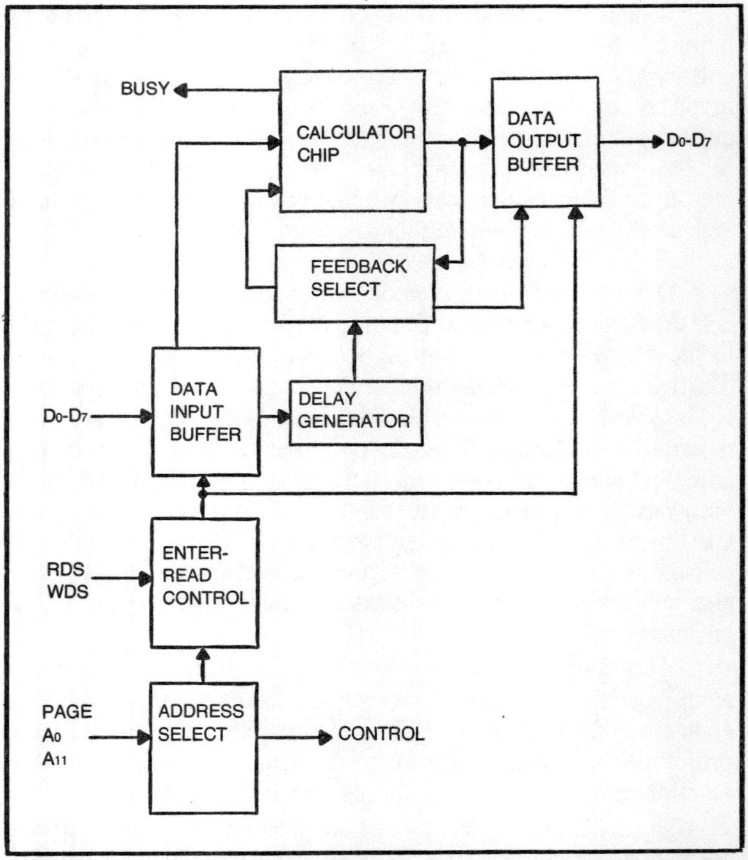

Fig. 8-5. Block diagram of a calculator board using a calculator chip for number processing. Even a simple chip can markedly increase computer throughput on arithmetic problems.

gates the read and write strobes to provide the enter and read signals at address time.

For the data signals it was decided to fully decode the lower four bits, giving up to 16 signals, primarily for digit selection, and to use the upper four bits for key line selection. However, to allow even greater flexibility, the circuit was arrayed to allow use of D_3 for key selection if needed. For chips that use BCD or octal input, the decoder package can be omitted and replaced by a set of jumpers on a header package. The prototype circuit is shown in Fig. 8-7. Type D 74173 registers are used to latch the input data and to provide for hold. The lower four bits are latched by either

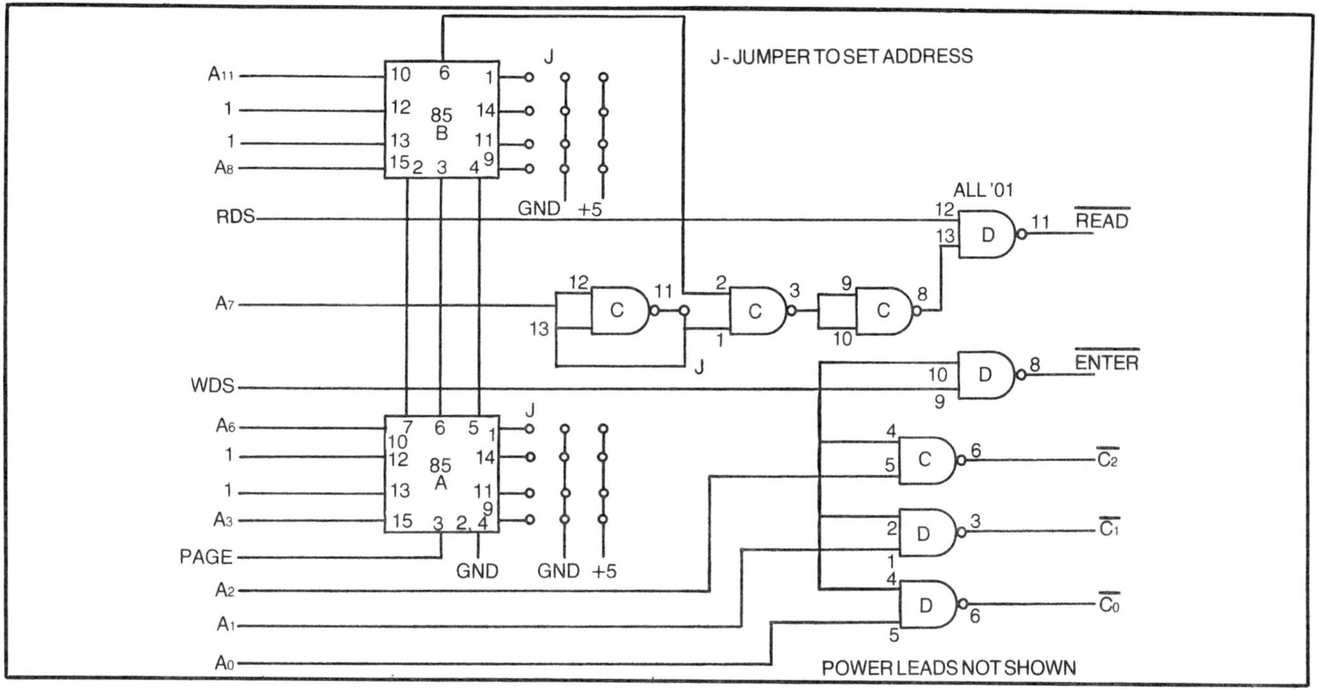

Fig. 8-6. Address decoding circuits for the calculator board. Three address lines are available for control. The address, in blocks of eight, is set by jumpers at the 7485 comparators. The write strobe is used to generate the enter command.

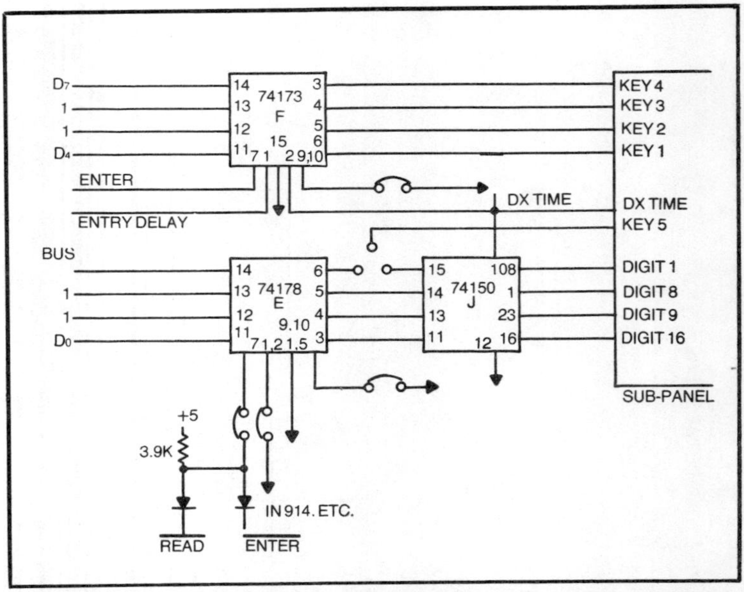

Fig. 8-7. Method of decoding a data input to provide for digit line-feedback, and for reading the desired output digit. Input lines D_0 to D_3 set up a strobe, D_x TIME, which is generated each time the digit occurs. For input, this is fed to the key line by digits D_4 to D_7.

the read or write signals. This provides output digit selection as well as input selection. The upper four bits are latched only by the write signal. This latch is isolated from the key lines at the end of the entry delay period by the signal generated. This signal may be needed to gate K5, if this option is used.

The lower four data bits are decoded into digit select by at 74150 (one of 16) decoder. Its output occurs at digit select time (D_x) to feed the key lines or to select the correct part of the segment signals. As is, this circuit allows selection of up to 16 data lines and four key lines. With very simple additions up to eight digit lines and five key lines can be handled, and up to 16 digit and 16 key lines are no problem.

The necessary wait and ready signals used in the prototype are generated by the circuit of Fig. 8-8. The entry delay and the following no-entry delay are generated by multivibrators, the periods are set to suit the chip. The two outputs are ORed together with another signal on the chip subpanel to creat a ready signal. They can be masked by an $\overline{S}$-$\overline{R}$ circuit fed from the address control bits; This allows use of the ready as a maskable interrupt.

The output circuits are shown in Fig. 8-9. Two more 74173 type D registers are used to hold the output data. The registers are clocked at D_x time. They hold data when an input occurs and when output is requested. If desired, the data at write can be locked out by control signals. The output can be energized directly by the control circuit provided, or can be energized by the read strobe, which is normal.

In this layout only two control latches are provided. More can be added to provide other control signals to the calculator chip. The output control mask can be used for one. Or, signals can be manual operations generated by jumpers or on-board switches, or from a front panel switch. The subpanel area shows provision for two such control leads. For further flexibility, a spare socket is shown. This is a 16-pin type, usable with any 14 to 16 DIP.

BOARD CONSTRUCTION

Don't build the board until you have the calculator chip on hand. The reason is that it may be possible, desirable or necessary to simplify some circuits, or to add to or modify others. However, I suggested that you build the board and check it out before starting the subpanel that carries the calculator chip. This makes it easier to concentrate on the calculator operation during the final check.

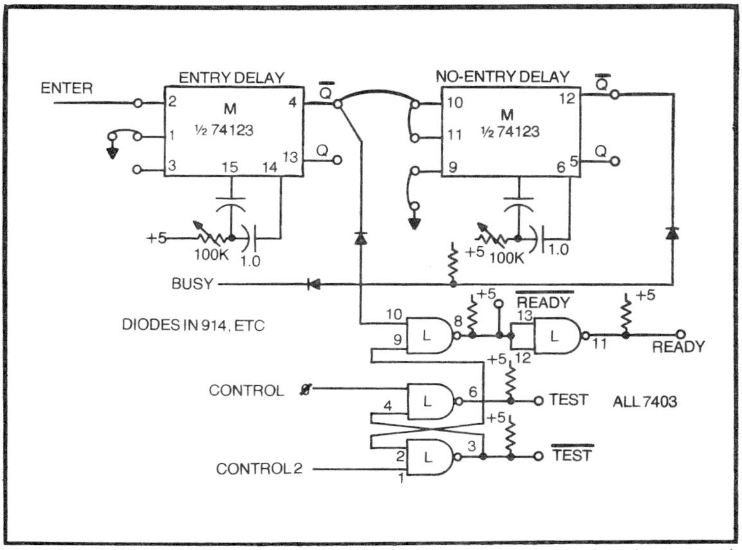

Fig. 8-8. Wait-ready circuits necessary to compensate for keyboard entry and no-entry delays, and for the calculator chip being busy on an operation.

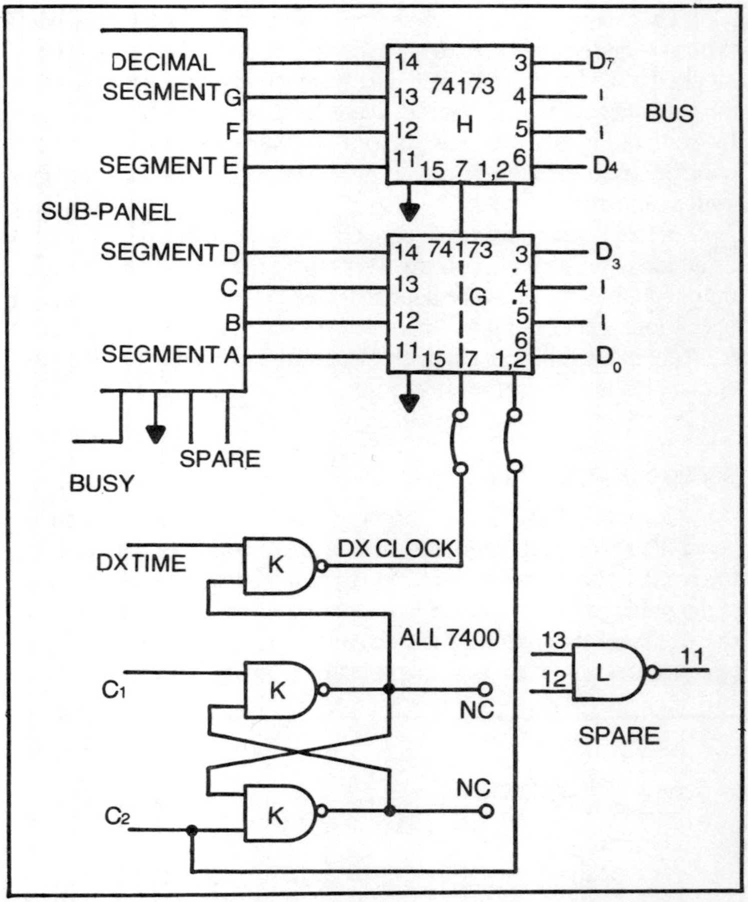

Fig. 8-9. Output circuit of the calculator board, with output being latched into the buffers at Dx time, under C_1 control. C_2 places output on the bus at read.

Board layout of the prototype is shown by Fig. 8-10. Layout is straightforward, following signal flow. Note the fairly large blank area for the subpanel. In the prototype this is left unetched and used as a ground bus. A better procedure cuts an opening to allow access to the subpanel.

The layout provides a number of connection points for signals, including back plane finger connections. These are for flexibility, and can be omitted if they are not needed by a particular chip.

The prototype uses elevated wirewrap sockets to allow soldering on both sides of the board. However, jogged socket construction is recommended because it is simpler, stronger, and needs less back plane space. A little rearrangement of the layout makes soldering on the top of the board easier.

CALCULATOR PANEL PRECHECK

With the board finished proceed to the preliminary check. The following steps are suggested:

1) Check for shorts and opens.

2) Install packages A and B. The B equals should be high, and go low when any of address lines A_3 to A_{11} (except A_7) are low, the page input is low, or if any of the B inputs are low.

3) Install C. Check the input high-low action, and the control line action.

4) Install D. Check the board selected, read, write, and control line action.

5) Install E. Input lines D_0 to D_3 should latch on read or write.

6) Install F. Input lines D_4 to D_7 should latch on write. Output should be controllable by the entry delay line, or the D_x line.

7) Install J. Digits D_0 to D_3 should control which of the digit lines feeds D_x.

8) Install M. Check that entry delay and no entry data are generated. Set these temporarily to 2 ms and 4 ms. Check the not-ready OR.

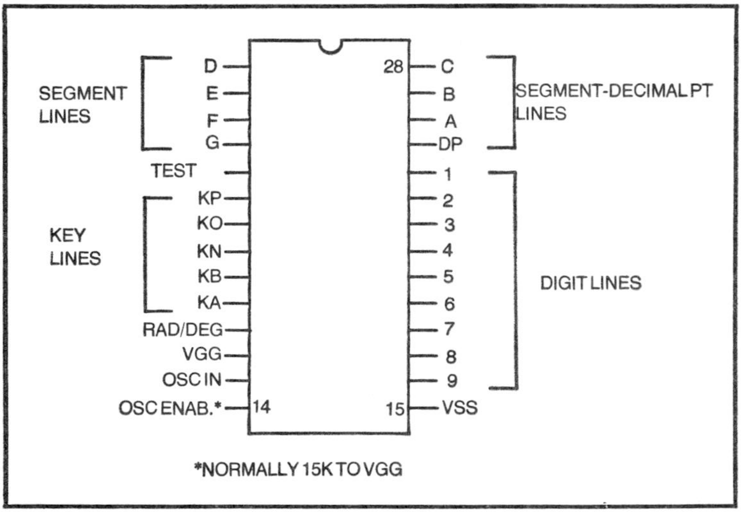

Fig. 8-10. Pin connections of the C-599 scientific calculator chip used in the prototype. See text and references for other chips.

9) Install L. Check control action and ready generation.

10) Install K. Check control action, D_x propagation.

11) Install G and H. Segment data should latch at D_x, time and appear on output (data) lines at read. With these checks complete, proceed to subpanel construction.

THE CALCULATOR SUBPANEL

The following is written for the General Instruments C-598 Scientific Chip used in the prototype. It is available from at least one of the hobby supply houses, and is used in several scientific calculators sold by department and mail order stores. The C-596 chip is similar but slightly simpler. Pin connections are shown in Fig. 8-10. (The 597 and 598 chips are nearly identical.) Except for details of pin connections and board layout, the following material applies to any chip.

The circuitry of the subpanel must provide the correct operating voltage to the calculator chip, and provide the necessary buffers and level changing between the chip inputs and outputs and the TTL circuits of the main panel. It must also route signals to the main panel connections.

Let's review a little. All the calculator chip outputs I know are open drain MOS-FET transistors, connected as open drain to V_{ss}. The equivalent circuit is shown in Fig. 8-11A with its resistance approximation. For proper operation the external circuit must sink current when the transistor is on the 1 state.

The usual input is a protected FET input with input resistor, shown in simplified form in Fig. 8-11B. Because of the internal resistor, the input rests in the 0 or no-key-pressed state. It must be pulled toward V_{dd} for the key-pressed condition. This must occur at the correct digit strobe. (Most chips interpret a continuous pull-down signal as an entry of one particular key.)

The C-598 chip is rated for V_{gg} equal to −15 V with −1.5 to 0 V at an input pin representing a logic 0, and −6 to V_{gg} representing a logic 1. You can set V_{ss} at +15 V and V_{gg} at 0. However, this requires low resistances in the output. Also, you have regulated +12 V available. So, for this chip V_{ss} is set at +12, and V_{gg} at −3 V.

The key line rests close to V_{gg}, now +12 V (high impedance), and must be pulled down to at least +6 V, preferably much lower. The upper voltage is beyond the rating of normal totem-pole TTL circuits, but several high-voltage open-collectors are available. A factor in choice is the number of key lines. As shown in Fig. 8-12,

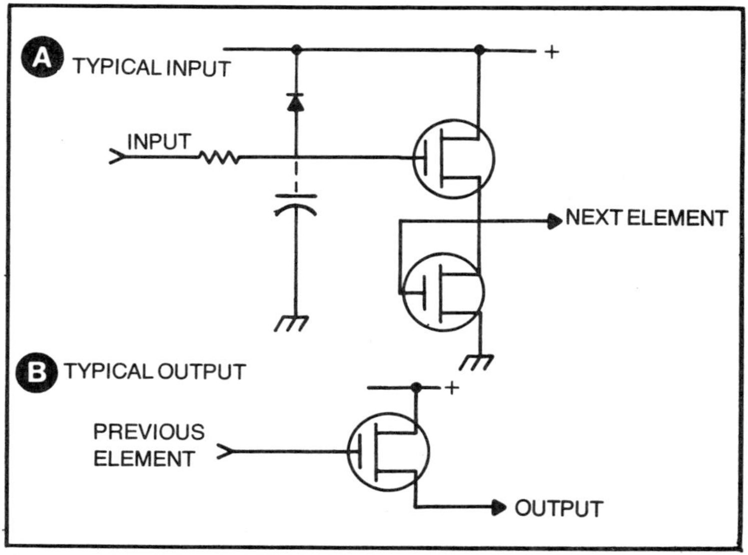

Fig. 8-11. Typical input and output circuits for calculator chips. Most chips operate at 9V, and must have interface circuitry at each point of connection to the calculator board.

the C-598 can be arrayed to operate with only two lines, by providing dual key inputs, selected by a function key. Or, the function can be omitted by providing five key line inputs, each key now being unique.

While you can use the fifth key-line provision of the basic board, a buffer package is still needed. A four to 10-line, high voltage, decoder buffer is available that reduces complexity in the digit select circuit. The elementary key-line circuit and the complete circuit become that of Fig. 8-13.

The standard MOS-output-TTL-input circuit is shown in Fig. 8-14 for the high and low conditions. The resistances can be calculated from the condition given. For the prototype, the standard values of 2.2K are used. (Note the change in resistance values if +15 and 0 V are used for the bus.) The complete output circuits are shown in Fig. 8-15.

You need a ready signal. With the C-599 chip this is fairly easy. The decimal point disappears from the display whenever the chip is in the internal calculate mode. Also, all data digits are suppressed. The sign digit remains on, flashing at intervals as the calculation progresses. Disappearance of a digit can be checked by noting that segments b and c must have disappeared at digit time.

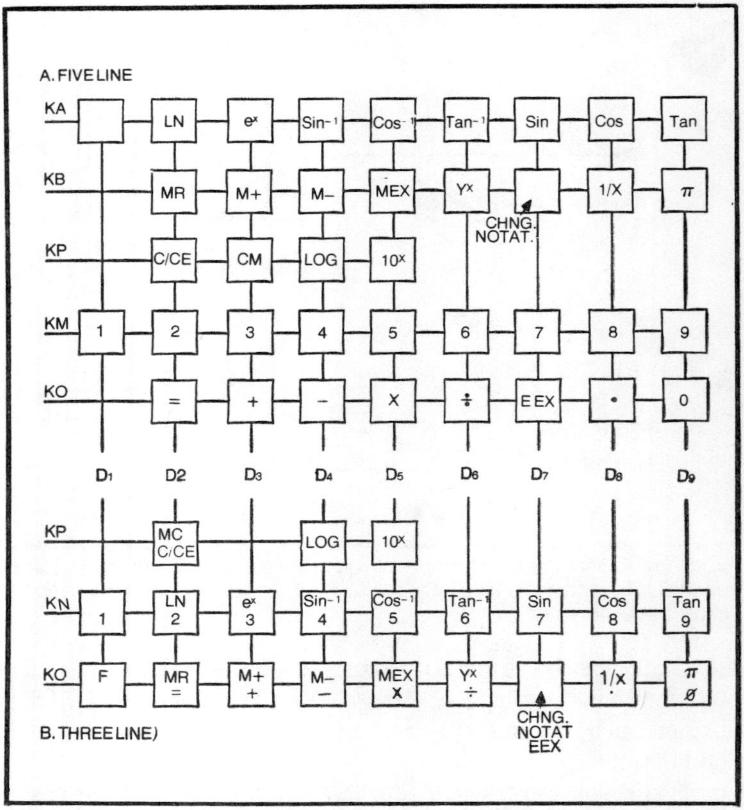

Fig. 8-12. Key line connections for the C-599 chip. These alternatives are rarely found.

Another signal may be useful. This is the error signal. It can be developed by a software routine, or by checking segments e and f. To give some flexibility in future use, the circuit of Fig. 8-16 was adopted. This provides for one or two inputs, and for gating. It also provides for direct or inverted output. Normally, it would serve as a simple buffer for the decimal point, which must be present or the calculator chip is busy. The fact that the DP signal is gated off during the interdigit interval is ignored, since its only effect is to increase the wait period a little. (This period can be filled out by adding a retriggerable multivibrator.)

The remaining connections that are needed are shown in Fig. 8-17. Since current demand is low, 20-30 mA, a simple zener circuit is used to secure the –3 V. The resistor to pin 14 energizes

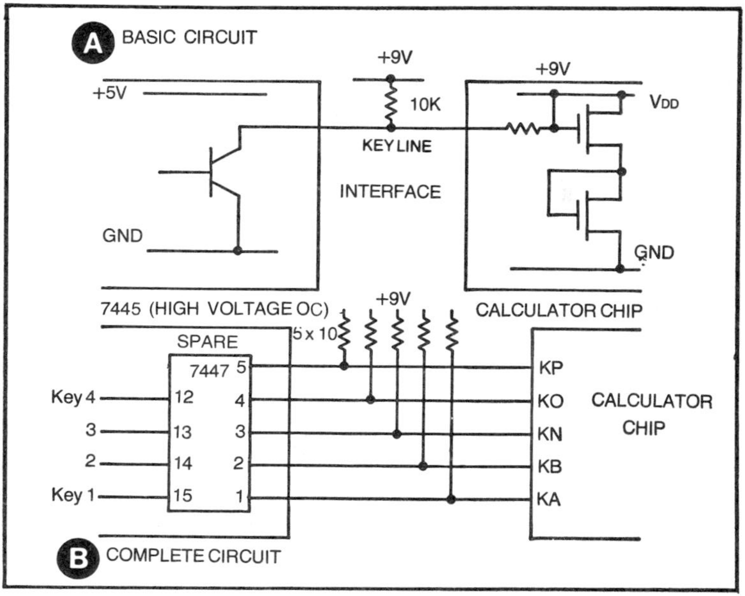

Fig. 8-13. Method of controlling key line input by use of a TTL high voltage open collector gate. A pull-down transistor could also be used.

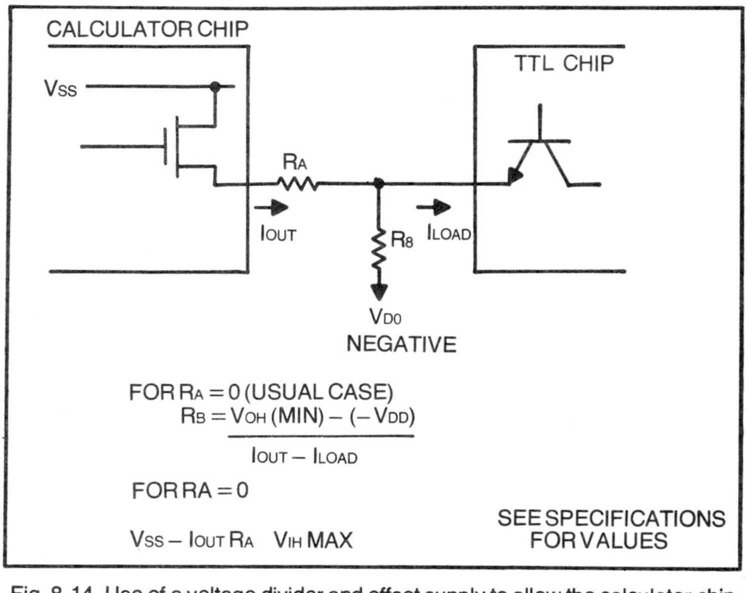

Fig. 8-14. Use of a voltage divider and offset supply to allow the calculator chip pull-up output to generate a TTL input.

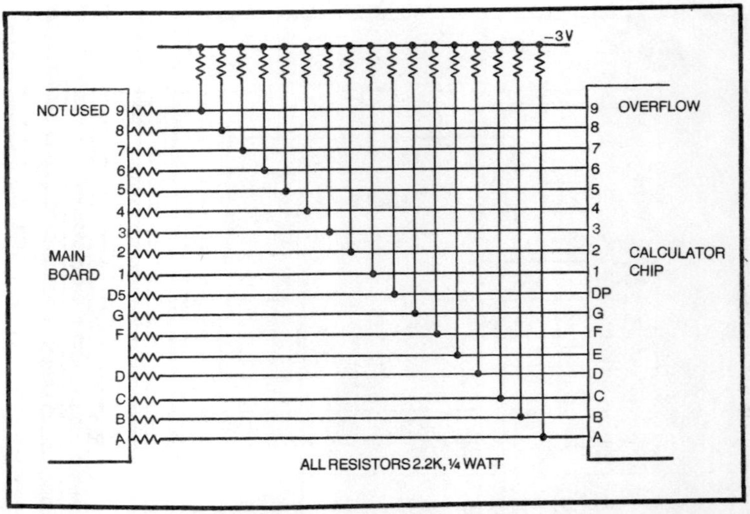

Fig. 8-15. Complete connection diagram for the digit and segment outputs of the calculator chip.

the internal oscillator. The degree-radian selection is left as an option, selected during construction by a jumper.

SUBPANEL CONSTRUCTION, TEST AND INSTALLATION

The layout used for the panel and subpanel is shown in Fig. 8-18. One-quarter watt resistors are used to help the size problem.

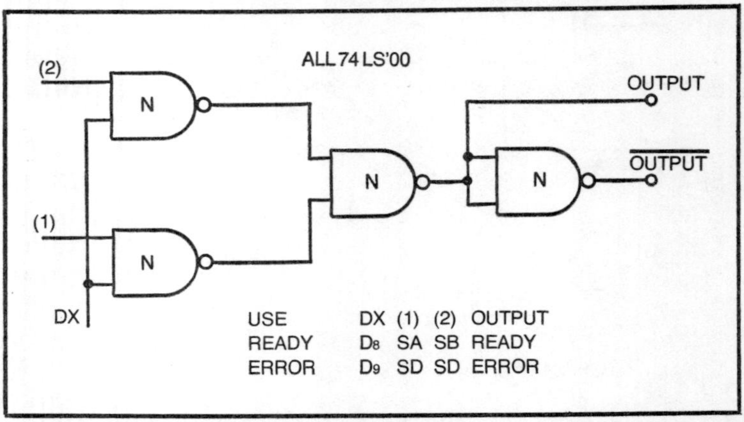

Fig. 8-16. Method of generation of error and ready (busy) signals by decoding. The segments and digits vary markedly from chip to chip, and may require an eight-input gate to decode them.

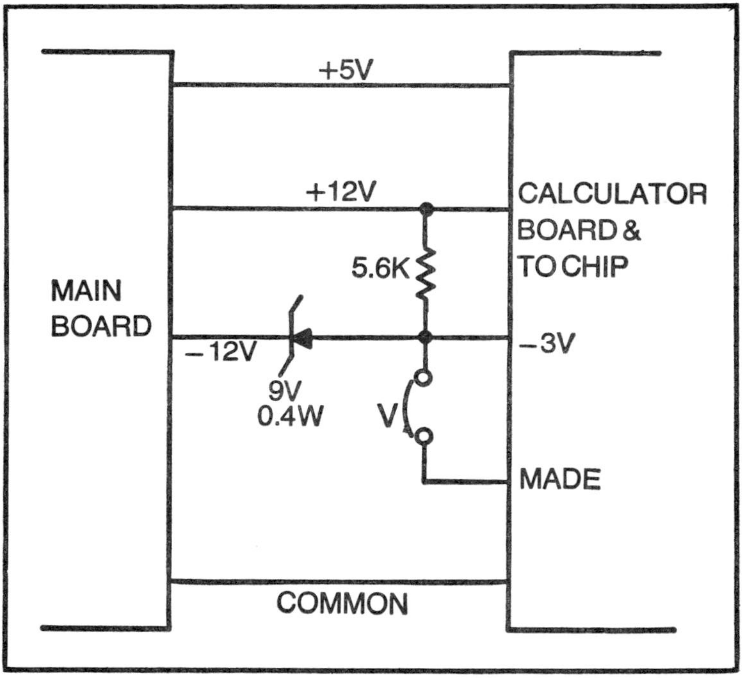

Fig. 8-17. Power circuits for the calculator chip, using the ±12 V leads to generate a −3 V offset supply. Mode sets degree or radian measure, and may not be present in some chips.

When complete the board can be checked adequately by applying power, and checking output signals with an oscilloscope. Temporarily jumpering a digit output to a key input allows calculation check, verification of connections, and measurement of timing for the particular chip.

The board is prepared for installation on the main board by inserting leads of number 26 bare, tinned wire, through the connection pads. These should be about one inch long, projecting through the component side of the board. They are then bent across the edge of the board in a sharp u. When installing this assembly on an uncut main board, use a piece of bare scrap board or other insulation between boards to prevent shorts. Figure 8-19 shows the completed board.

If it should be necessary to remove the subpanel, it is probably easiest to clip the interconnection leads. The stubs can serve as test points, if necessary. They would be removed, and new jumpers inserted for reassembly.

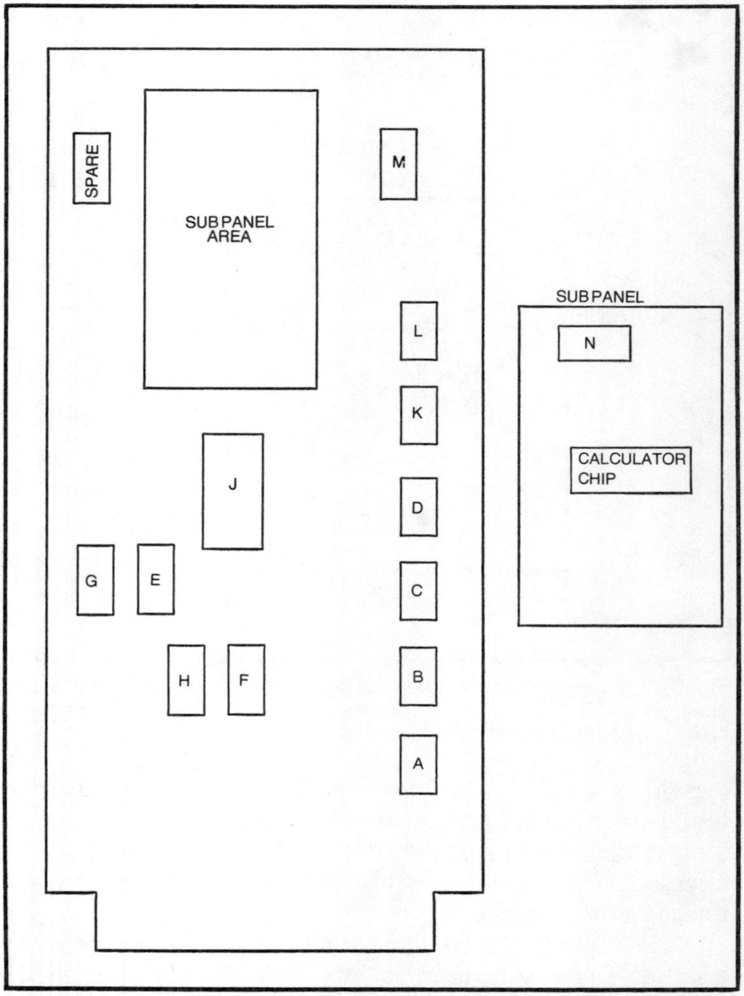

Fig. 8-18. Layout of the calculator board and the subpanel that carries the interface circuits and calculator chip. It is recommended that the calculator area be cut out.

FINAL CHECKOUT

Before starting final check, pause and make a table of data line inputs that correspond to calculator inputs, and the data line outputs that correspond to segment outputs. You need this data often in programmmg, so it is worth making a smooth copy. This can be in the form of 16 x 16 array for the input, and a 7 x 12 array for output (10 digits, sign and error signal). A 5 x 12 output array, of segments a, b, e, f, g can be used also.

For final bench check, place the board in the addressed mode by jumpers. Now put in trial data or operation signals, check for

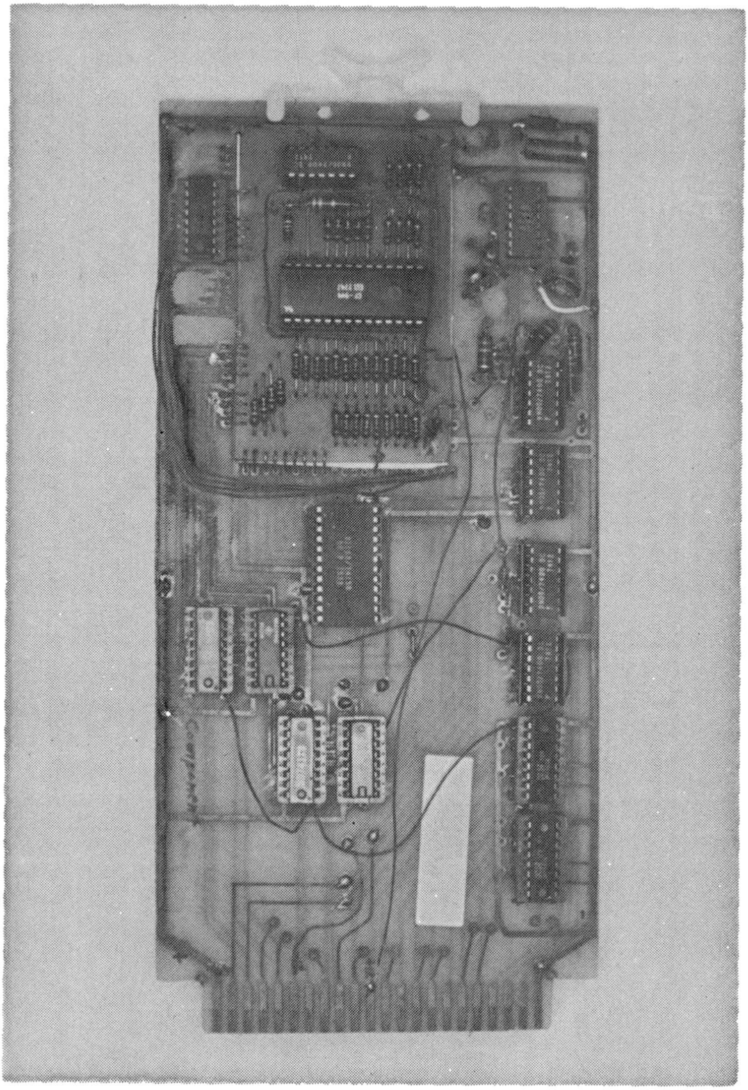

Fig. 8-19. The complete calculator board. An early design using wire-wrap sockets ¼ inch above the board to allow soldering on both sides. Use of jogged sockets is recommended.

correct results. Check for output digit selection and for ready-signal performance. Set the delay multivibrator to give reliable input and positive assurance that the chip is ready to accept new input.

Final checking is in the computer. This must be under software control, either from the front panel, or by a program. Check the major features such as digit and key selection and output, under panel control; then proceed to the program check.

SOFTWARE FOR THE COMPUTER PANEL

To take full advantage of the computational ability, you need a number of software routines. These can be written as callable subroutines or can be incorporated in other programs. Some of the routines to create are:

- Enter an eight-digit number with sign and decimal.
- Enter an eight-digit number with sign and decimal, and a two-digit exponent with sign.
- Check for error signal and set a flag if found.
- Read out an eight-digit number with sign and decimal.
- Read out a five-digit number with sign and decimal, plus exponent and sign.
- Determine, by multiplying two large numbers and subtracting the most significant digits, the number of digits used internally in calculation. (This is usually two or three digits more than are normally displayed, and may be as many as 15 digits.) Write a routine to recover all significant digits (ultimate precision).
- Write a routine for extra precision data entry.
- Write a routine to check that the decimal point follows the first significant figure, adjusting location as necessary.

In doing these routines, you need to adopt conventions for storage, entry, readout format, and operation symbols. See the various Tiny BASIC, BASIC, and Fortran conventions.

In addition to these I/O routines, you need to write the routine or routines for problem solving. There are several possibilities. One is relatively easy, but it places a large load on the user. It consists of writing a series of data words corresponding to the needed keystrokes in sequence, then following this with a link instruction in the form of a DO, keystrokes i to n. A subsequent instruction reads digits i to n plus sign.

If you think about this, you see that there is a simple map relation between the symbols 1, 2, SIN, +, etc. and the keystrokes, and the data/address used by the link instruction. Since this is true, the load on the user can be reduced greatly by a look-up table.

Table look up can be handled in two ways. One uses a sequential sort, like the IL table of NIBL. The second generates a hash function to use for look up. The algorithim for this is not too difficult:

1) If an ASCII letter is followed by a blank, use it as a NIBL variable to determine the look-up input.

2) If an ASCII number or period is followed by another number and a blank, use them in sequence as look-up input.

3) If an ASCII symbol is followed by a blank, use it as the look-up input.

4) If an ASCII letter is followed by a letter (SIN, EXP, etc.) generate a hash, and use it as the look-up input.

Often the trick of using the sum of the value of the first letter plus the second letter shifted one left and the third letter shifted two left will give a unique hash on the first trial. If not, a little search will show an easy rehash.

Whichever technique you use, retain the symbols and short forms of standard BASIC. This simplifies program portability.

You might wish to add another item to the algorithim. If a letter is followed by a number, use the number as an offset from the address given by the letter read as a NIBL variable. This gives a set of 260 variables of any length desired. Obviously, it uses a lot of memory.

If you are rewriting NIBL, it's no problem to insert the added symbols needed in the IL Table. The specific codes that must be entered to the calculator board latches vary with the chip used. For the CF-598, the codes for data input and output digit select are shown in Table 8-3, and those for seven-segment to BCD conversion are shown in Table 8-4. These are the items to be placed in ROM.

If one of the segment scan chips is used, say that from the TI-30, the output conversion is a little more complex. The output is by active segments. These must be sorted to the correct digit, then data for another segment secured and sorted. For most displays, five sorts are sufficient to establish the characters, but this should be checked by preparing a table for the characters used by the calculator designer. The usual unique segments are a, b, e, f, g.

In writing routines, remember the decimal points. Common practice in floating point routines is to normalize these, that is, to

Table 8-3. Data Line Codes for Enter and for Digit-Select at Readout. The Operating System Must Translate the Arithmetic Symbols Into the Key Stroke Data Codes in the Correct Sequence, and Must Read the Output Digits in Order. Two Output Modes are Suggested, One With Exponent and One in Fixed Mode. Normalize the Exponential Form to the Form N.NNNNNbEE, the B Being a Blank.

Entry	Data Line 0	1	2	3	4	5	6	7
0				1				1
1							1	
2	1						1	
3		1					1	
4	1	1					1	
5			1				1	
6	1		1				1	
7		1	1				1	
8	1	1	1				1	
9				1			1	
=	1							1
+		1						1
−	1	1						1
x			1					1
÷	1		1					1
EEX		1	1					1
•	1	1	1					1
√					1			

Entry	Data Line 0	1	2	3	4	5	6	7
LN	1				1			
e^x		1			1			
Sin^{-1}	1	1			1			
Cos^{-1}			1		1			
Tan^{-1}	1		1		1			
Sin		1	1		1			
Cos	1	1	1		1			
Tan				1	1			
MR	1					1		
M+		1				1		
M−	1	1				1		
MEx			1			1		
Not.		1	1			1		
1/x	1	1	1			1		
π				1		1		
CE	1			1				
CM		1		1				

Digit	Data Line 0	1	2	3	4	5	6	7
1								
2	1							
3		1						
4	1	1						
5			1					
6	1		1					
7		1	1					
8	1	1	1					
9				1				

place the decimal point to the left of the first significant nonzero figure. This is easily done by multiplying or dividing by 10 as needed. Some calculator chips can do this by a special key. Others automatically shift the decimal to a standard position on any operation. Don't forget to include a routine to check for automatic fixed-to-floating conversion.

SPECIAL NUMBER PROCESSOR DESIGNS

It is evident that a considerable amount of hardware and software is required because of the characteristics of calculator chips. This includes feed of output to input for key selection, generation of a form of WAIT to compensate for key debounce, output digit selection and output conversion.

Essentially all of these can be eliminated by going to one of the special chips intended for number processing. The cost of this approach may actually be less than the cost of using a calculator chip, the savings in auxiliary parts being greater than the calculator chip cost differential.

An example of a design specifically intended for use with the SC/MP is shown in Fig. 8-20. This is an extension of a circuit given in the manufacturer's data sheet. The simplicity of the design is evident.

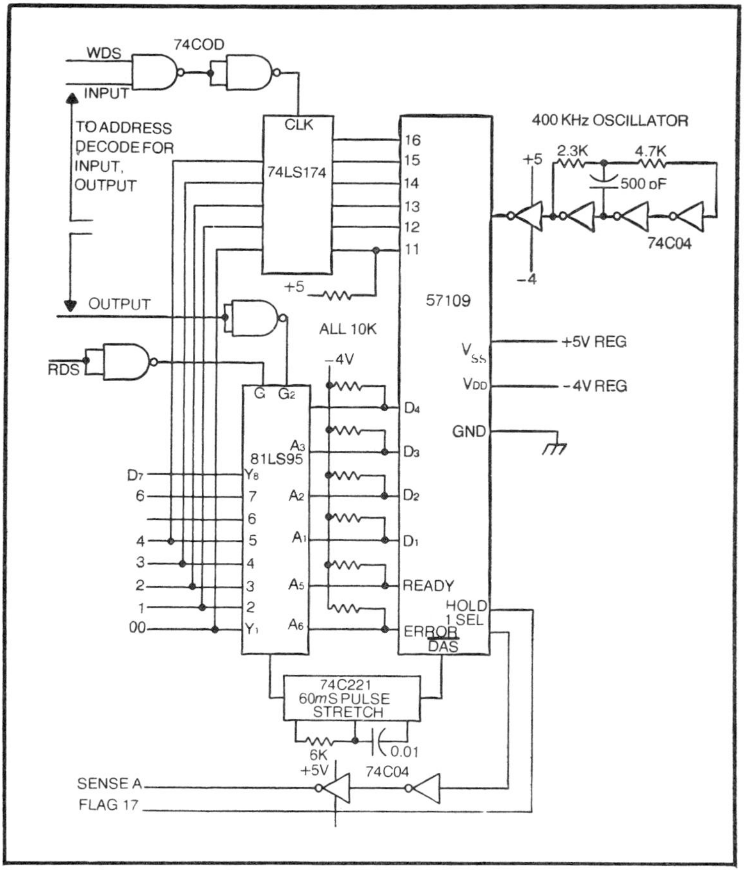

Fig. 8-20. A number processor using the MM57109 chip, and intended for SC/MP applications. Since the chip is designed for microprocessor use, interfacing is much simpler. Address selection can be the same as in Fig. 8-11.

Table 8-4. Seven-Segment Plus Decimal Code as Eight-Digit Binary Words. The Code Varies With Chip Manufacturers. The Operating System Must Translate This Code Into Hex With Provisions for Decimal and Errors.

	Segment							
Character	a	b	c	d	e	f	g	DP
0	1	1	1	1	1	1		
1		1	1					
2	1	1		1	1		1	
3	1	1	1	1			1	
4		1	1			1	1	
5	1		1	1		1	1	
6			1	1	1	1	1	
7	1	1	1					
8	1	1	1	1	1	1	1	
9	1	1	1			1	1	
							1	
C	1			1	1	1		
E	1			1	1	1	1	
F	1				1	1	1	
H		1	1		1	1	1	
H			1		1	1	1	
	1				1	1		
Blank								
								1

Of course, one of these special chips can be used with the basic processor board with proper configuration of the subpanel. This could be done if a calculator chip fails in use, and an exact replacement is not available. Routine rewriting is necessary, of course.

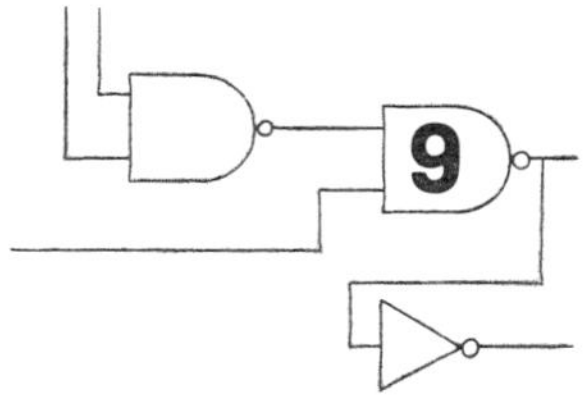

Using Other Microprocessors

When I began working on *How to Design, Build and Program Your Own Working Computer*, I studied the microprocessors that were available to determine which one seemed best. I considered their expected availability, cost, the simplicity of their circuitry, the availability of higher languages, and their support in terms of data and program availability.

The SC/MP was one of the simplest, yet retained a good balance of features. It also offered some unique features, such as a design adapted to control, and a higher language that offered immediate execution.

While this microprocessor has performed as expected, it's reasonable to look at more complex chip designs. Keep in mind the possibility of substituting for the SC/MP, and also the possibility of operating in conjunction with it using multitask and multiprocessing modes. This again raises the questions of chip selection and general design.

While single-board computers definitely reduce cost, construction and testing are complex, and it is sometimes difficult to keep them flexible. Keep in mind the original design concepts:

- Simple layout and construction
- Single function per board
- Add-on approach with a standard bus
- Full bus buffering.

Each board is a separate project.

With experience, a change toward more computing power is in order. But this does not necessarily require an increase in construction or test complexity. You can increase computing power by expanding the instruction set. This is the approach for the remaining material in this chapter.

There is one chip in particular that is a natural choice for a new design. This is the INS-8073 by the makers of the SC/MP. It uses the same concept of internal buses, a related instruction set, and has nearly the same capability for control. It also has 16-bit arithmetic, plus some search and match capability. The 8073 has 2.5K of on-chip memory mask programmed at the factory in a second version of NIBL. (This chip is discussed in Chapter 4.)

One benefit of a standard approach is the availability of programs. This leads to the chips that can use the 8080 family of instructions. There are now a number that can, their instruction set being a super-set of the 8080 instructions.

Because of its need for auxiliary chips, the 8080 itself does not fit well into the SCAMP:Y bus-board structure. Several of the newer chips do, the Z80 for example. It is powerful, fast, inexpensive, and readily available. It's a likely candidate for replacement or multiprocessor use.

Let's look at these two chips, the INS-8073 and the Z80, in some detail, then consider their use as the CPU in a home-built system.

THE INS 8073 ARCHITECTURE

The general structure of the INS is shown by the block diagram of Fig. 9-1. The central part resembles the SC/MP very closely, but differs in three major respects: (1) The address bus is 16-bits, rather than 12-bits, with four bits multiplexed on to the data bus; (2) there is an added temporary register associated with the arithmetic unit, and with the high and low address buses; and (3) in addition there are differences and extensions in use of registers.

A major change is the operation of the four registers holding addresses. Designated as program counter (PC) and as pointers 1, 2, and 3 (P1, P2, P3) in the SC/MP, these are instead designated as PC, SP, P2 and P3 in the 8073 (the SP indicating stack pointer). As before, stacks can be maintained on P2 and P3. However, the interrupt line SA no longer causes an interchange between PC and P3. Instead SA causes a designated address to be loaded into PC, forming an interrupt call. SB calls a second address.

Another major extension is the association of the extension register and the accummulator, which can form an extended accumulator (EA) of 16-bits for double precision arithmetic. The temporary register is primarily for these operations. The SC/MP serial input-output capability of the extension register has been eliminated.

Outside the central area, the largest extension is the addition of internal RAM, 64 bytes in all units of the series, and internal ROM, 2.5K bytes in the 8071, 72 and 73.

Many features of the INS-807X series remain the same as the SC/MP. Bus access signals are identical to work with the microbus. (An IEEE standard on this is expected.) Address strobes are similar, except that NADS is no longer needed, so the strobe structure is simpler. The control provisions, three flags and two sense lines are identical.

THE INS-8073 INSTRUCTION SET

The resemblance of the 8073 instruction set to that of the SC/MP is due mainly to the architecture. There are one, two and three-byte instructions, as shown in Fig. 9-2. Microcoding of the first byte is used, much like the SC/MP, with bits D_0 and D_1 used to designate one of the pointers. D_2, or D_0 to D_3, may contain data.

The instruction set and the mnemonics for these are shown in Table 9-1. This form does not include the first and second operand, nor indicate the presence of data. The DELAY instruction is eliminated, and that there are several important added instructions. You can call one of the 16 subroutines with a single-byte instruction. PUSH and POP handle accumulator and pointer designated LOAD-READ instructions. BRANCH is added to extend JUMP capabilities.

The complete instruction set is shown in the first byte map of Fig. 9-3. This shows the addressing by microcoding, as in the CALL and ADD instructions. Microcoding causes the similarity to many of the SC/MP instructions, for example, ADD, AND and XOR. But note that others are changed, such as the location of ILD and DLD. It's best to regard the SC/MP and 807X series instruction sets as completely independent.

Figure 9-3 is coded to show the one-byte, two-byte and three-byte instructions. Note that the one-byte instructions are all in the upper half of the table, and those that are three-byte either have X02 in the first half-byte, or a low multiple of X04 in the second half, showing the use of microcoding in generating them.

All in all, the instruction set of the 807X series represents a considerable extension of SC/MP. Of course, this also means more

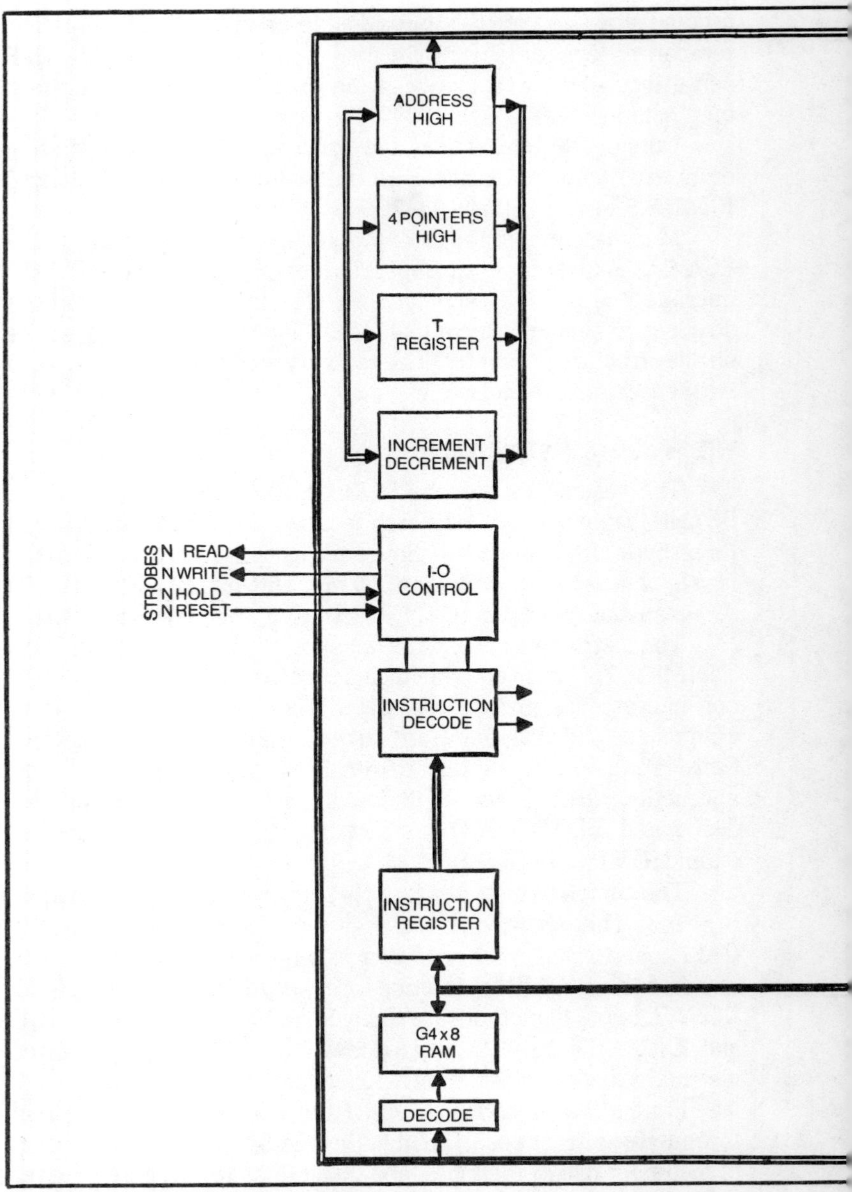

Fig. 9-1. Block diagram of the INS-8073 microprocessor that includes a 2.5K version of NIBL on the chip. Contrasted to the SC/MP, the address outputs are 16-bit, rather than 12-bit plus four multiplexed bits. Two registers have been added, which permits 16-bit arithmetic. However, there is no serial input or output.

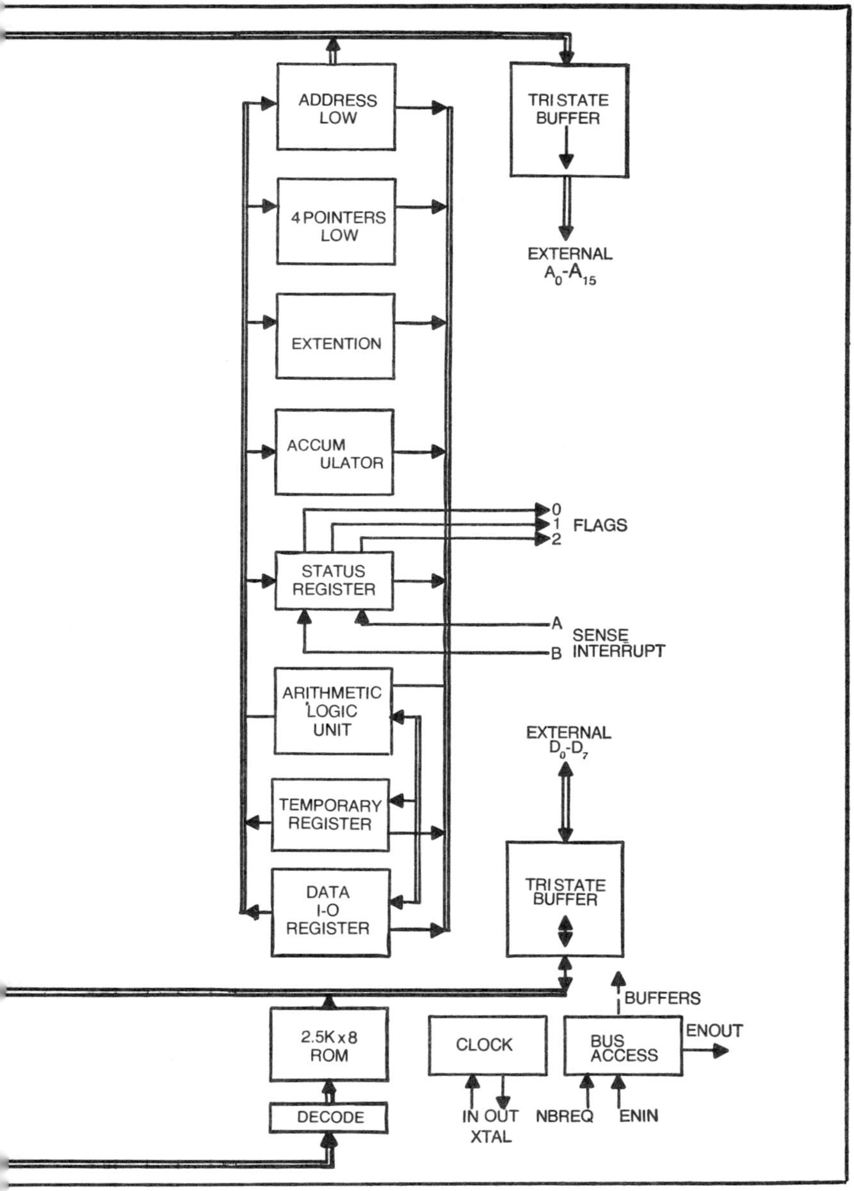
ADDRESS LOW
4 POINTERS LOW
EXTENTION
ACCUM ULATOR
STATUS REGISTER
ARITHMETIC LOGIC UNIT
TEMPORARY REGISTER
DATA I-O REGISTER
TRI STATE BUFFER
EXTERNAL A_0-A_{15}
0
1
2
FLAGS
A
B
SENSE
INTERRUPT
EXTERNAL D_0-D_7
TRI STATE BUFFER
BUFFERS
2.5K x 8 ROM
DECODE
CLOCK
IN OUT
XTAL
BUS ACCESS
ENOUT
NBREQ
ENIN

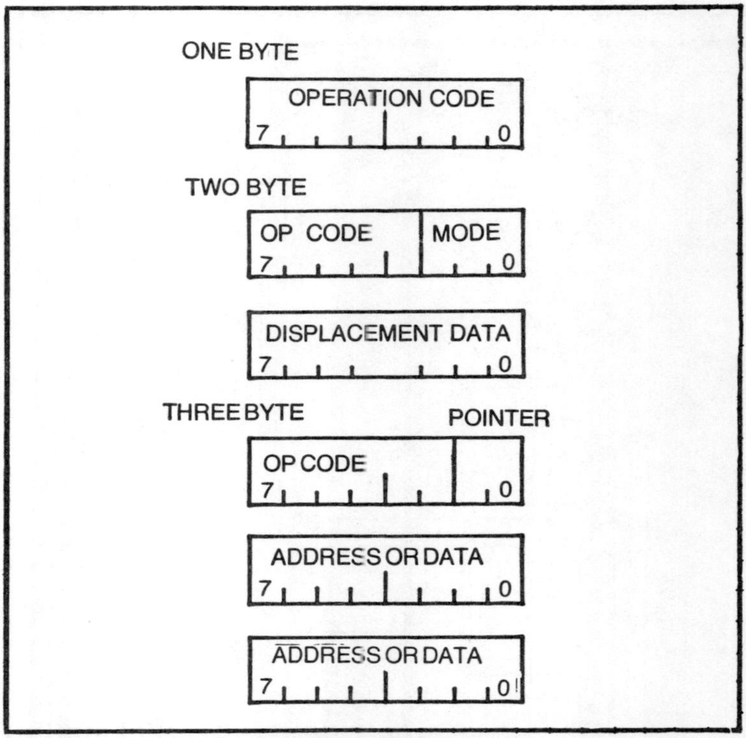

Fig. 9-2. Form of instruction codes in the INS-807X series.

to remember and watch when you program in assembly code or machine language.

THE Z80 ARCHITECTURE

In general block form the architecture of the Z80 is very simple. See Fig. 9-4. An interesting exercise is to lay this block diagram beside one of the large microphotos of the Z80 chip. (These are commonly available in computer and book stores.) The true complexity of the design then becomes apparent.

Much of the power of the Z80 lies in its register set, shown in Fig. 9-5. It is an extension of the set of the 8080, differing in the addition of the primed index registers. The instruction set of the Z80 is set up to use all instructions of the 8080 exactly. These factors allow the Z80 to run programs originally created for the 8080.

The accumulator performs all arithmetic and logic functions, using register or memory data. The flag register records the status

Table 9-1. INS-807X Instruction Set. Numbers in Parentheses Show Whether the Instruction is One, Two or Three Bits, and Whether it has Variable Length.

ADD	(1,2,3)
AND	(1,2)
BRANCH	(2)
CALL	(1)
DECREMENT/LOAD	(2)
DIVIDE	(1)
EXCLUSIVE OR	(1,2)
EXCHANGE	(1)
INCREMENT/LOAD	(2)
JUMP	(3)
LOAD	(1,2,3)
MULTIPLY	(1)
NO OPERATION	(1)
OR	(1,2)
POP	(1)
PUSH	(1)
PUSH/LOAD	(3)
RETURN	(1)
ROTATE	(1)
SEARCH	(1)
SHIFT	(1)
STORE	(2)
SUBTRACT	(1,2,3)

of operations and conditions. The execution detail of many instructions depends on status, and needs careful study to use correctly.

The register pairs BC, DE and HL may be used individually for eight-bit operations or as a pair for 16-bits; usually for addressing, but also for 16-bit arithmetic. All of these registers are duplicated in the alternative register set. A group of exchange instructions make both sets available to the programmer.

The program counter performs the normal function of keeping track of location in execution. The stack pointer plus register exchanges give simple use of multiple level interrupts, subroutine nesting, and various forms of data handling. The last is further simplified by the two 16-bit index registers that allow tabular data manipulation and fast code relocation, often by a single instruction.

The interrupt vector register is used to form the upper eight bits of a service address table, with the lower eight bits provided by the interrupting device. Several modes of interrupt operation are offered.

The memory refresh register is designed to give refresh addresses to dynamic memory. It generates row addresses for all such memory during the second half of each instruction fetch cycle (designated as the MI cycle) and is incremented each fetch. At the worst, the resulting refresh with a 2.5 MHz system clock is over

2nd HALF BYTE

1st HALF BYTE

	0	1	2	3	4	5	6	7
0	NOP	XCH A, E					LD A,S	LS S,A
1	← CALLS → 0	1	2	3	4	5	6	7
2	JSR	PUSH P2	PUSH P3	JMP		LD SP	LD P2	LD P3
3	← LD → EA, PC	EA, SP	EA, P2	EA, P3				
4	LD A,E				← LD → PC, EA	SP, EA	P2, EA	P3, EA
5	AND A,E				PUSH PC		PUSH P2	PUSH P3
6	XOR A,E				BP		BP P2	BP P3
7	ADD A,E				BRA		BRA P2	BRA P3
8	← LD → EA,PC	EA, SP	EA, P2	EA, P3	EA	EA-D	EA@P2	EA@P3
9	← ILD → A,PC	A,SP	A,P2	A,P3		← ILD → A-D	A@P2	A@P3
A	← LD → T,PC	T,SP	T, P2	T,P3	T,I	T-D	T@P2	T@P3
B	← ADD → EA,PC	EA,SP	EA,P2	EA,P3	EA,I	EA-D	EA@P2	EA@P3
C	←→ A_1(PC)	A_2(SP)	A_1(P2)	A_3(P3)	A,I	A-D	A@(P2)	A@(P3)
D	← AND → A, PC	A,SP	A,P2	A,P3	A,I		A@P2	A@P3
E	← XOR → A,PC	A,SP		A,P3	A,I	A-D	A@P2	A@P3
	← ADD → A,PC	A,SP	A,P2	A,P3	A,I	A-D	A@P2	A@P3

TWO BYTES
IN UPPER HALF •

Fig. 9-3. Instruction set of the INS-807X series as a half-byte matrix. There are many resemblances to the SC/MP set, but also many changes and extensions.

8	9	A	B	C	D	E	F
PUSH EA	LD T,EA	PUSH A	LD EA	SR EA	DIV EA,T	SL A	SL EA
———	———	———	———	———	———	———	———→
8	9	10	11	12	13	14	15
				MPY EA,T	BND	SSM P_2	SSM P_3
POP A	AND 5	POP EA	OR S	SR A	SRL A	RR A	RRL A
——→ E,A				←— XCH EA,PC	EA,SP	EA,P2	EA,P3 —→
OR AE				RET		POP P_2	POP P_3
				BZ		BZ P_2	BZ P_3
SUB A,E				BNZ		BNZ P_2	PN P_3
←— ST EA,PC	EA,SP	EA,P_2	EA,P_3 —→		←— EA-D	EA@P_2	EA@P_3 —→
←— DLD A,PC	A,SP	A,P_2	A,P_3 —→		←— DLD A-D	A@P_2	A@P_3 —→
←— EA,PC	EA,SP	EA,P_2	EA,P_3	EA,I	EA-0	EA@P2	EA@P3 —→
←— A,PC	A,SP		A,P_3 —→		←— A	A@P_2	A@P_3 —→
←— OR A,PC	A,SP	A P_2	A,P_3	A,I	A-D	A@P_2	A@P_3 —→
←— A,PC	A,SP	A,P_2	A,P_3	A,I	A-D	A@P_2	A@P_3 —→

THREE BYTES
ALL

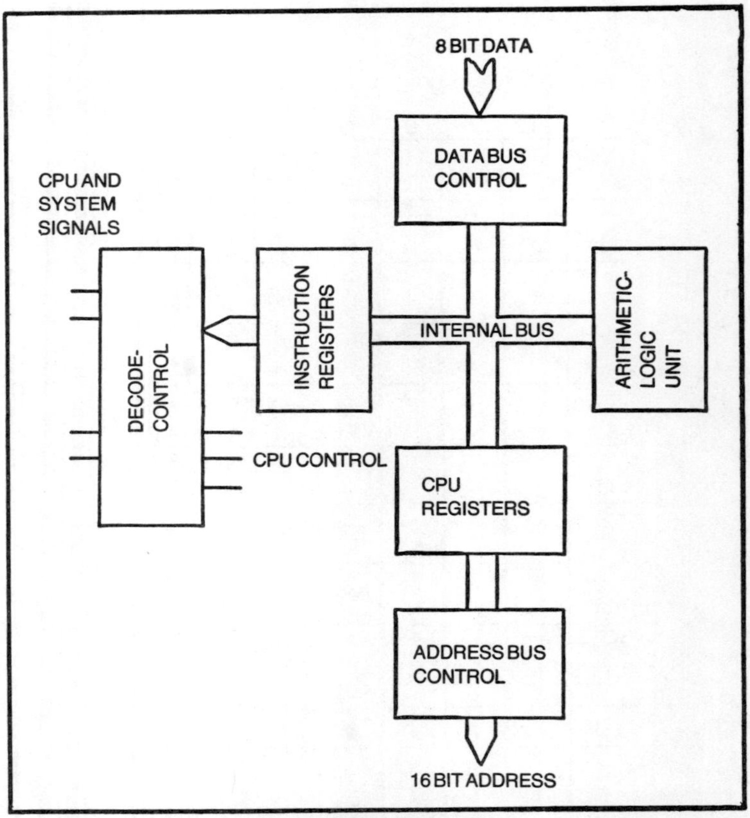

Fig. 9-4. Basic block diagram of the Z80 microprocessor.

twice the required refresh rate of common dynamic memory chips.

The simplified block diagram does not show the control signals that are covered later. They include address strobes, interrupts and a family of bus control signals. In addition to bus request and bus request acknowledge, the bus is used by the CPU for data, I/O and refresh.

THE Z80 INSTRUCTION SET

The instructions of the Z80 may be one, two, three or four bytes, with the additional bytes being either amplification of the basic instruction, or data. Microcoding the instruction is used extensively to designate registers and register pairs.

The basic instructions are given in Table 9-2. The table does not distinguish the first and/or second operands, nor indicate the presence of data, but concentrates on the basic nature of the instruction. Most of the mnemonics are clear from standard usage, with a few exceptions such as compare (CP) for block search, and complement (CPL). The block search, block transfer, bit and exchange groups are not found in the 8080.

The general arrangement of the instructions can be seen from the first byte instruction map of Fig. 9-6. For all but four, the first byte completely determines the nature of the operation, and whether the instruction is one, two or three bytes. For the four, the second byte is required to determine the nature of the instruction. These instructions may be two, three or four bytes.

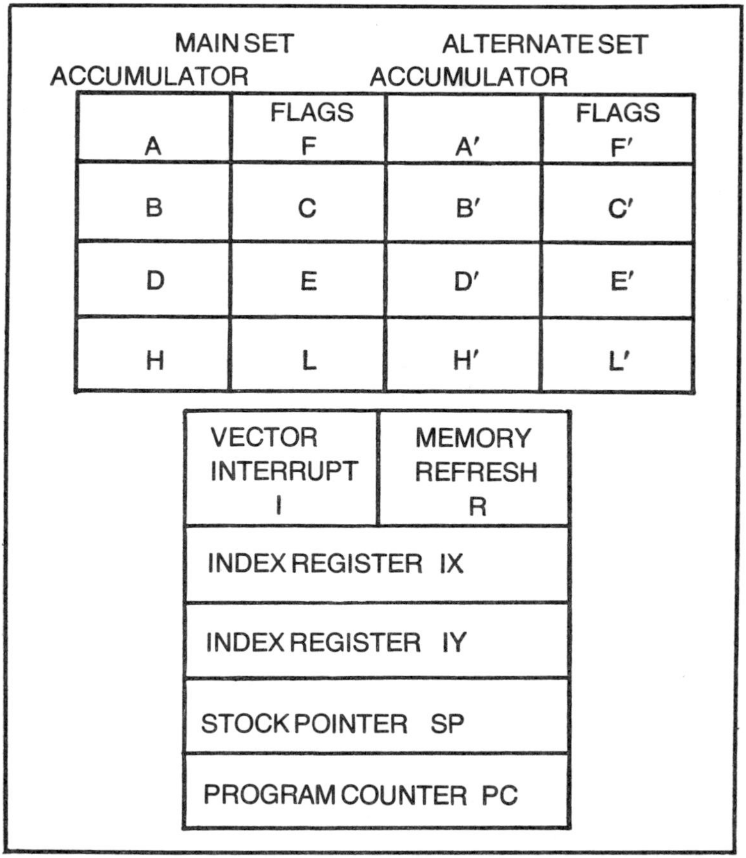

Fig. 9-5. Register set of the Z80. Much of the power of this processor comes from the family of register interchange instructions. The set is an extension of those in the 8080 processor.

Microcoding is used in all instruction bytes. For the first byte, the codes are indicated in Fig. 9-6. Nine coding tables are needed to keep track of the various register and data combinations. Additionally, some of the operations and operation results affect the flag registers. Four of the flags are used for conditional jumps, two for decimal arithmetic.

Because of the length of the instruction list, the number of the internal registers, and the complex interaction of instructions and flags, it is not practical to memorize the instruction set (as can be done with the SC/MP). Aids are necessary. Some useful ones are prepared by the chip manufacturer, and others can be found in the references. Extensions of the instruction map of Fig. 9-6 to include bytes one and two are suggested.

COMPARISON OF ADDRESSING MODES

A good part of the power of a microprocessor lies in the addressing available. A rough indication of the power is the number of different ways addressing is possible.

Table 9-3 lists addressing modes for the SC/MP, the 807X series and the Z80. In all cases but one, the nomenclature and list order is taken from manufacturer's publications. The word "substitute" is added to the SC/MP and INS-807X series list to indicate the effect of a displacement of −128. Possibly there should be another word added, "implied", to indicate such actions as addressing the I/O ports.

Once a threshold is reached, extensions of the addressing modes does not increase the ability to solve problems. Usually, however, the additional modes make solutions possible with fewer steps and in less time. Standardized notation of address modes is helpful.

In addressing the location of preassigned addresses that are implied by an instruction or invoked as a response are important. The SC/MP did not include any in its normally addressable space. However, both the 807x series and the Z80 do. Figure 9-7 shows their locations.

In the 8072 and 8073, the first 2.5K of address space is internal, preassigned, and not directly available to the programmer. In the 8070 these locations are external but partially preassigned. The preassigned locations are used for power up, interrupt jumps, and calls. All of the family place 64 bytes of RAM at the top of memory. In addition, the 8073 (the NIBL2) assumes that the locations FD_{00} to FD_{04} contain information about I/O

Table 9-2. The Z80 Instruction Set, With the Number of Bytes Indicated in Parentheses. Nearly All Have Complex Alternatives.

ADD	(1,2,3)
AND	(1,2,3)
BIT	(2,4)
BLOCK SEARCH	(2)
CALL	(3)
CLEAR	(1)
COMPLEMENT	(1,2,3)
DECIMAL ADJUST	(1)
DECREMENT	(1,2,3)
EXCHANGE	(1,2)
EXCLUSIVE OR	(1,2,3)
HALT	(1)
INCREMENT	(1,2,3)
INPUT	(2)
JUMP	(2,3)
LOAD	(1,2,3)
NEGATE	(2)
NO OPERATION	(1)
OR	(1,2,3)
OUTPUT	(2)
POP	(1,2)
PUSH	(1,2)
RESET	(1,2,4)
RETURN	(1,2)
ROTATE	(1,2,4)
SET	(1,2,4)
SHIFT	(1,2,4)
SUBTRACT	(1,2,3)

routines, and automatically jump to the start of page 8 (if it is ROM) and excute the program there.

In the Z80, calls and interrupts are also assigned to low memory. Further, if N vectored interrupts are used, 2N bytes of addresses are needed for jumps. These can be located anywhere in memory not otherwise preassigned.

BUS ASSIGNMENTS AND CHARACTERISTICS

To use either of the two newer chips, you must be sure that the necessary bus signals are present or can be developed. This involves the polarity and timing of the signals. The last is particularly important for the relation between the memory strobes and data valid periods. In addition, the new chip may have signals that allow simpler use of some of the add-on boards.

Table 9-4 shows the signals of the SCAMP:Y bus, the SC/MP, 8073 and Z80 chips. These are grouped by function, for the closest equivalent signal.

For the newer chips some change in addressing is necessary. SC/MP and SCAMP:Y are page oriented, with page lines needed

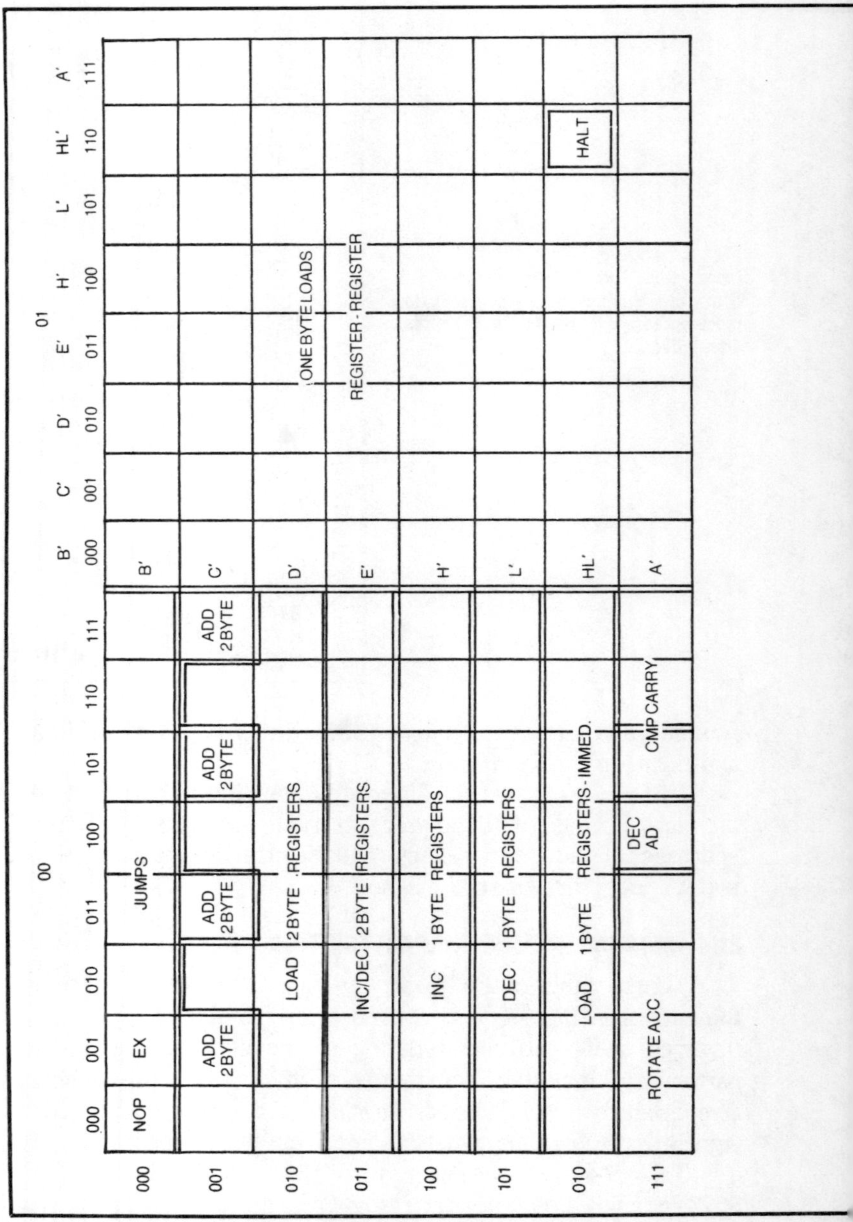

Fig. 9-6. Simplified map of the first byte of the Z80 instruction set. With four exceptions this byte specifies the function of the instruction, and, often, the register or register sets involved. The most common microcoding is indicated.

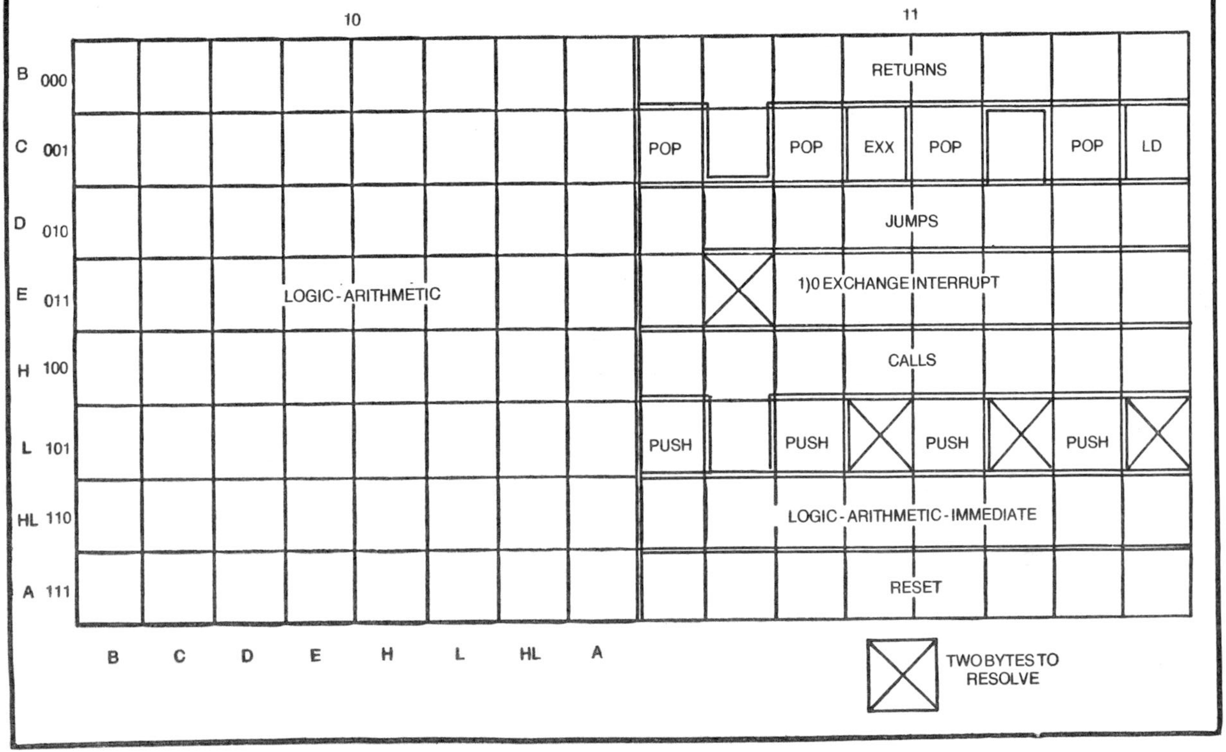
10
11
B 000
C 001
D 010
E 011
H 100
L 101
HL 110
A 111
LOGIC - ARITHMETIC
RETURNS
POP
POP
EXX
POP
POP
LD
JUMPS
1)0 EXCHANGE INTERRUPT
CALLS
PUSH
PUSH
PUSH
PUSH
LOGIC - ARITHMETIC - IMMEDIATE
RESET
B
C
D
E
H
L
HL
A
TWO BYTES TO RESOLVE

Table 9-3. Addressing Modes of the SC/MP, INS-807X and Z80 Chips, as Used By the Manufacturers. Computing Power Generally Increases as Addressing Freedom is Increased.

SC/MP	807X	Z80
PC Relative	Implied	Immediate
Pointer Relative	PC Relative	Immediate Extended
Immediate	Pointer Relative	Modified Page Zero
Indexed	Immediate	Relative
Auto Indexed	Immediate Extended	Extended
Substitute	Direct	Indexed
	Auto Indexed	Register
	Absolute	Implied
	Indirect	Bit
	Substitute	

for many boards. It is necessary to modify the control board to allow the lines A_{12} to A_{15} to control the page-select circuit or to provide an additional decoder. A simple (but limiting) change is to remove the 74100 latch used to hold the bus data after the address strobe, and to use a header to connect the address lines A_{12} to A_{15} to socket terminals 9, 8, 5 and 4 respectively. The data lines present no additional problem. The new chips do not have multiplexing, and the change made for addressing eliminates the latch.

SC/MP and 807X series have on-board oscillators, but the Z80 requires a timing input. The SCAMP:Y bus has a timing output. This can feed a Z80 in a multiprocessor system, but an oscillator is needed if the Z80 is used alone. Any oscillator can serve as an input buffer, an output buffer, or an oscillator. This can be standardized for use with the 8073, or even for addition to an SC/MP CPU board.

The sense lines of the SC/MP and 8073 correspond closely to the interrupts of the Z80, except for the nonmaskable feature. Necessary action can be by hardware. For the Z80 the three-flag outputs if needed, are generated, as for teleprinter use. With some restrictions, they can be generated by an AND of a data line and the IORQ input-output signal. Similar provisions can be made for serial input and output. Or a parallel-serial board can be used. (Note that the prototype serial board has the necessary gating provision for IORQ.) For the 807X series, a flag can be used for S_{out} and sense B for S_{in}.

Reset is essentially identical in all three chips. Bus control is different in the Z80, but its two bus signals can provide the necessary functions, including those for multiprocessor use. (See the next chapter.)

The hold and wait functions serve the same purpose, extending read and write cycles as long as the line is active. The major difference is that the SC/MP and 807X are static devices with internal operations (except clock) suspended; whereas the Z80, being dynamic, produces the hold by generating a wait cycle of one clock cycle duration as long as the line is active.

In the SC/MP, the continue line causes the CPU operations to be suspended after completing the current cycle (without loss in internal status). This is useful in many multiprocessor applications. The line does not exist in the 807X. It serves approximately the same purpose as the software controlled HALT of the Z80, the line being an output that indicates the CPU is waiting for an interrupt. The address and refresh lines remain active.

The read and write strobes of the three chips are identical in purpose, and nearly so in detail. The address strobe of the SC/MP

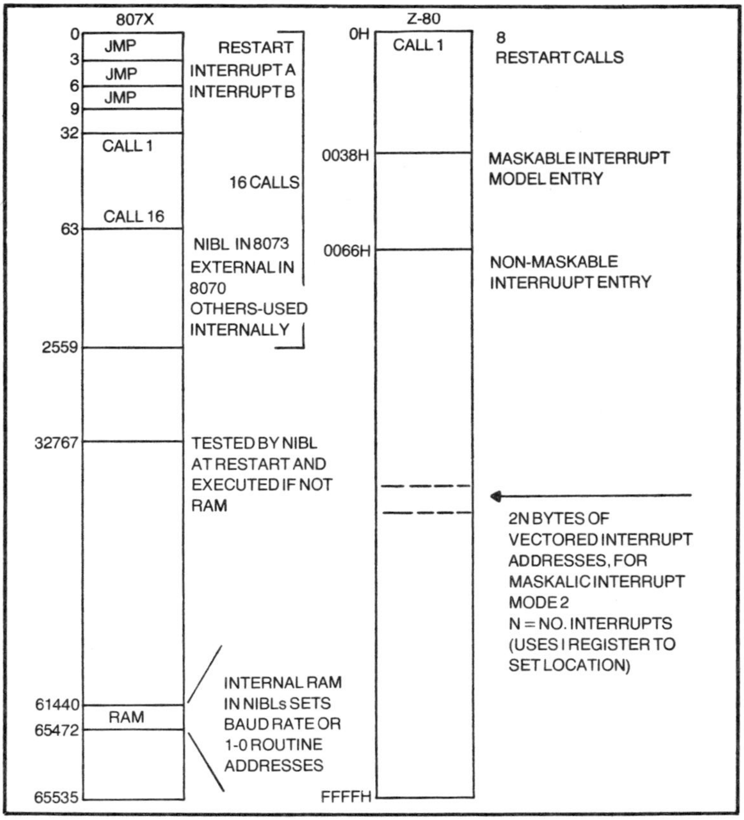

Fig. 9-7. Locations of preassigned memory in the INS-807X and Z80. The preassigned calls are a great programming convenience.

Table 9-4. Current Bus Signal Assignment in the SCAMP:Y Using the SC/MP Chip, and Possible Assignment Using the INS-807X or Z80 Chips.

	Bus Signal Source-Use		
SCAMP:Y BUS	SC/MP	807X	Z80
A_0-A_{11}	A_0-A_{11}	A_0-A_{11}	A_0-A_{11}
Page	D_0'	A_{12}-A_{15}	A_{12}-A_{15}
D_0-D_7	D_0-D_7	D_0-D_7	D_0-D_7
Xout	Xout	Xout	
X_{in}	X_{in}	X_{in}	ØIN
S_A-S_B	S_A-S_B	S_A,S_B	$\overline{NMI}$, $\overline{INT}$
F_0-F_1-F_2	F_0-F_1-F_2	F_1, F_2, F_3	
S_{out}-S_{in}	S_{out}-S_{in}		
RST	NRST	NRST	$\overline{RESET}$
ENOUT	NENOUT	NENOUT	$\overline{BUSAK}$
ENIN	NENIN	NENIN	
BREQ	N-BREQ	N-BREQ	$\overline{BUSRQ}$
HOLD	NHOLD	NHOLD	$\overline{WAIT}$
CONT	CONT		$\overline{HALT}$
ADS	NADS		
RDS	NRDS	NRDS	$\overline{RD}$
WDS	NWDS	NWDS	$\overline{WR}$
			$\overline{MREQ}$
			$\overline{IOREQ}$
			$\overline{RFSH}$
			$\overline{MI}$

is not needed by the other chips, but it is used by the single-step function of the SCAMP:Y control panel. The circuit could be disabled or omitted when the new chips are used, but single-step capability is a great aid when a new panel is being debugged. The design problem is considered later.

The Z80 has several signals that are not part of the other chips (although one, the instruction fetch indication, is almost the same as the I-flag of the SC/MP). The refresh is not available on the other chips, which require memory board refresh when dynamic memory is used.

The signals $\overline{MREQ}$ and $\overline{IORQ}$ are intended for bus control during memory and I/O operations. Very nearly, their function is equated by the use of page assignment signals for all memory mapped operations.

TIMING SIGNAL

The characteristics of the bus and control signals for the three chips are shown in Figs. 9-8, 9-9 and 9-10. The major difference is the duplex use of the data bus in the SC/MP with consequent changes in timing of the strobes. There are, of course, many differences in detail, but they are almost entirely self-compensating. This and the use of fast TTL devices on most boards, keep timing problems to a minimum. They are not much harder than for the original SCAMP:Y design.

BOARD USE OF SIGNALS

The signals used by the various boards of SCAMP:Y are charted in Table 9-5. Except for the serial board, the address and data buses and the read or write strobes are used by all boards. This is also true of the page signals.

Of the other signals, the SIN and SOUT of the SC/MP are used in one or two boards. The function is easily equated by changing them to SB and FO.

BUS ASSIGNMENT FOR THREE CPU BOARD DESIGNS

A simple way of handling the use of any of these chips in the SCAMP:Y bus system is given in Fig. 9-11. A few changes are needed on other boards and the bus leads. This is no problem with the wire-wrapped bus of the original design.

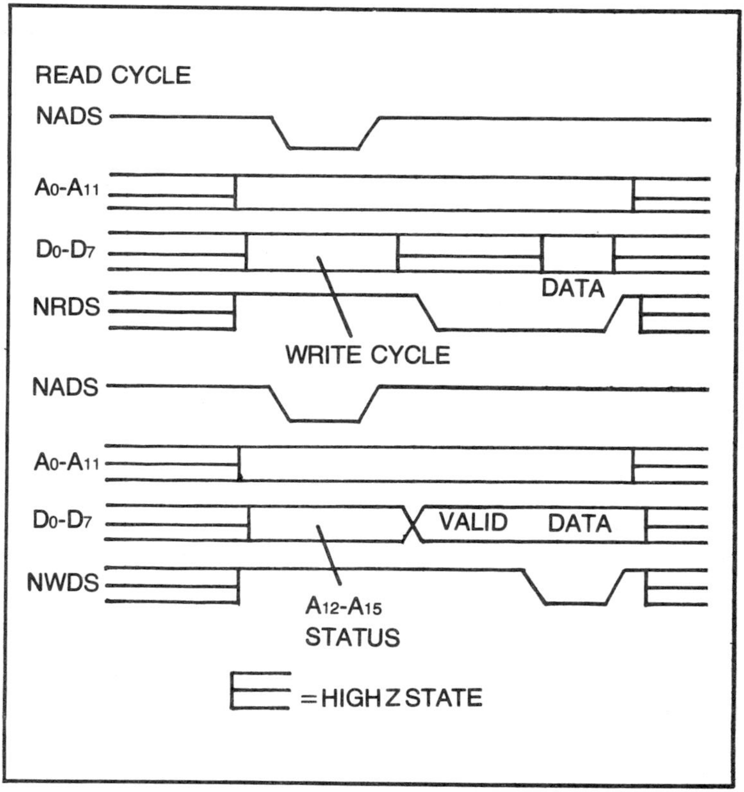

Fig. 9-8. Form of read and write signals for the SC/MP. See specifications for timing.

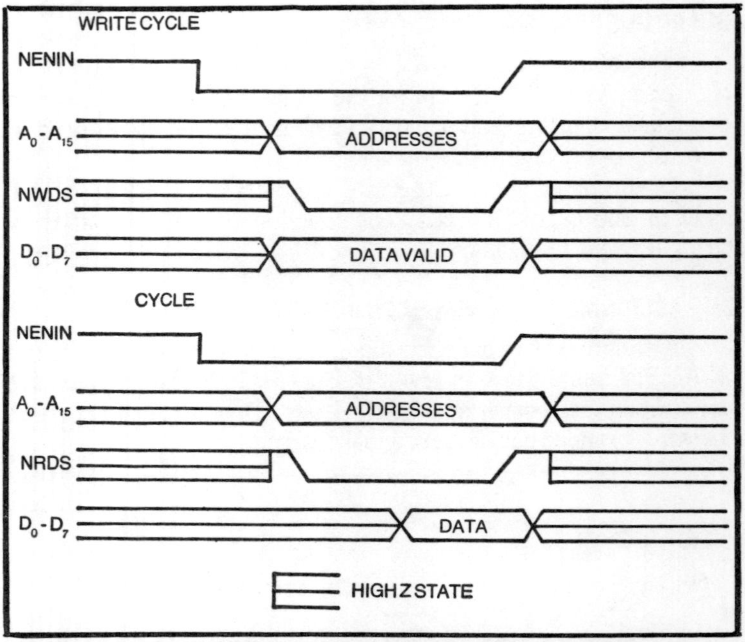

Fig. 9-9. Form of read and write signals for the IN-S807X. See specifications for timing.

The design takes full advantage of the SC/MP and 807X series CPUs, doesn't use all of the potential of the Z80. To add Z80 capability, use a board with on-board buffers with jumper provisions so the buffered lines can be connected to one of the three unused lines of the CPUs.

BOARD DESIGN FOR THE 8073

The circuit for a replacement CPU board using the 8073/NIBL2 chip is shown in Fig. 9-12. Following the chip and bus similarity, this is very close in design to the prototype (Fig. 1-8). It includes some changes for better performance and greater flexibility.

The data bus buffers are modified to use the bidirectional chips to cut costs. As in the original CPU, this buffer rests in the read state, going to the output or write mode under WDS control.

The address bus buffers are changed to tristate to simplify use of the bus for DMA, multiprocessing, etc. The bus can be left active, or placed in the high state by ENOUT, or by external control.

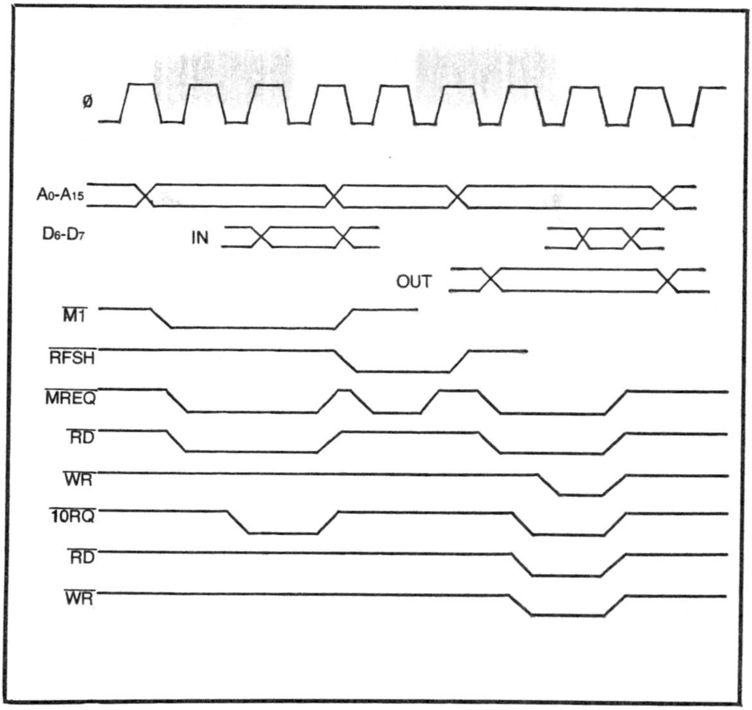

Fig. 9-10. Simplified form of read and write signals for the Z80 with the I/O also shown. See specifications for further data, and for timing.

Table 9-5. Board Use of Bus Signals For the Individual Prototype Boards of This Series. Since the Back Plane is Wire-Wrapped, Location Freedom is High.

Signals	CPU	Control	RAMI	S1O	CROM	INT	P1O	ARITH	RAM-2	PROM	PROM BURN	TAPE CONT
A_0-A_{11}	•	•	•		•	•	•	•	•	•	•	•
D_0-D_g	•	•	•		•	•	•	•	•	•	•	•
X_{out}	•											
S_A,S_B	•	•		•	•		•					
F_0, F_1, F_2	•	•		•								
RST	•	•										
ENOUT	•											
ENIN	•											
BREQ	•	•										
HOLD	•	•									•	
CONT	•											
S_{out}	•	•		•								•
S_{in}	•	•		•								•
WDS	•		•				•	•	•		•	•
RDS	•		•		•		•	•	•	•	•	•
ADS	•	•										
Page	•	•	•		•	•	•	•	•	•	•	•

PIN	ORIGINAL SC/MP	807X	Z-80$
1	GND	→	→
2	A_0	→	→
3	A_1	→	→
4	A_2	→	→
5	A_3	→	→
6	A_4	→	→
7	A_5	→	→
8	A_6	→	→
9	A_7	→	→
10	A_8	→	→
11	A_9	→	→
12	A_{10}	→	→
13	A_{11}	→	→
14	D_7	→	→
15	D_6	→	→
16	D_5	→	→
17	D_4	→	→
18	D_3	→	→
19	D_3	→	→
20	D_1	→	→
21	D_0	→	→
22	+5	→	→

PIN	ORIGINAL SC/MP	807X	Z-80
23	GND	→	→
24	ADS	A_{12}	A_{12}
25	RST	RST	RESET
26	N CONT	A_{13}	A_{13}
27	XTAL-0	XTAL $^{1-0}$	XTAL $^{1-0}$
28			
29	N SIN	A_{14}	A_{14}
30	N SOUT	A_{15}	A_{15}
31	NFO	NFO	*
32	NF1	→	*
33	NF2	→	*
34			
35	N SENA	N SENA	$\overline{\text{NM1}}$
36	NSENB	NSENB	$\overline{\text{INT}}$
37	RNIN	ENIN	
38	ENOUT	ENOUT	BUSAK
39	NBRQ	NBRQ	$\overline{\text{BUS RQ}}$
40	HOLD	HOLD	WAIT
41			
42	RDS	RDS	RD
43	WDS	WDS	WR
44	+5	→	→

* NFO = 10 RQ AND D_0 LATCHED

Fig. 9-11. Possible extension of the SC/MP and SCAMP:Y signals to provide for use of the INS-807X or Z80 microprocessors. These can be used for replacement or multiprocessing.

A small modification made to the oscillator allows the buffer to be connected as an output buffer or as an input driver. This allows use of the one-chip oscillator as the system master clock or slave. Omit the crystal if the slave state is desired.

The Schmidt trigger-buffer is omitted from the reset line NRST. The trigger is included within the chip, so it is not sensitive to rise time, as is the SC/MP. The simple integrator delay is retained to give reliable power-on reset. A zener diode gives overvoltage protection.

The strobe modified buffers reduce the delay between chip output and the bus, and also simplify use of the bus lines by other strobes. The output buffers are tristate; two gates in parallel are retained to ensure the good noise margin in the low state. The strobe lines can be left active, or placed in high by ENOUT or by external control. Since they are pulled up by the input current of other boards, there is a possibility of the read and write strobes going high if the timing of the signals on the strobe lines is not correct. The other circuits remain the same as the prototype.

Incidentally, if there is ever a reason to make a new SC/MP type of CPU board, the above changes should be considered. They will simplify future designs in some cases.

A Z80 CPU BOARD DESIGN

A design for a SCAMP:Y CPU board using the Z80 as the central processor is shown in Fig. 9-13. Many things are the same as the 8073 board.

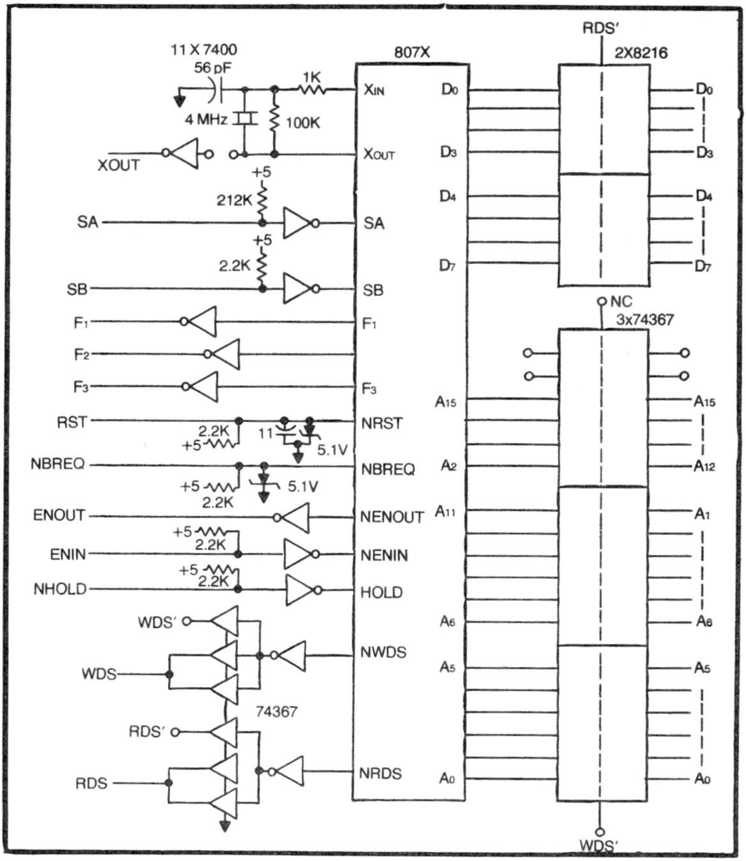

Fig. 9-12. Preliminary design of a CPU board using an INS-807X chip, and intended for the SCAMP:Y system. This circuit has not been tested.

The data and address bus buffering is identical to that of Fig. 9-12. This is also true for the read and write strobe lines. Freedom in connection is provided by jumpers. However, since the signals are not precisely the same as for the SC/MP and 8073, a spare set of gates is provided, that can be used to generate combination strobes, inverted signals, etc. The gates may be AND, OR or their negatives.

Since the Z80 chip does not include an oscillator, you must include one on the board. This is a simple multivibrator. A separate buffer with jumper provisions for connection as input or output is included. For use as an input, the crystal is omitted from the MV circuit.

The creation of four flags by latching signals from the data lines D_0 to D_3 by the I/O request signal is a special feature. This simple approach blocks further use of one of the features of the Z80 design. However, except for the Teletype provisions of SCAMP:Y, its bus is oriented to memory-mapped I/O, so the loss is negligible. There may even be a speed gain when the Z80 is used in the memory-mapped mode, rather than in the I/O bus change.

The reset circuit of the Z80 and the 8073 chips operate in the same fashion, so the circuit is the same. Other circuits are provided with simple buffers, input or output, and can be jumpered to unused terminals if they are needed externally.

For convenience, the pin connections of the three CPUs are reproduced here as Fig. 9-14.

ALTERNATIVES—PURCHASED DESIGNS

The beginning home builder may find the idea of starting from scratch more than he is willing to undertake. A way to handle this is to purchase one of the small single-board computers, and to use it as the core of his own designs of expansion boards. This course is becoming attractive for even experienced builders. As noted in Chapter 1, the effects of competition have reduced prices to the point where it is difficult for the home builder to get components for the price of a complete kit or board. This is true for any item used in quantity.

There are two single-board systems that fit into the SCAMP:Y concept very well. The first is the NIBBLER, an SC/MP based board that includes a NIBL ROM, 2K of RAM, and a 110 baud I/O port. All external connections are buffered with respect to the CPU, but not to memory.

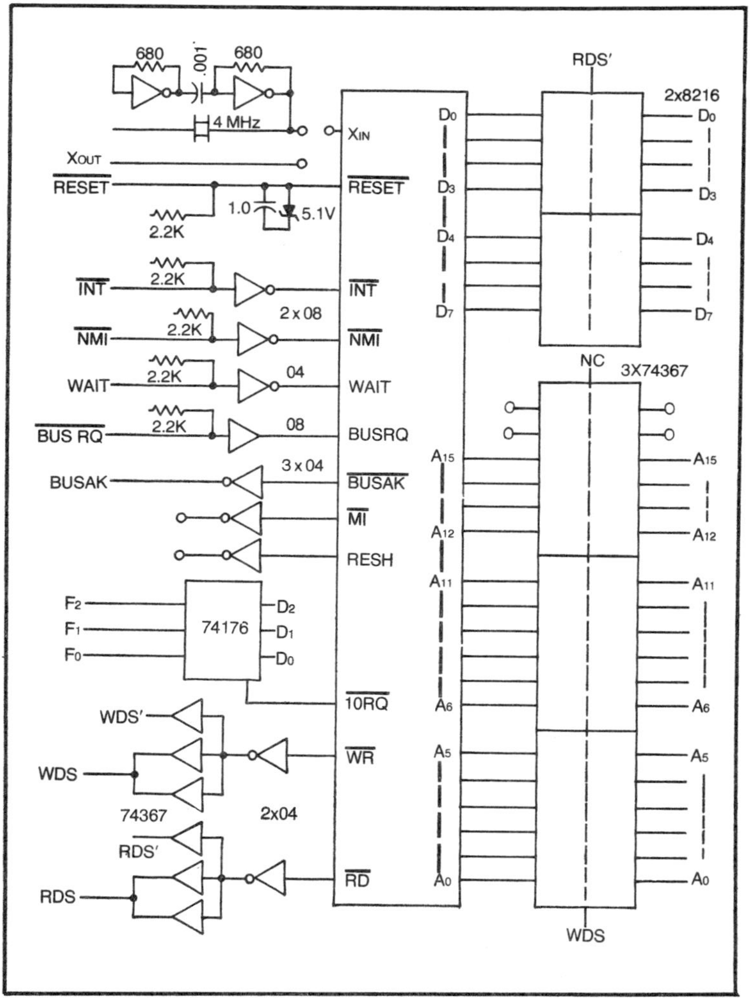

Fig. 9-13. Preliminary design of a CPU board using a Z80 chip, and intended for the SCAMP:Y system. This circuit has not been tested.

The bus structure of the NIBBLER is more complex than for the SCAMP:Y with two unidirectional data buses. There are also some special signals. These are mainly for memory expansion. There is provision for deselect of on-board memory. The full bus assignment is shown by Fig. 9-15.

In addition to the CPU board, a back-plane board, A/D board, memory expansion and power supply boards are available. Wire-

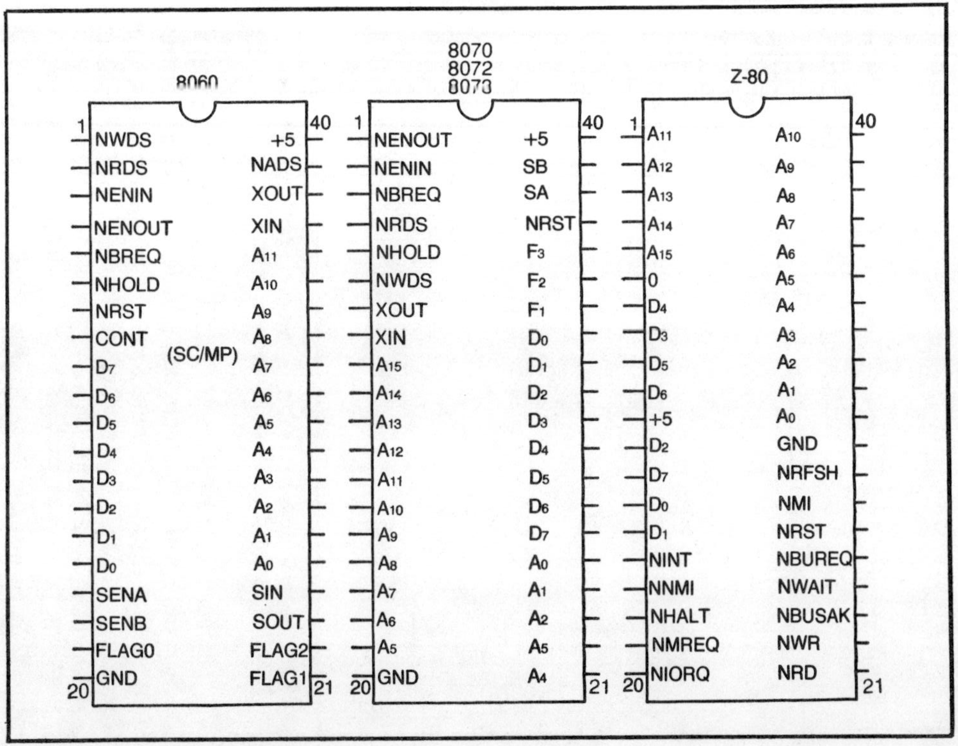

Fig. 9-14. Pin connections for the SC/MP (INS8060), the INS-807X series, and the Z80.

wrap bus interconnections of any of these and the SCAMP: Y panels is no great problem.

Another single-board CPU unit is the MICROACE, nearly a single package computer. It uses a Z80 CPU, a special ROM with single character BASIC, and can accommodate 2K of RAM on board. Output is modulated rf to a TV set. A cassette I/O is provided.

As can be seen from Fig. 9-16, the bus of the MICROACE is reasonably close to that of the SCAMP: Y. However, many of the lines are directly connected to the CPU, and buffering is minimal. The unit provides a simple and low cost method of adding a TV-typewriter to a system, and, of course, can serve as the CPU or a multiprocessor.

1	+5	25		49	RESET-N_1
2	+5	26	DOUT1-PI	50	NENIN-N_1
3	DIN7-P_1	27	DESELECT-PI	51	NBREQ-N_1
4	I-O OUT-P_1	28	D OUTØ-OP_1	52	N HOLD-N_1
5	DING-P_1	29	SEL G-N_1	53	CONT-P_1
6	1-0 IN-P_1	30	SEL 7-N_1	54	NDS-N_2
7	DINS-P_1	31	−12	55	
8	1-0 OUT-N_1	32	−12	56	NWDS-N_2
9	DIN4-P_1	33	+12	57	
10	1-0 IN-N_1	34	+12	58	CLOCK-P_2
11	DIN3-P_1	35	AD14-P_1	59	AD11-P_2
12	RDRCONT-N_1	36	AD15-P_1	60	AD1Ø-P_2
13	DIN2-P_1	37	AD12-P_1	61	AD9-P_2
14	DOUT7-P_1	38	AD13-P_1	62	AD8-P_2
15	DIN1-P_1	39	NRDS-N_2	63	AD7-P_2
16	DOUTG-P_1	40	SIN-P_1	64	AD6-P_2
17	DINØ-P_1	41		65	AD5-P_2
18	DOUT5-P_1	42	SOUT-P_1	66	AD4-P_2
19		43		67	AD3-P_2
20	DOUT4-P_1	44	FLG2-P_1	68	AD2-P_2
21		45		69	AD1-P_2
22	DOUT3-P_1	46	SENA-P_1	70	ADØ-P_2
23		47		71	GND
24	DOUT2-P_1	48	NENOUT-N_1	72	GND

BLANKS ARE SPARE PINS

Fig. 9-15. External bus connections for the Nibbler, a single-board microprocessor using the SC/MP with NIBL in ROM. The unit is widely advertised.

	A-SIDE	B-SIDE
1	D_7	+5V
2	RAM CS1	+9V
3	SLOT	SLOT
4	DØ	GND
5	D_1	GND
6	D_2	$\overline{Ø}$
7	D_6	AØ
8	D_5	A_1
9	D_3	A_2
10	D_4	A_3
11	INT	A_{15}
12	NMI	A_{14}
13	HALT	A_{13}
14	MREQ	A_{12}
15	IDRQ	A_{11}
16	RD	A_{10}
17	WR	A_9
18	BUSAK	A_8
19	WAIT	A_7
20	BUSREQ	A_6
21	RESET	A_5
22	M_1	A_4
23	REFSH	(NC)

Fig. 9-16. External bus connections for the Microace, a single-package processor using a Z80 and a video display with BASIC in ROM.

Instruction books are available separately for these designs. The NIBBLER book is a worthwhile addition to the SCAMP:Y users library, since it includes a full listing of NIBL in page format.

There are many other single-board designs on the market, but I haven't reviewed them for conformity to the SCAMP:Y structure.

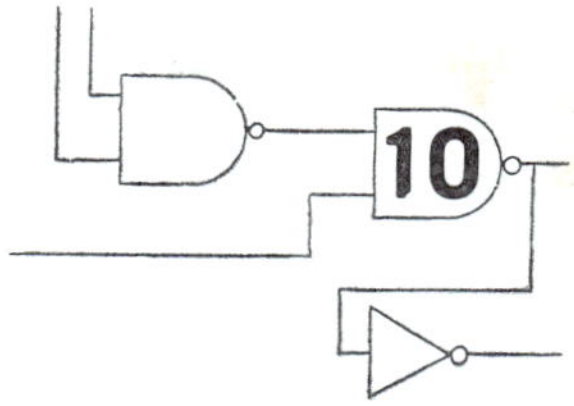

Multitasking, Multiprogramming and Multiprocessing

The three words in the title of this chapter are industry buzz words. Despite this, the concepts they are intended to convey are important and useful, to the home computer builder or user.

TIME-SHARED SYSTEMS

The easiest way to see the basis of these concepts is to look at a large *time-sharing system*. See Fig. 10-1.

The heart of the system is not the large central processor at the top of the drawing, but rather the group of smaller network processors at the center. They serve as interfaces, between the system users (represented by the teleprinters at the bottom) and the processor at the top. In this particular form, the net processors don't communicate directly with the central processor, but instead, via a disk file. This is a file-buffered system.

The way the typical system works is shown in Table 10-1. The Table shows an event at the user's terminal, the program called by the network computer, file reference and CPU action. Programs and names are typical, but simplified. A program name followed by a question mark asks for something. A period indicates that a statement is printed.

The network computer continuously scans its 10 to 20 incoming lines for a user request. When detected, it calls a sequence from a common file to identify the user, set up a charge, and request the type of service desired. A large installation may

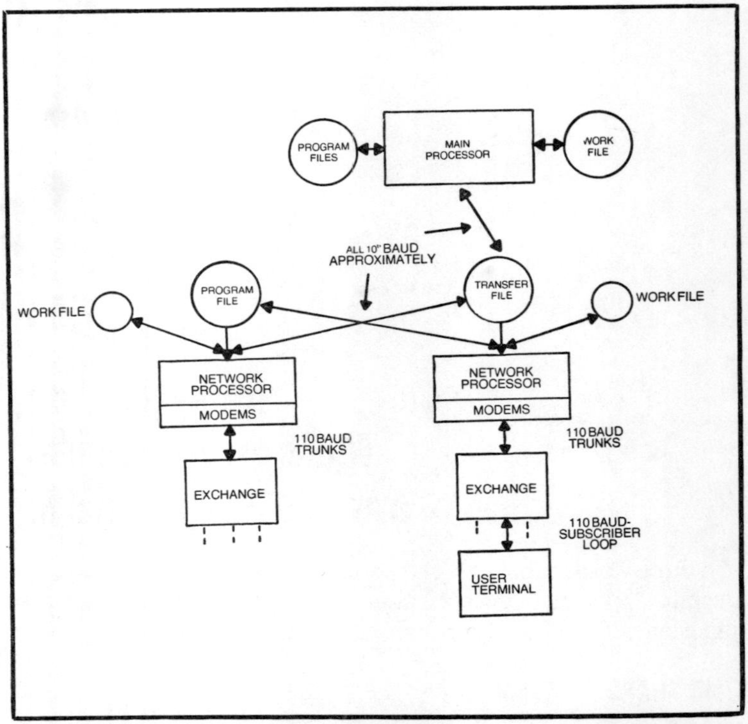

Fig. 10-1. Schematic of a large time-sharing computer system. Users communicate with one of several auxiliary or network processors, which transfer the users' problems to the central processor at run time. The central processor works on the problem for a short period, then places the problem and intermediate results in temporary storage, and proceeds to another problem. Complete results are transferred to the input processor for relay to the user.

allow time-sharing or *delayed-batch* operation, one of several high-level languages, teleprinter of TV-typewriter terminals, and several special features, such as page composition, library search and so on. This table assumes Fortran is requested and is available.

The network computer now asks for the type of run. NEW is the response if the program has not been previously entered, debugged, and saved. In Table 10-1 the response is the name of a file. When it is found, a request for instructions follow; in this case a RUN. On most systems other possible instructions at this point are LIST, APPEND and KILL.

So far all operations have been between the user terminals, the network computer, and its common files. RUN starts a new mode; the Fortran program is first sent to a transfer file. The central processor is now told that it has a new job. It completes its current round, then picks up the new problem. Typically, one second is devoted to each user in turn. If the work element is simple, it may not even take this long. If the program being run is long, it may take several seconds to compile it.

In this case the CPU runs for its allocated time, transfers the partially completed job to storage, goes to the next users problem, loads it and starts to work on it. The time spent in transferring the programs in and out of files represents an overhead: the cost of time sharing. Disk files allow low overhead.

In Table 10-1 the compiled program requests input data. It is loaded into the transfer file, and relayed to the user. This can be done automatically by starting the teleprinter tape reader relay. The input data travels back to the central processor, and execution resumes, again in short blocks. When execution is complete, the results are collated following the program direction and transferred to the network computer, which relays them and goes through a more-work-or-sign-off routine.

Despite the overhead of transferring files of partially executed programs, this system can be very cost effective. The central processor, being relieved of slow input and output operations,

Table 10-1. Typical Flow in a Time-Sharing System. The Input, File Search, and Other Service Programs are Necessary, and Represent an Overhead.

Terminal Event	Program At Network Computer	File Reference	Main Processor
Dial Up	Line Scan	W	
	PASSWORD?	T	
Code In	Verify	T	
	Time Start	W	
	SYSTEM?	T	
Fortran	FILE?	T	
Name	WAIT.	T,W	
	Search/Found	T	
	READY.	W	
RUN	Load	T	x
	DATA?	T	x
Input	Load	T	x
	WAIT.	W	x
	Collate	T	x
(Printout)	Output	T, W	
	READY.	W	
Name 2	WAIT.	T, W	
	Search/Found	T	
	READY.	W	
KILL	Delete	T, W	
	READY.	W	
STOP	Time Get	W	
(Time and Charge)	BYE.	W	

doesn't waste time in wait loops. During busy parts of the day, it can, and often does, run at maximum through-out (overhead considered). As work falls off, remote batch processing is picked up. Typically, one shift is devoted to contract processing.

A good system, partly loaded, is *transparent* to the user. That is, it appears that he is linked directly to the central processor. As the load increases, WAIT messages become more common. The user is charged only for actual processor time.

HOMENCLATURE AND TIME-SHARING

It is easy to see why the words at the top of this chapter were developed for time-sharing. The entire concept is based on multiple programs in execution at the same time. As concepts developed, people realized there were many *tasks* (locating customers, finding files, keeping track of charges, etc.). This led to the division into programs for control and programs for computation, i.e., multitasking. A further development was the use of separate processors with divided task and program assignments. (This form of multiprocessing started with the time-sharing system at Dartmouth and developed in parallel with BASIC.)

Let's look at the application of these concepts to home computing. Obviously, we must look at the scale involved and keep it within reach of the home computer user.

MULTIPROGRAMMING AND THE HOME COMPUTER

One of the easiest time-sharing concepts to apply in the home system is multiprogramming. The major reason is that no changes are required in the computer, and only minimum additions outside of it.

Go back to the system in Fig. 10-1. At a particular time three terminals might be using the system. The program flow is shown in Fig. 10-2. At the end of time allocated for execution of Program A, a time-driven interrupt stores work in progress and increments operation to Program B. The method of this process depends on the goals. It is no problem to insert or delete additional programs, nor to provide for the unequal length of programs.

As an example of what could be done, assume Program A is a home energy system control program scanning thermostats and thermometers, and that Program B is a program to scan the home security locks, fire alarms, etc. Program C could be one that began earlier, but wasn't finish; equivalent to a delayed batch problem in the time-sharing system.

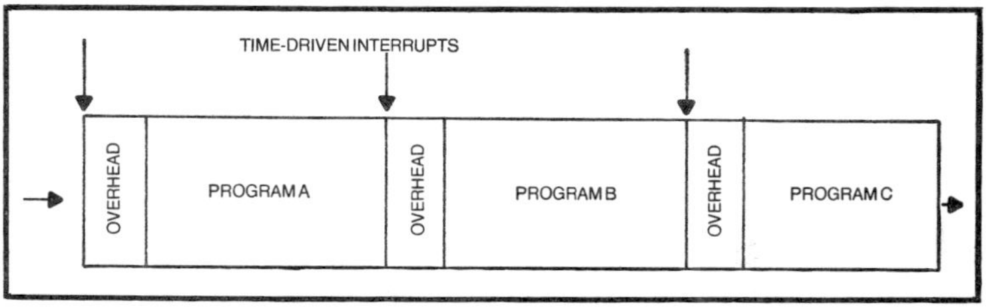

Fig. 10-2. Typical flow in a time-driven interrupt system. Again, an operations overhead is involved.

A simple way to allocate time to these programs is to use a clock chip for the interrupt driver. Figure 10-3 shows a simple driving timer, to give one interrupt per second, one per 10 seconds or one per minute. The interrupts can be developed at varying time intervals, as shown in Fig. 10-3. The overhead program can keep track of programs by their duration, or can read the clock output to determine the time block. (This circuit is not included here.)

Of course, it is possible to set up a complete time-sharing system for the home. Several requirements must be met for the central processor. It requires a supervisory program, including executive, monitor and routines; an interrupt system, usually time interval driven; assignable memory with protection; and dynamic data and status relocation (plus program relocation, if possible).

Data, status and program relocation can be simplified if work space is arranged by pages. This can be extended to provide automatic data and status relocation. For example, in NIBL you can create Pages 1A, 1B, 1C . . . by a computed AND of Page (n) and SELECT (m). Page 1A is assigned to user A, and so on.

With disk systems and their high transfer rate, it is possible to eliminate the hardware, and do the relocation under overhead-program control. If somewhat longer delays are allowable, tape systems can be used.

Protected programming is important. Accidents happen that can destroy data or programs.

Such a complex home system is not easy to build, nor is it inexpensive. But it can be worthwhile.

Time sharing is not the only way to set up multiprogramming. Another way, which is better in some circumstances, is interrupt driven multiprogramming. See Fig. 10-4. The computer is assigned a base operation and works on a base program until an interrupt occurs. Using a home energy system example, this method could scan and record all sensor readings. Suppose the

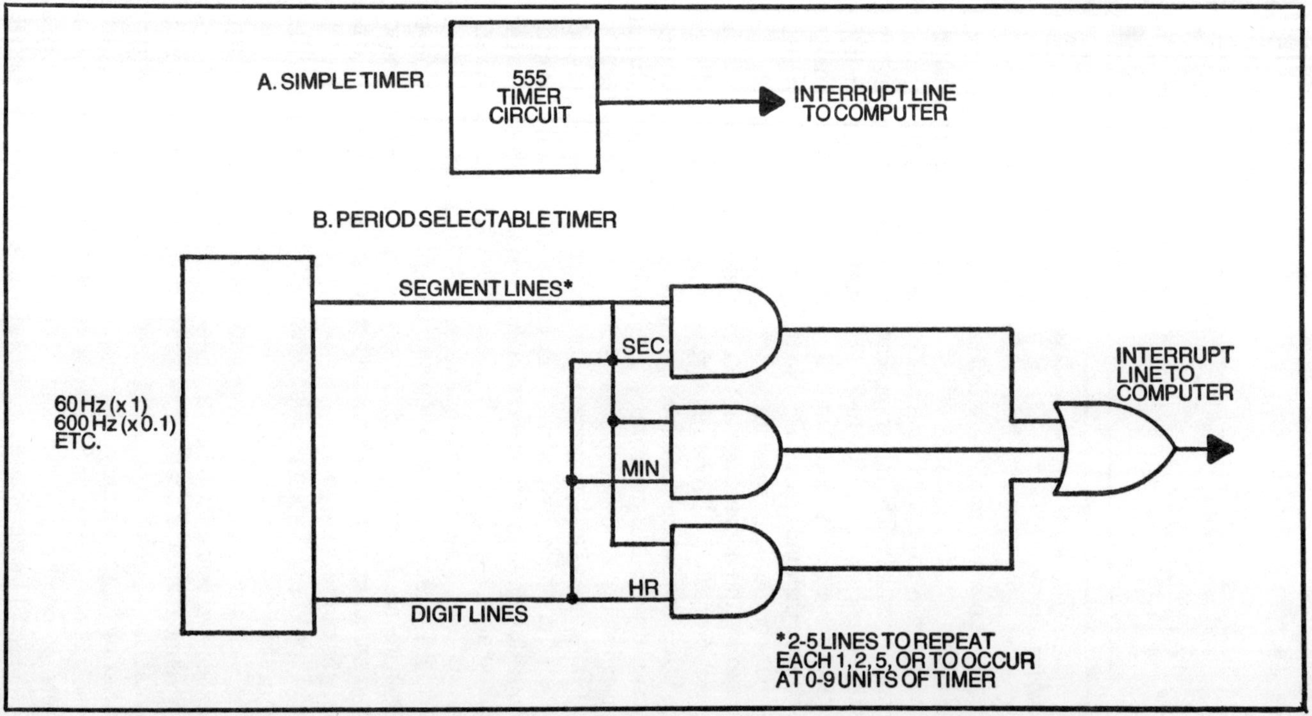

Fig. 10-3. Simple timers for home use in a time-sharing or time-driven-interrupt systems. The clock method is most accurate.

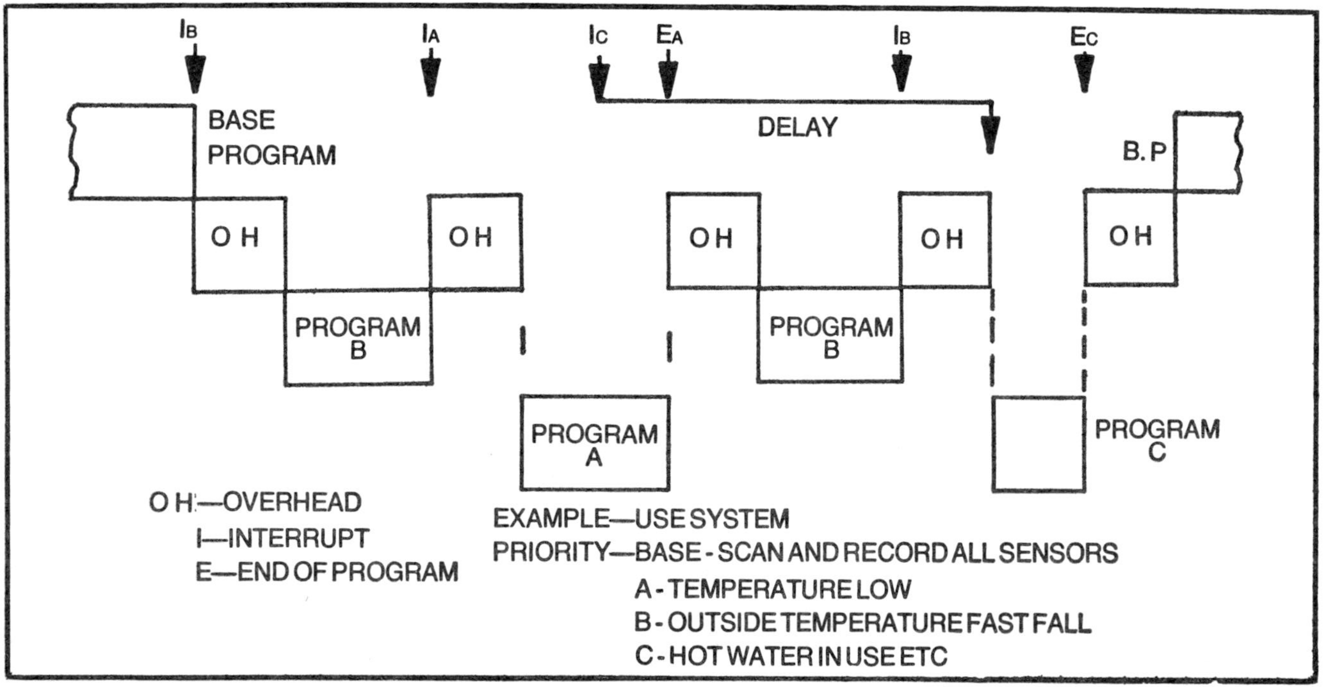

Fig. 10-4. Example of a priority driven interrupt system. When not interrupted, the processor operates on the base program. Interrupt A has higher priority than B, C, etc. and B has higher priority than C.

outside temperature sensor indicates that the temperature is falling at an unusual rate and generates an interrupt that calls Program B to rebalance the energy flow settings. Before this work is complete, the internal temperature of the hot water tank drops below limits. Since this is a lower priority than Program B, it is temporarily ignored. However, another event, falling inside temperature, of higher priority, interrupts the rebalance, starting the heating furnace and its flow pumps by Program A. Then, Program B execution resumes, and when it is finished, Program C is picked up and executed. If there are no more interrupts, execution of the base program resumes.

While this interrupt driven system handles a control problem, the technique is equally valid for a straight computational system. For example, in the household system above, the highest priority might be the den, the second priority the study centers in the childrens' rooms, and the lowest priority might be the game room. Or, the highest priority might be the control console, the second the magnetic tape recorder, the third the magnetic tape reader, and the lowest, the paper tape reader. In some, a combination of time and priority might be indicated.

The hardware for the priority interrupt need not be complex. Figure 10-5 shows the basics. The priority encoder generates the interrupt signal, and also shows the highest priority as output. This can be buffered for the bus, or encoded as BCD and then placed on the bus. In either case it is in response to a read request. In a small system with a full page devoted to interrupt, the decoding can be simplified by eliminating address decoding. OR can encoding several signals of equal priority. Provisions for determining which one caused the interrupt may be added by jumpers to the identify buffer.

MULTIPROCESSING AND THE HOME COMPUTER

The two basic ways of interconnecting two processors are shown in Fig. 10-6. The first, variously called *loose coupling, Serial Connection* and *Slow Transfer*, uses relatively few wire connections. Currently, it is likely to use ASCII as the exchange code to either the RS232 standard or the Teletype 20 mA loop. This is the connection commonly used with smart terminals (the TV-Typewriters that include their own microprocessor).

The second basic interconnection is sometimes called *tight coupling, parallel access* or *bus access*, and uses the data bus, the address bus, and at least some of the control buses. Transfer may be at CPU speeds, or, altered if a buffering device is used.

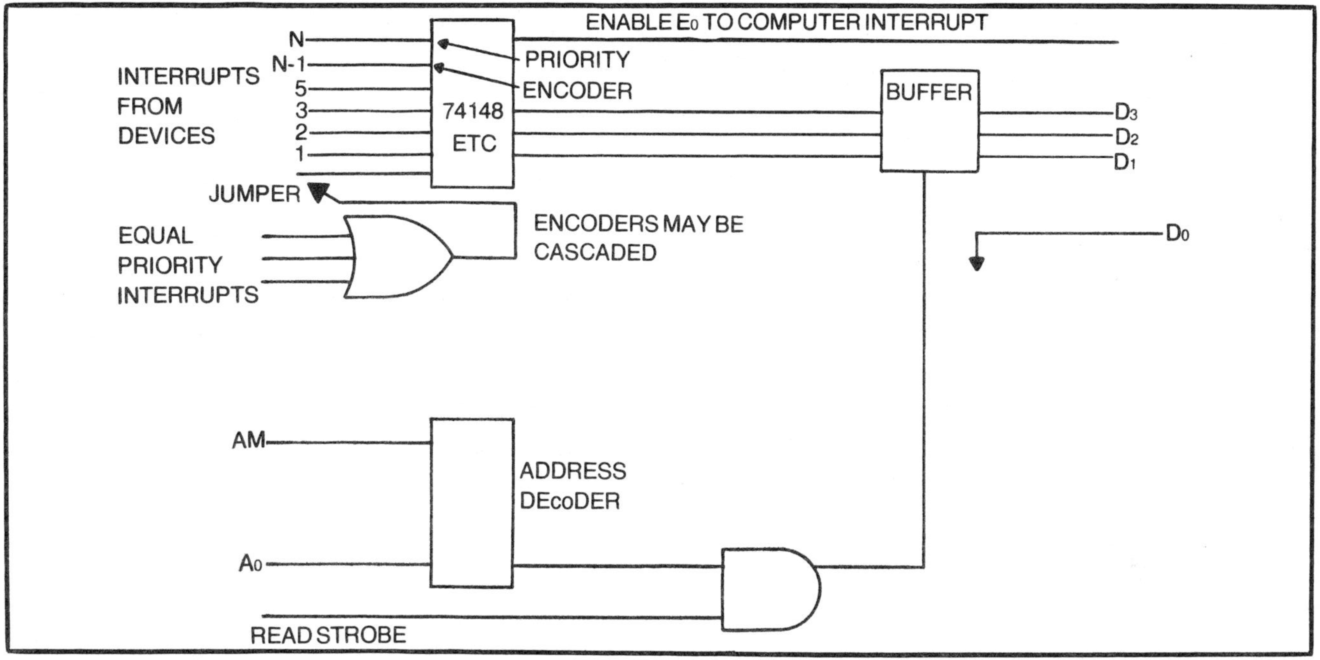

Fig. 10-5. Using a priority encoder to generate an interrupt, and to provide data to form the address of the highest priority interrupt present at the time. The schematic sets D_0 to low, to allow for a two-byte vector to the service routine. Equal priority inputs may be assembled by the OR, but the service address vector must be determined otherwise.

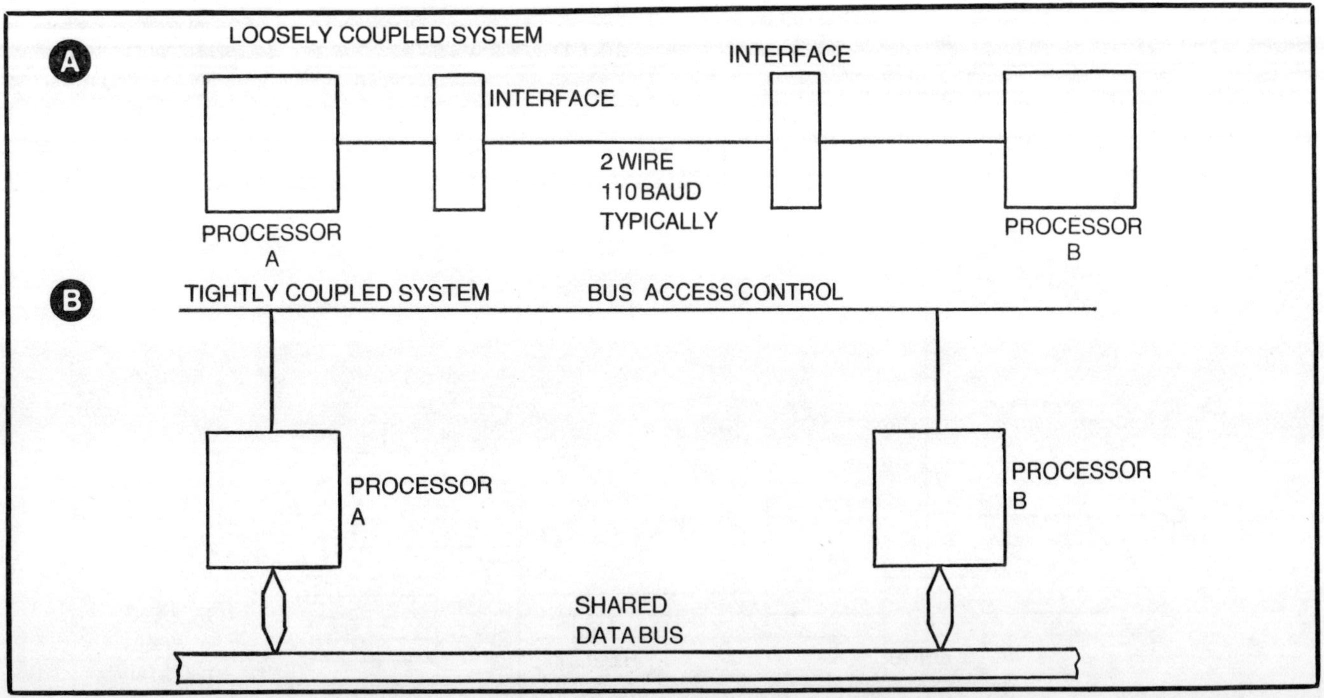

Fig. 10-6. Schematic of two systems of computer interconnection. The loosely coupled system often uses telephone circuits with data flow at 110 baud. The tightly coupled system will often use the data bus at the normal processor speed.

When considering multiprocessor techniques for home building, the main limits are cost and time. The three techniques covered were chosen with this in mind; they are not the only possibilities.

Loose Coupling

The serial or loosely coupled technique of Fig. 10-7 expands a little on the multiprocessor discussed earlier. It is a small scale version of a time-sharing system, using "somewhat" smart terminals as the interface between the users and the main or central processor. The user terminals can be dumb, with only a hex keypad, an LED display, and very little memory. The KIM, the ELF and terminals are representative of this class.

Slightly more complex terminals can use one of the keyboard-TV receivers. The AIM65 and several other single-board computers can serve here. Smart terminals are more expensive, but good. Don't forget to look at the TV games—several of these have typewriter keyboards, TV-video, and an internal processor. However, the central processor does all of the actual computation.

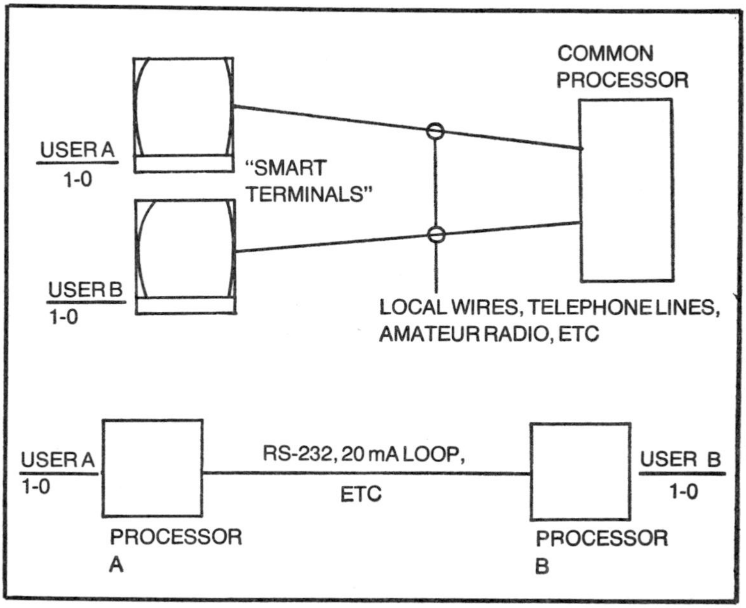

Fig. 10-7. Elementary multiprocessor systems, both loosely coupled. In A the users normally have no intercommunication.

Another approach is to base the terminal on hard copy, which is the usual practice in remote time sharing systems. While this introduces a noise source, it has some advantages where students or home offices are concerned. For one thing, it encourages learning good typing speed. It also encourages the good practice of knowing what you want to accomplish before tying up the processor.

This approach need not be expensive. Some of the very small, single boards can be used; the SC/MP based LCDs, for example, or the later design prepared for the 8073. The printer can be a surplus Baudot machine, a surplus Teletype, or even a keyboard and an electric typewriter with a set of solenoids. This can be simulated if several parallel output boards are constructed with feedback to the printer. The printer doesn't operate when a key is pressed unless the CPU is actually accepting data. Some programming is necessary to avoid confusion between programs in execution and I/O operations from the several terminals.

If you want straight interconnection between two computers, as shown in the lower half of the figure, the problems in programming become simpler, since time-sharing is reduced or eliminated. However, some time-sharing may be desirable, if one CPU has more memory, a larger instruction set, and so on. Program exchange only requires some handshake operations, which can be manual. For true multiprocessing new OS features are added to both CPUs.

Tight Coupling

Figure 10-8 shows one version of parallel-fed or tightly-coupled processors, which we have already used in simple form. This is usually called a *master-slave* system, since the auxiliary processor(s) are completely under the control of the central processor. In this system it is usually understood that the slave processors initiate operations only when they get instructions from the master unit. The slaves can then complete an instruction series without further supervision.

There is, however, a variation in which the slave processor only executes a single, fixed routine, inserting new data into the routine on central processor instructions. This is the operation in the upper slave unit of the figure, a graphics generator. In this simpler version, the routine can be a digital form of raster generator keyed to a bank of memory, with memory written or cleared by the central processor. This is the type used in the

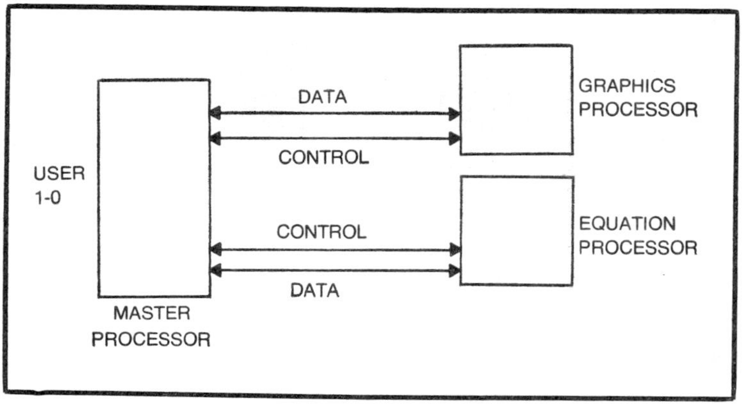

Fig. 10-8. Elementary master-slave multiprocessor. The graphics of the Heathkit H-19 computer are in this form. The calculator board of Chapter 8 is an elementary form of this system, with operations being controlled by the master CPU, and by chip-internal ROM.

TRS-80. More complex units, such as the H19, may use a microprocessor here and assign it other duties.

The second slave unit shown is an equation or algebraic processor, which can be the general type considered in Chapter 8. An alternative is to use a full microprocessor, with a floating point package in memory and sufficient RAM to handle data needs.

Reasons for this master-slave set up may be design convenience, or a desire to increase processing speed. With reasonable control software, the arithmetic processor can double the system output for algebraic equations. In some cases the gain is substantial. Look at the example of an energy system generating a family of control signals, each based on instrument readings and their rate of change. A programmable calculator chip could handle all of the calculations, leaving the main processor free for data gathering, control and alarm functions.

It's likely that this form of multiprocessing will grow and become more important in the future. Several companies now offer arithmetic chips, disk controllers and graphic generators. Peripheral interface units, data encryption units and bus protocol controllers are also available. Sometimes it's difficult to decide whether a particular unit is actually a processor, or whether it is simply a complex buffer.

Shared-Resource Multiprocessing

Another multiprocessor that can be used in home computers is the shared-resource type sketched in Fig. 10-9. Here two or more

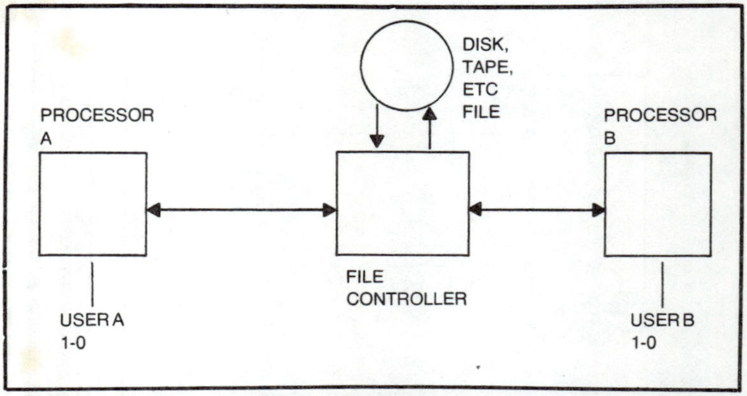

Fig. 10-9. The shared resource form of multiprocessor with a master file available to two or more processors. While any resource may be shared, it is not worthwhile if both processors need to use the resource extensively.

processors are completely independent in programming or function, but share one or more peripheral units. The shared unit may be memory, as shown in the figure, and used with the terminal processors of Fig. 10-1. Of course, resource sharing does not need to be limited to memory. Table 10-2 suggests some multiprocessing applications.

Probably the easiest way to increase execution speed is to work with the I/O execution first, then with one of the processors for math, and so on. A TV-typewriter expansion can be a great convenience. As the system grows, a file controller becomes a great time saver.

Summing Up Multiprocessing

In multiprocessing, possibly even more than in other expansions, keeping track of time delays and loading is very important. Without special buffers, about 20 feet of twisted pair or 10 feet of coax is the maximum cable run. With special high power buffer-drivers, coax can be driven in lengths up to about 2,000 feet (but this is not recommended for the home computer use). The best way to go is serial connection. The constant-current 20 mA Teletype loop is good, since it was designed to be immune to many cable troubles. Even for a multiroom installation with several smart terminals and a central resource system, the loosely coupled serial data system is easiest.

Home use of multiprocessing, distributed computers and time-sharing is just starting. There is a lot of knowledge about large commercial, industrial and military systems, but little of it applies to home building. Anyone who tries home multiprocessing is pioneering.

MULTITASK OPERATION AND THE SMALL COMPUTER

To understand the basic concept of multitask operation, assume that you have two independent programs to run. As shown in Fig. 10-10A, each of these is composed of four related parts: Program input, data input, execution, and result output. In normal home operation these programs are run serially. That is, the four parts of the first program would be completed in sequence, then operation would proceed to the second program. As indicated by the flow line, the second program is completed in the same order.

If you meet certain requirements, you don't need to follow this sequence. In fact, you may want to depart from it. Suppose you load the first program, then find that you have mislaid the data sheet. Rather than letting the processor set idle, you decide to load another program while you search for the data sheet. Then you can feed in the data for the first program. The operation sequence is changed to that of Fig. 10-10.

As the number of pending programs or operations increase, and as the computer capability becomes greater, multitasking becomes more and more attractive. For one thing it keeps the

Table 10-2. Examples of Multiprocessor Possibilities. All Are Used in Large Systems, and Several Are Becoming More Common in Home and Business Microprocessors.

Thru-Put Enhancement
Dual Speed Computation
IO Execution
TVT Presentation Generation
Disk or Tape Control
Math Processing
Statistics Processing
Conversion Factor Generation
Time-Interval Generators
Data Conversion (Digital, Analog, etc.)
File Controller
Control System "Normal-Fault Location" Pairs
Shared Resource Systems (e.g., disk file)
Distributed Processing (e.g., partial local preprocessing)
"Smart" Terminals
"Smart" Instrumentation
"Smart" Data Buffer (e.g. for speed standardization)
"Smart" Security System
"Smart" Memory

computer at work, instead of sitting idle. It reduces the cost per unit of execution time, and may even be necessary to keep up with your programming load.

In this elementary example, there only are two programs with the same number of steps or tasks, and an operator decision. A full system places events under control of a program structured to allow a variable number of programs, a variable number of tasks for each, and a method of determining event sequence, probably based on a priority.

The collection of routines that handle these problems of resource allocation, organization of data, and communications between devices is called the control program. It must have data available to allow it to identify, select, execute or modify routines, records, files, operations or data. The most important elements of the control data to the home computer are those that allow programs to be broken into tasks, and those that assign and keep track of resources (especially memory).

There are very definite requirements for multitasking. The control program is just one. Adequate data capacity is another. The ability to transfer program segments from standby memory to execution memory is a practical necessity. In most cases this requires a disk system.

For most users multitasking under program control is not warranted. It is much better to remember the technique, occasionally using it in the way it is outlined in Fig. 10-10. Multitasking is simplified by good programming techniques. Organize programs as group of tasks capable of independent execution.

Multitask operation of small computers is just starting. Again, if you choose to use it, that you are going to be a pioneer.

INTERRUPTS

Because interrupts and bus control are so important to the concepts of multiprocessing, let's complete this chapter with a review of the techniques used by the three CPUs considered, starting with interrupts.

In the SC/MP there is only one interrupt, which can be masked. This is the SENSE A line, which always controls bit 4 of the status register, and which also becomes the interrupt when the interrupt enable flag, bit 3, is set under program control. The action of an interrupt line is shown in Fig. 10-11. A test is performed prior to each instruction fetch. Note that there is a single response to an interrupt, the interchange of the program

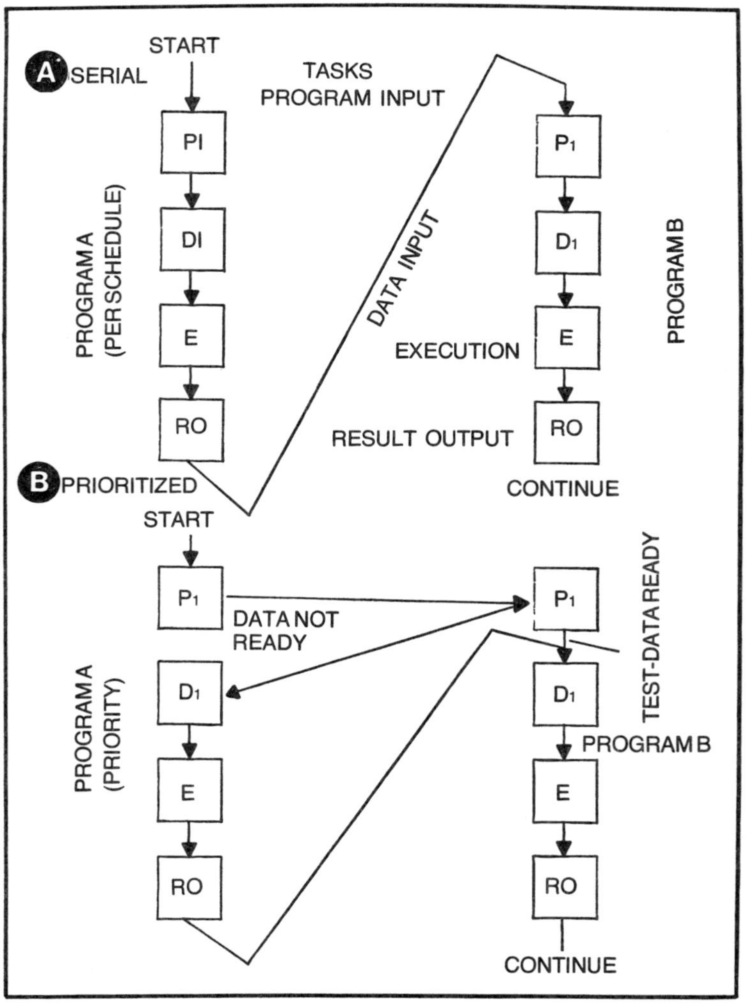

Fig. 10-10. Elementary example of multitask operation. At A events are sequenced within a program, and at B they are interrupt-driven by a priority scheme. One or more processors may be involved.

counter, and the P3 register. P3 must point to the interrupt service routine.

Since there is only one interrupt, a combiner must be used if more are desired. This may be the type shown in Fig. 10-5. Interrupt and return routines are given in Appendix C of the *SC/MP Application Handbook*.

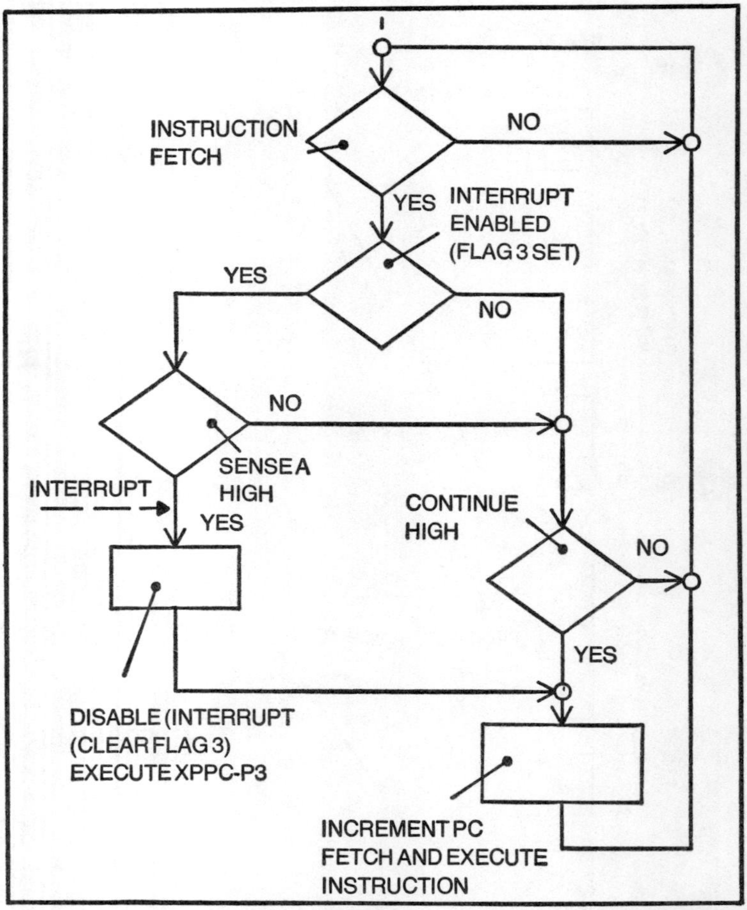

Fig. 10-11. Logic flow for the SC/MP interrupt. Sense A serves as a maskable interrupt.

NIBL uses P3 as the subroutine pointer, and makes no specific provision for interrupts. However, an interrupt is possible since the Teletype input, sense B, if tested during program execution with program control returning to the keyboard monitor if a signal is present. Interrupts can be inserted into any input program by using the statement LINK (interrupt routine address). The first part of this routine must save the P3 return address. This is followed by the interrupt routine address, and the interrupt mask is removed. If no interrupt is present, the return reverses the process.

The 807X interrupt system is similar, but has considerably greater flexibility. Both SENSE A and SENSE B can serve as interrupts. They are both maskable, but not independently. The A line has priority if both are active. Each interrupt has a predetermined location in memory, described by Fig. 9-7.

While interrupts can be masked, the flags are edge triggered, which simplifies interrupt design and allows receipt of a second request while one is being processed. However, the second request is ignored until the interrupt is re-enabled.

Interrupt service routines should conclude with OR S = 01_{16} followed by RET. This causes processing to resume at the point where it was interrupted. The processing sequence is shown in Fig. 10-12.

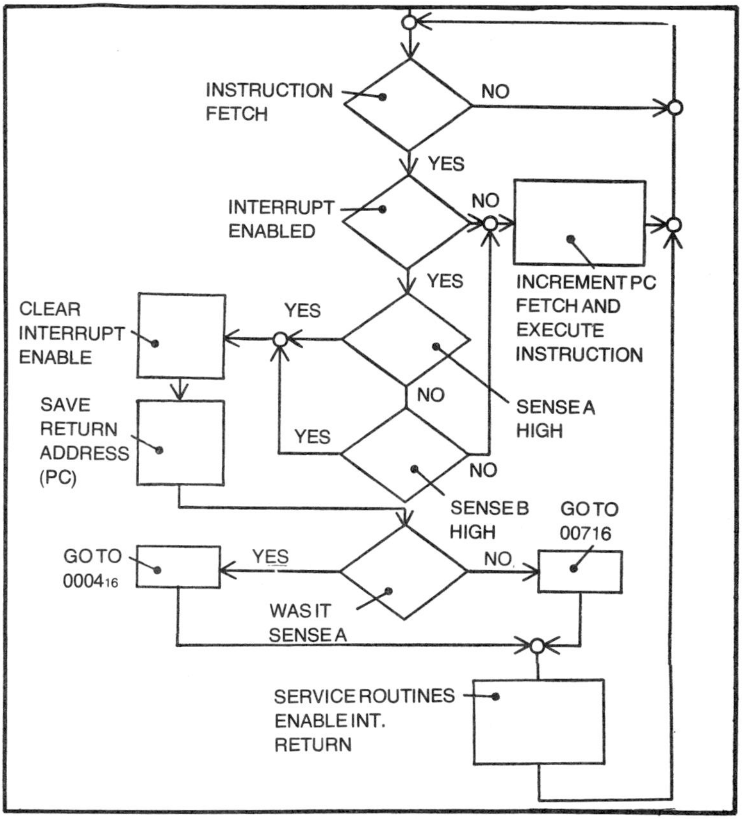

Fig. 10-12. Interrupt logic flow in the INS-807X series. There is only one mask, but it affects both interrupt sense lines.

The RESET line offers a nonmaskable interrupt. It causes a restart at the location specified by memory locations 02_{16} and 03_{16}. However, it clears the PC, SP and all status register bits exempt S_A and S_B. For that reason must be used with extra caution.

The NIBL 2 in the 8073 executes any designated GOSUB when either interrupt 1 or 2 is present. This is done by a special instruction ON n, X, where X is the gosub line number. The LINK construct can also be used.

The interrupt scheme for the Z80 has still more freedom, and is accordingly more complex. The interrupt jump assignments are shown in Fig. 9-7. The logic flow is shown in Fig. 10-13. This includes the test for bus request. The interrupt flow is shown in Fig. 10-14. A family of instructions allows the interrupts to be enabled or disabled, and sets the mode. Another family handles the returns, including a conditional return.

It seems that none of the higher level languages for this chip provide an instruction for interrupt control. Dependence is placed on the automatic save associated with the interrupts. A few languages provide for linkage to machine language routines, and those with a POKE function can provide a construct to test the maskable interrupts. Probably the easiest method to use is a priority interrupt to the NMI input.

BUS CONTROL

One of the easy ways to set up a multiprocessor system is to arrange the processors in a chain, as shown in Fig. 10-15. This is equivalent to establishing a priority to their requests for bus access, since the left-most processor can deny access to the ones at the right, and so on down the line.

For the case of the SC/MP, a simulated test of this method of control gives the throughput results of Fig. 10-16. A second processor nearly doubles the throughput, even under worst-case conditions. However, while adding more processors does give some further increase, at about four, the throughput tends to become constant, at about two and a half times that of a single processor. The reason for this is shown by the graph in Fig. 10-17. The left-most processor first takes equal time. Then, as the number of processors increases, it takes greater than equal time. While this happens, its throughput decreases. Other processors get less time, and thus have relatively small throughput.

While this test shows what happens when the processor only is considered, the system gain can be spectacular. For example,

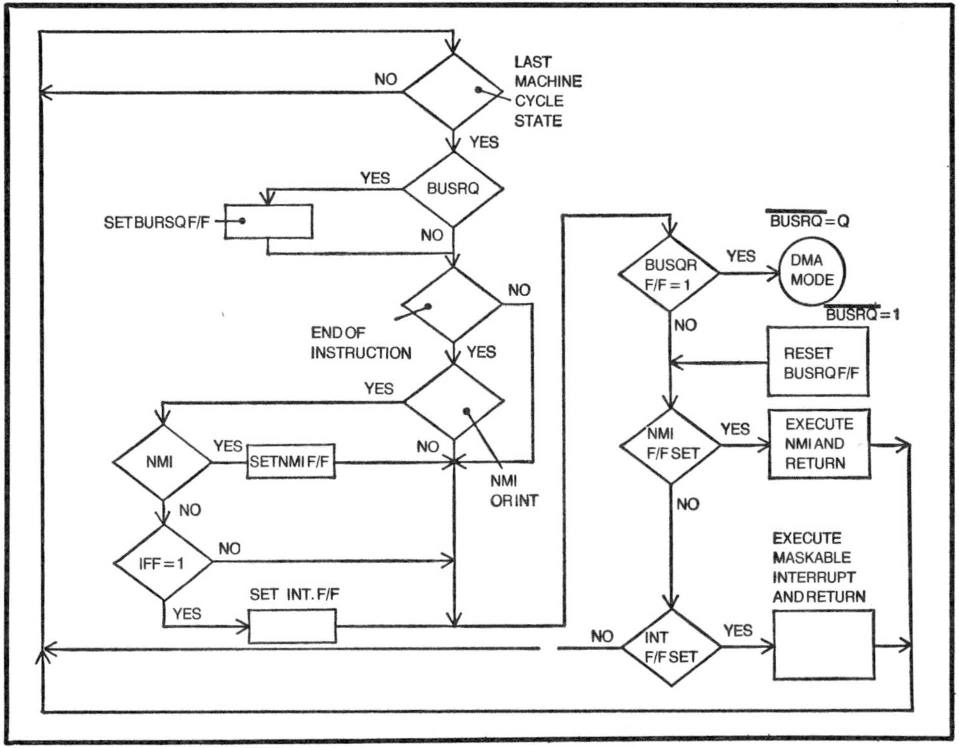

Fig. 10-13. Interrupt logic flow in the Z80. One interrupt can be masked, the other can't.

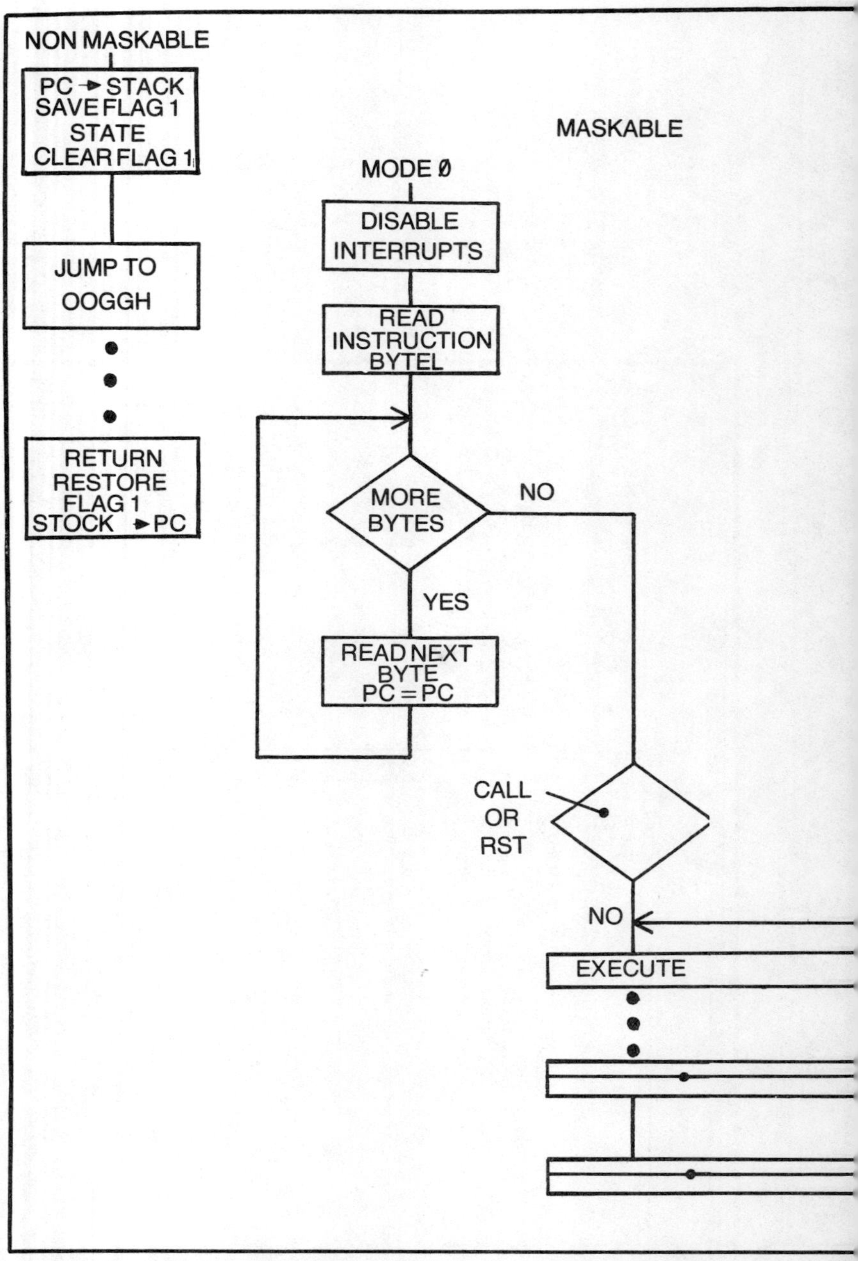

Fig. 10-14. Execution details in the Z80. The service routines depend on the mode set.

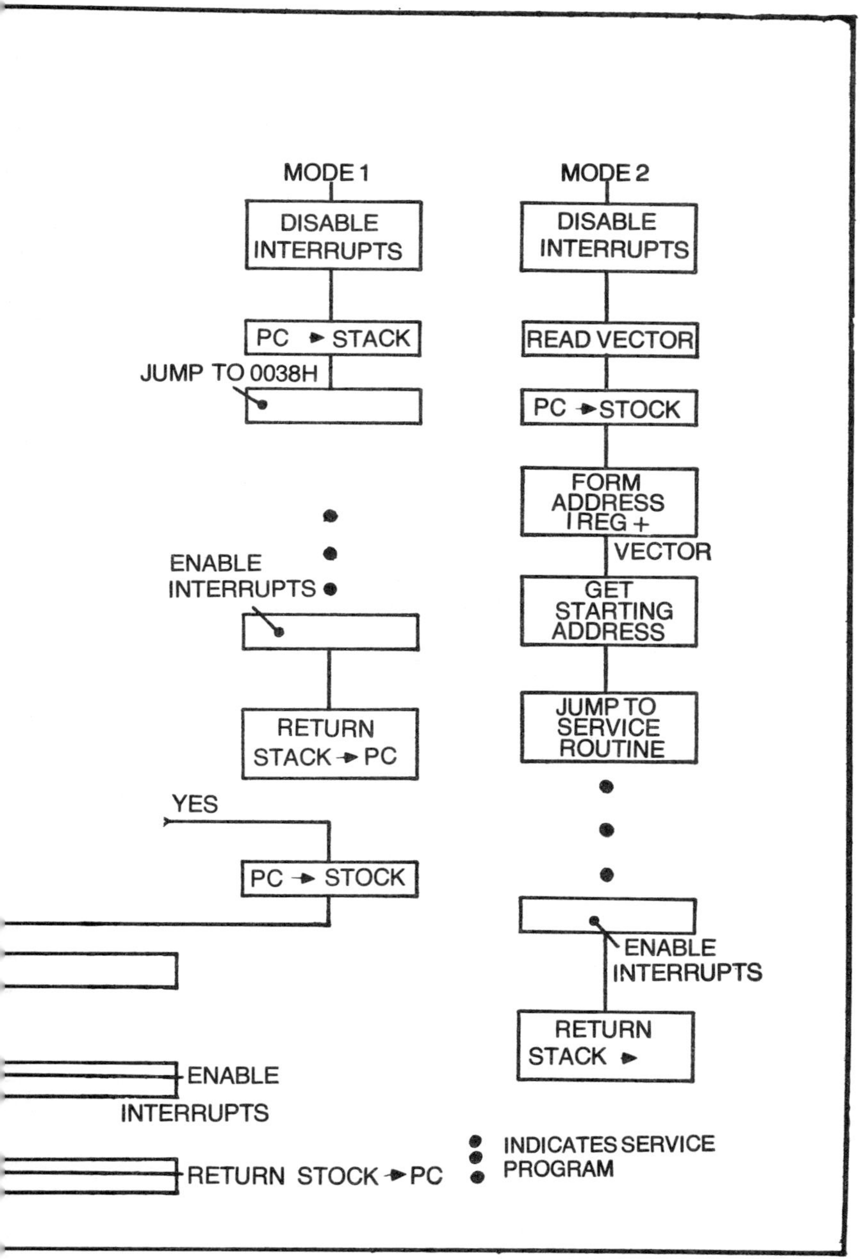

MODE 1
DISABLE INTERRUPTS
PC ► STACK
JUMP TO 0038H
ENABLE INTERRUPTS
RETURN STACK ► PC
YES
PC ► STOCK
ENABLE
INTERRUPTS
RETURN STOCK ► PC
MODE 2
DISABLE INTERRUPTS
READ VECTOR
PC ► STOCK
FORM ADDRESS I REG +
VECTOR
GET STARTING ADDRESS
JUMP TO SERVICE ROUTINE
ENABLE INTERRUPTS
RETURN STACK ►
INDICATES SERVICE PROGRAM

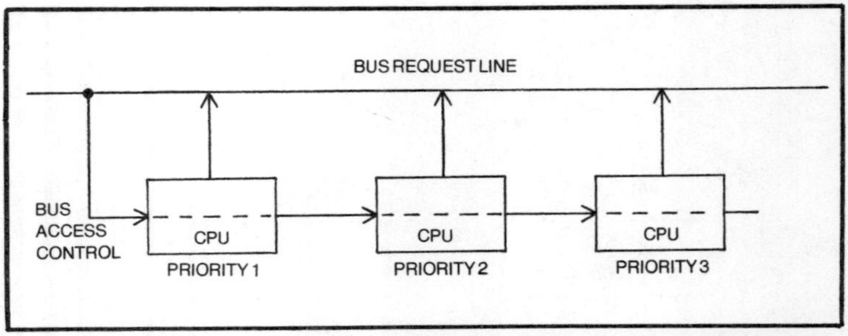

Fig. 10-15. Principle of daisy-chain operation, with the left-most CPU having the highest priority. This mode is very useful in shared resource systems.

consider the common event of Teletype input, where the processor may execute 50 instructions on receipt of a character, then go into a loop of 4,000 instructions, waiting for the next character. Assuming the Teletype input is handled by one processor, and allowing a second to do regular processing, can increase system throughput 80 to one. The gain is less marked with a high-speed reader, but still very much worthwhile.

For the SC/MP and the 8087, the connections for this form of multiprocessing is shown in Fig. 10-18. The daisy chain effect is also shown. This form of connection is possible because NBREQ is both an input and an output. However, the logic flow is different in

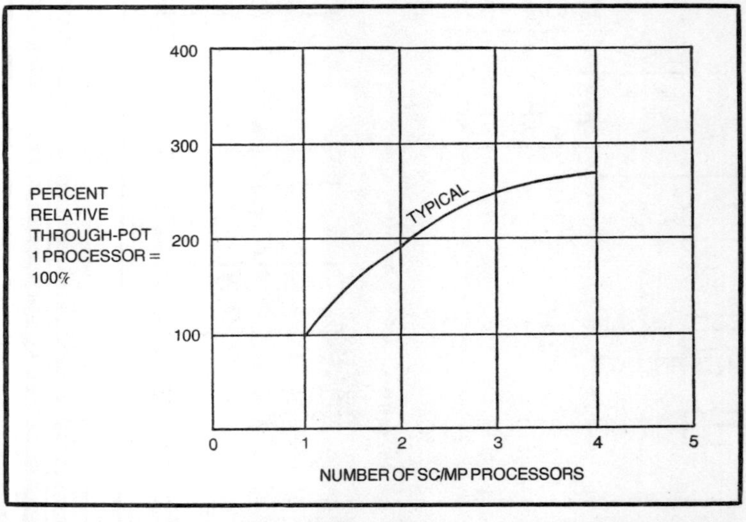

Fig. 10-16. Example of relative gain in throughput in a multiprocessor system. The data was taken with SC/MP processors with a test program.

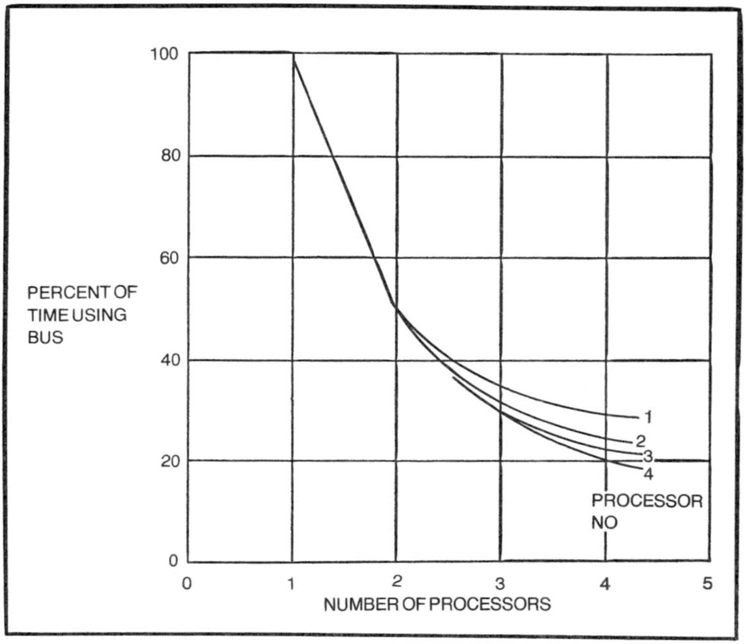

Fig. 10-17. Percentage bus usage in the SC/MP multiprocessor system. With two in the system, the second is essentially using free bus time. With three or more present, there is an increasing probability that one or more will be waiting for the bus.

the two chips, as shown in Fig. 10-19. For these two chips, the connection to a system bus will be of the form shown in Fig. 10-20. Obviously, polarity must be observed.

Daisy chain operation with the Z80 is also possible. The connections are given in Fig. 10-21. The logic flow is shown in Fig. 10-13 in connection with interrupts.

There are, of course, other connections that can be used, such as nonmaskable bus-use request, etc. It's unlikely that a home builder will need more than the simple chain, since interrupt drivers are so powerful.

The gain in throughput results from the one simple rule—keep the slow devices to the right of the chain.

For those who want to go further, the number of possibilities is very large, especially when multitasking and multiprocessing techniques are combined. Figures 10-22 and 10-23 show a beginning. The first is oriented to calculations used in engineering, but can also serve for scientific, statistical or business purposes.

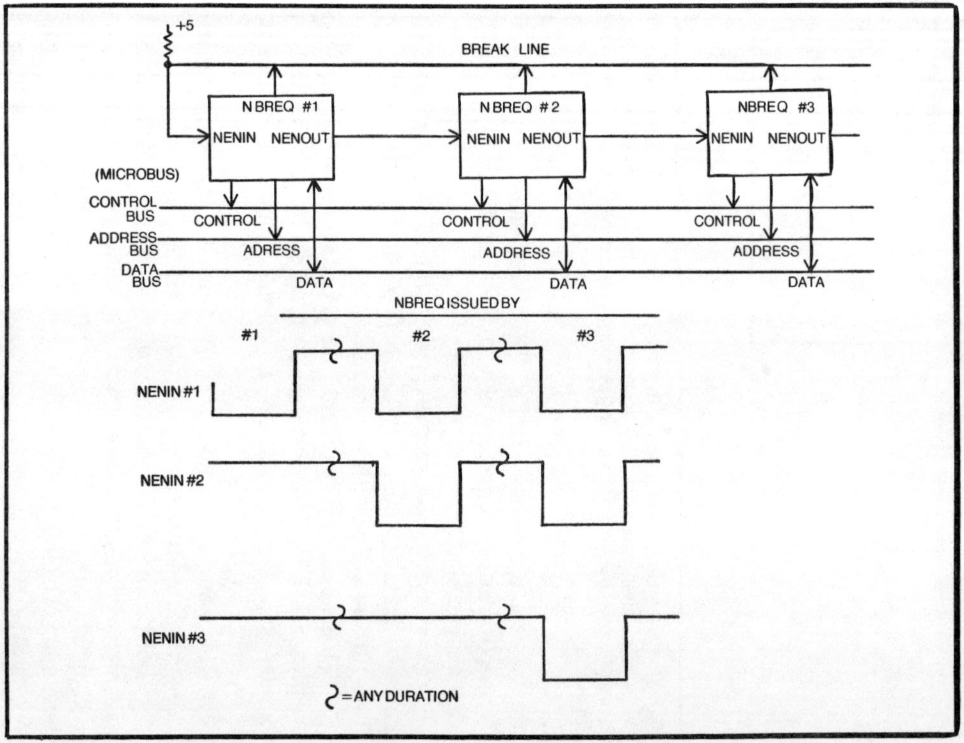

Fig. 10-18. Specific connections for SC/MP and INS-807X daisy-chain operation.

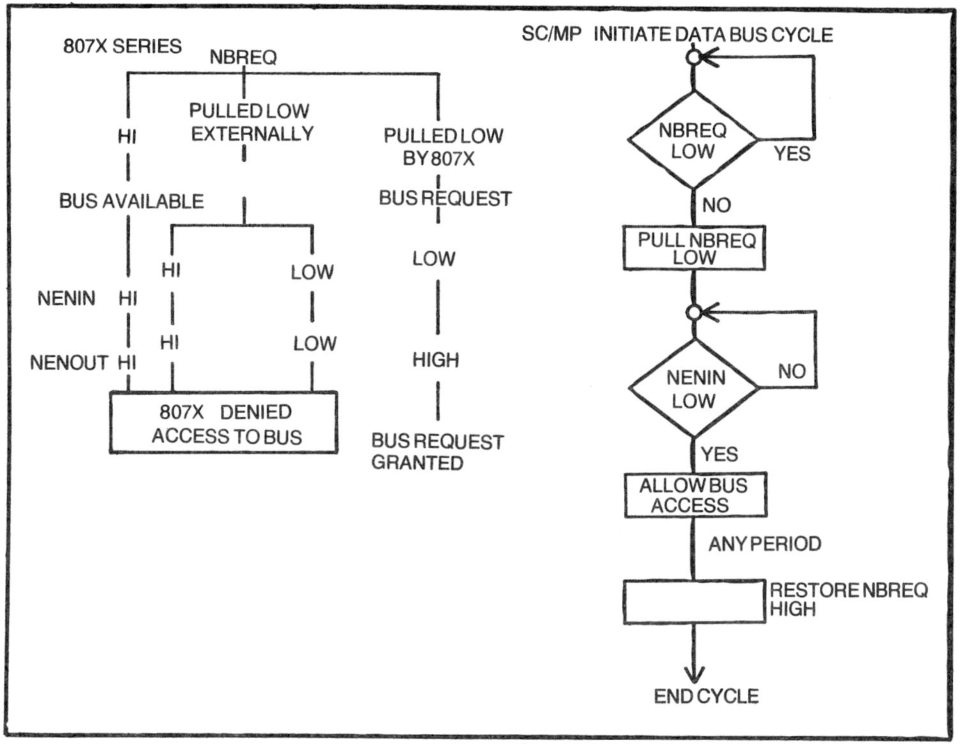

Fig. 10-19. Bus access logic for the SC/MP and INS-807X CPUs.

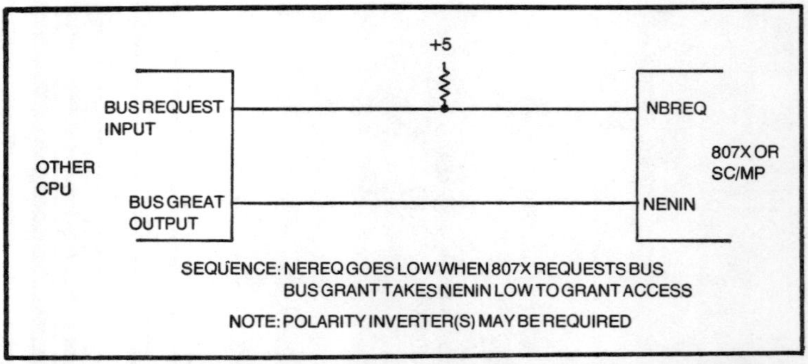

Fig. 10-20. Generalized bus control connections between an SC/MP or INS-807X CPU and other processors (This is another area where standardization is needed.)

The second is oriented to the home, with general purpose computational power plus energy and security management.

While these are quite simple in concept, remember that much creative work will be necessary in their execution. Design guides for some of the hardware are available, but construction and programming will not be trivial jobs.

WHY A HOME SYSTEM?

One of the traps in home use of microprocessors lies in the difference between "having a computer" and "using a computer". I have observed friends and neighbors become interested in the field, buy a minimum system, expand it for a time, and then lose interest. They haven't stopped to answer the question, "Why am I doing this?"

Consider the pie chart of Fig. 10-24. This shows the expected uses in the home of a low-cost, well-programmed computer

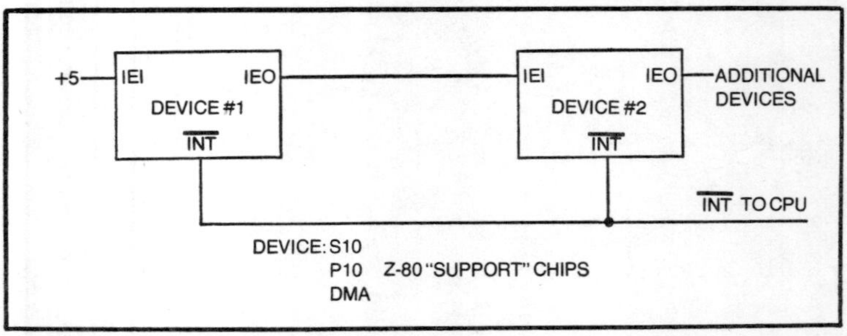

Fig. 10-21. Method of establishing a priority daisy chain in a system using Z80 CPUs.

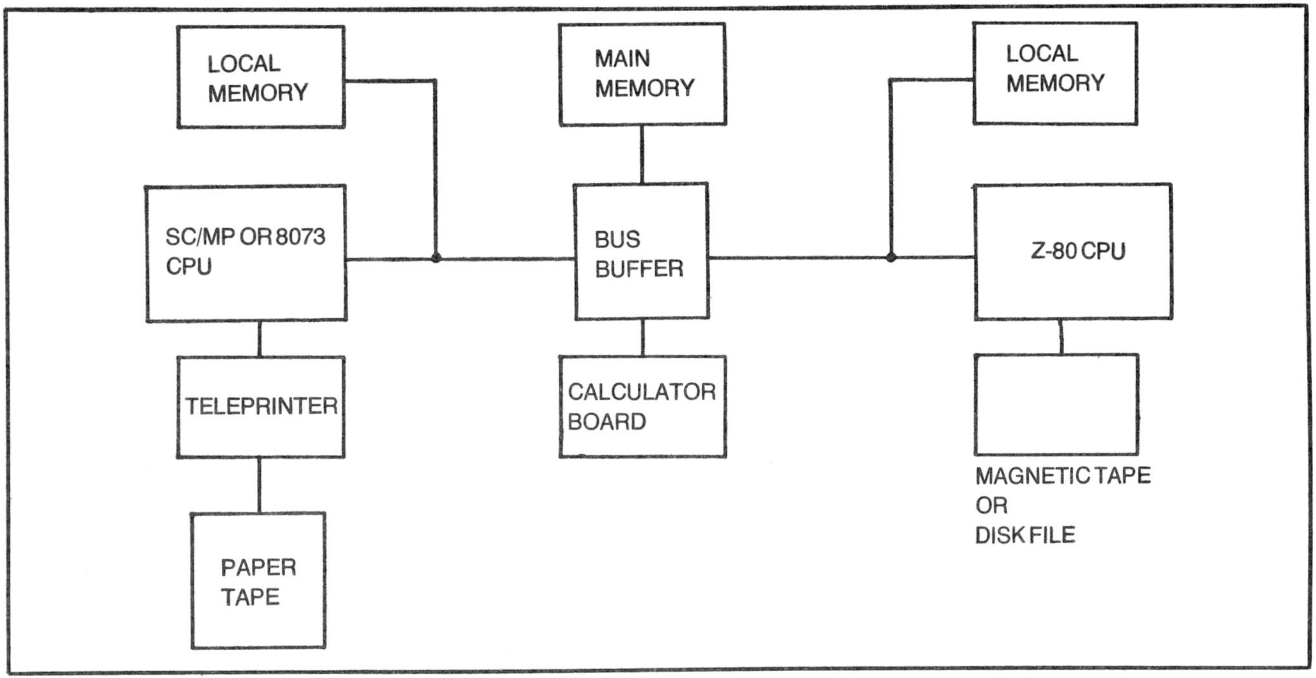

Fig. 10-22. Schematic of a calculation oriented multiprocessor system. Basically, an SC/MP or INS-8073 is used for I/O, a calculator board for computation, and a Z80 for data transfer, logic and binary arithmetic. The main memory is operated like a post office.

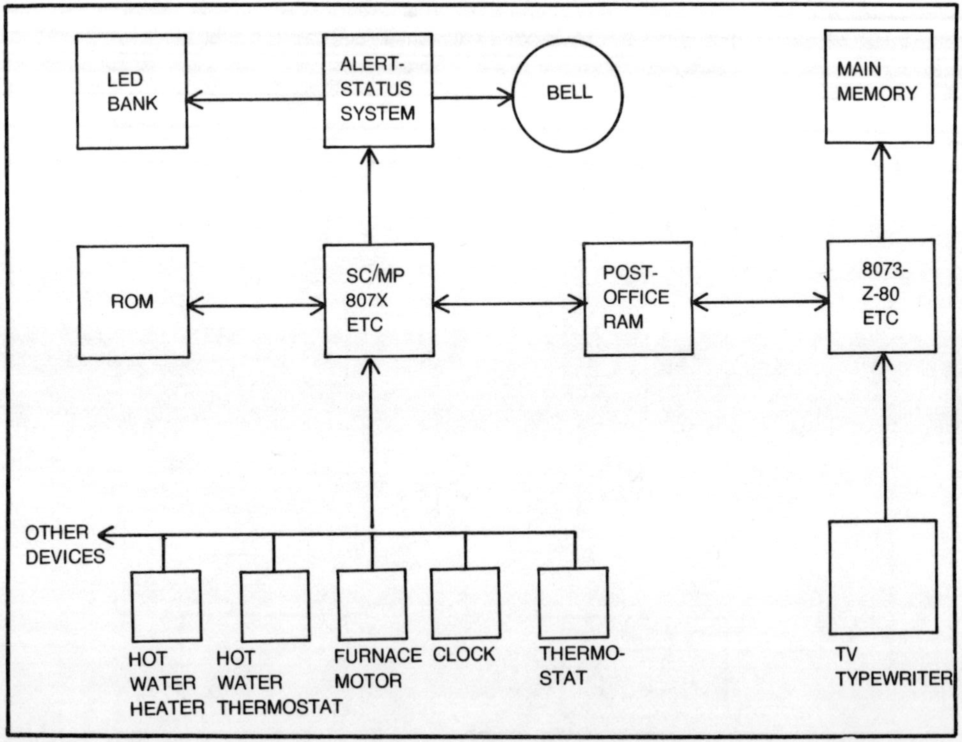

Fig. 10-23. Schematic of a computerized-home multiprocessor. An SC/MP or INS-8073 is used primarily as the controller-supervisor-threshold system for device and security operations, with another CPU system used as a smart terminal to establish operating conditions for home operation. Many variations are possible, of course.

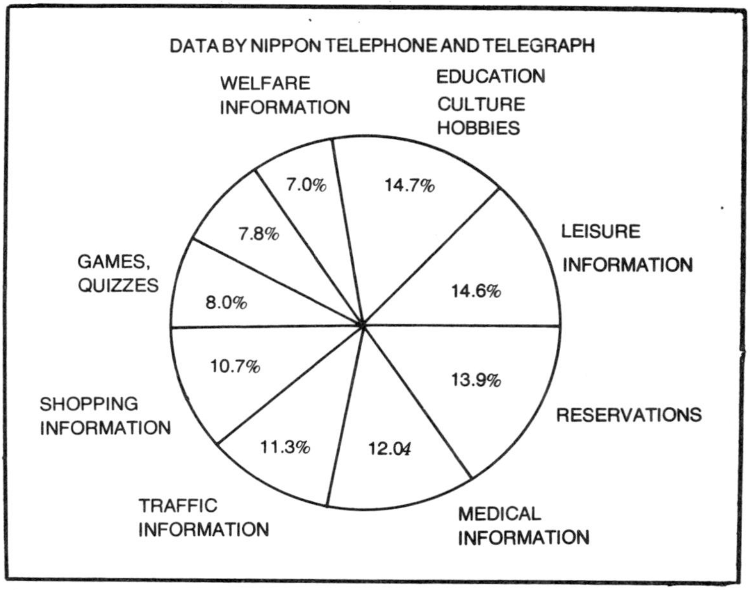

Fig. 10-24. Results of a survey of expected home use of a home computer system having access to a large information system. The data should be applicable to all affluent countries. Note that the system supposes that a good information utility exists.

system. Note that games (currently the most common use for package home computers) are anticipated to account for only eight percent of home computer use. The largest block, education, culture and hobbies, is expected to have twice as much use.

This survey was prepared by a telephone company, and may have slanted questions. Over seventy-five percent of home computer use will be for information, according to the survey. This may result from slanted questions, since information is likely to be provided by the telephone lines. The bias may also be reflected by the absence of such important areas as home-energy management and home security. And important areas of information are not included, such as the weather forecast.

Incomplete though this analysis may be, it does emphasize the fact that people expect the home computer to be a part of tomorrow's living. You can be sure that the commercial computer companies are working toward that goal. But for now, and for a long time to come, it's going to be ideal for the home builder and home programmer—a chance to pioneer and to work out approaches to a whole new field.

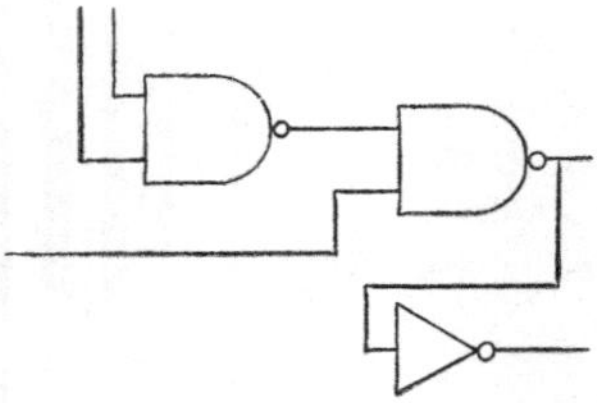

Appendix

The following programs, published in *Computer Newsletter*, have been selected as useful to SC/MP microprocessor users, SCAMP:Y, or others. They are reprinted with permission of National Semiconductor, publishers of *Compute*.

Program Name and Author	**Issue**
Fig. A-1. "Bootstrap Loader," R.P. Haviland	v.5,n.1
Fig. A-2. "Subroutine Supervisor," Erik Skovgaard	v.4,n.3
Fig. A-3. "Block Transfer," J.F. Verrij	v.4,n.3
Fig. A-4. "Hex Memory Dump," Thomas M. Farr, Jr.	v.5,n.3
Fig. A-5. "Reverse Polish Tape Loader," L.J. Stempnik	v.4,n.3
Fig. A-6. "Punch Complemented Binary Tape," Bernie Kute	v.3,n.1
Fig. A-7. "Quad Square Root," Jon Kleiser	v.5,n.1
Fig. A-8. "Double Unsigned Multiply," Jon Kleiser	v.5,n.1
Fig. A-9. "Fifteen-Bit Square Root," Ed Schoell	v.3,n.4
Fig. A-10. "SC/MP Editor," Jamieson Rowe	v.4,n.1
Fig. A-11. "PLOT," Jermyn Microcomputer Center	v.3,n.6
Fig. A-12. "SC/MP Dissembler (in NIBL)," Jim Derosa	v.4,n.2
Fig. A-13. "Word Game," Soo Nam Kim	v.3,n.10

Note: Also see references.

SC/MP BOOTSTRAP LOADER

by Robert P. Haviland
2100 S. Nova Road, Box 45
Daytona Beach, Florida 32019

The short bootstrap loader program shown below is for beginning SC/MP users who work with Baudot tape or equivalent. It should also be of interest to others who need such a loader.

Assumptions:

- Baudot tape, including start bit, five data bits, two stop bits; start is bus bit 0.
- Serial (SI, SO) input and output SENSE B = Character Ready.
- Two successive Baudot characters per eight-bit word.
- Baudot bit 5 indicates a control character (only rubout used).
- Baudot bits 1-4 entered in binary coded hex (Baudot A = Hex 3, etc.)
- Single step used to set starting address, to escape loader, and to jump to program execute.
- Register use as shown in program.
- For ASCII, can omit combining operations, and add checksum and more control characters.

NOTES:

1. To execute, single step to 11, then enter +51 (Hex 33). This jumps to 3E.
2. To escape, single step to 11, then enter +53 (Hex 35) and proceed, or use reset.

Fig. A-1. "Bootstrap Loader."

Address	Data (Hex.)	Mnemonic	Comment
00	08	NOP	Single Step
01	C4	LDI	
02	--	--	Manual enter Hi Start Address
03	36	XPAH-P2	
04	C4	LDI	
05	--	--	Manual enter Lo Start Address
06	32	XPAL-P2	
07	06	CSA	Auto Step: Wait Loop: Check Status
08	D4	ANI	
09	20	20	
0A	98	JZ-PC	Skip if character started
0B	FB	-5	
0C	8F	DLY	Delay 11 ms
0D	0D	13	(for 2.5 MHz clock)
0E	19	SIO	Get a bit
0F	06	CSA	Recheck Status
10	D4	ANI	
11	20	20	
12	98•	JZ-PC	Skip if end of character
13	04	+4	
14	8F	DLY	To End of Bit
15	0D	13	
16	90	JMP-PC	Skip if another bit
17	F4	-12	
18	19	SIO	End of character, add Stop bit
19	C4	LDI	Check control character
1A	FE	255	
1B	60	XRE	
1C	98	JZ-PC	Skip if not control
1D	E9	-23	
1E	06	CSA	Check which half
1F	D4	ANI	
20	01	01	
21	9C	JNZ-PC	Skip if second half
22	0E	+14	
23	C4	LDI	Set a mask
24	1E	31	
25	50	ANE	Get 1/2 byte
26	1E	RR	Shift
27	1E	RR	
28	1E	RR	
29	1E	RR	
2A	CA	ST-PZ	Store Temporarily
2B	00	0	
2C	C4	LDI	Set a flag
2D	01	01	
2E	07	CAS	
2F	90	JMP-PC	Go to Wait Loop
30	D6	-42	
31	C4	LDI	Set a mask
32	1E	31	
33	50	ANE	Get 1/2 byte
34	DA	OR-P2	Combine
35	00	00	
36	1E	RR	Normalize
37	CE	ST@P2	Store, increment
38	01	+1	
39	C4	LDI	Clear flag
3A	00	00	
3B	07	CAS	
3C	90	JMP	Go to wait
3D	C9	-55	
3E	92	JMP-P2	End of load -2
3F	FE	-2	Then to Start of Load

Fig. A-1. "Bookstrap Loader" (continued from page 267.)

SC/MP Subroutine Supervisor or The Long Arm of P3

by Erik Skovgaard
Carlsberg Research Center
10 GL. Carlsbergvej
DK-2500 Valby
Denmark

The limited number of registers in the SC/MP makes it a pain in the neck to write large programs with several subroutines and I/O. I finally got tired of setting up pointer registers (my cross assembler does not have the JS macro). Instead, this subroutine supervisor was written which makes programming a lot less boring.

```
                                ;CALL XPPC P3 (P3=SSUP)
                                ;THE FOLLOWING 3 BYTES MUST CONTAIN
                                ;1. FUNCTION CODE
                                ;2. ADDRESS HI
                                ;3. ADDRESS LO
                                ;THE FUNCTION CODE IS AS FOLLOWS:
                                ;   00=RETURN (NO ADDR NEEDED)
                                ;   01=READ (RESULT IN AC)
                                ;   02=WRITE FROM AC
                                ;   03=GOSUB
                                ;THE FOLLOWING ARE INDIRECT ADDR MODES
                                ;I.E. P3 POINTS TO THE ADDR TO BE USED
                                ;P3+0 IS HI,P3+1 IS LO ADDR BYTE
                                ;   04=READ INDIECT
                                ;   05=WRITE INDIRECT
                                ;   06=GOSUB INDIRECT
                                ;THE DEPTH OF NESTING MAY BE CONTINUED
                                ;BUT ONLY FIRST RETURN ADDR IS SAVED
                                ;
                                ;P2-11, -13, AND -14 MUST NOT BE
                                ;DESTROYED BY THE SUBROUTINE
                                ;P2-12, AND -15 ARE ALSO USED.
                                ;TIMING: 341 FOR GOTO, 153 FOR RET.
                                ;+190 MICRCYCLES EXTRA FOR INDIRECT.
                                ;
0200 C2F3    RET:   LD -13(2)          ;RESTORE P3
0202 37             XPAH P3
0203 C2F2           LD -14(2)
0205 33             XPAL P3
0206 C703           LD @3(3)           ;ADJUST FOR RETURN
0208 C2F5           LD -11(2)          ;GET AC
020A 3F             XPPC P3
                    ;*** ENTRY POINT
020B CAF5    SSUP:  ST -11(2)          ;STORE AC
020D C301           LD 1(3)            ;GET FNS WORD
020F 98EF           JZ "RET"
0211 CAF4           ST -12(2)          ;STORE FNS WORD
0213 C302           LD 2(3)            ;SET UP NEW P3
0215 CAF3           ST -13(2)
0217 C303           LD 3(3)            ;H=P2-13
0219 33             XPAL P3            ;L=P2-14
021A CAF2           ST -14(2)          ;SAVE OLD P3
021C C2F3           LD -13(2)
021E 37             XPAH P3
021F CAF3           ST -13(2)
0221 BAF4    STST:  DLD -12(2)
0223 9814           JZ "READ"
0225 BAF4           DLD -12(2)
0227 9816           JZ "WRT"
0229 BAF4           DLD -12(2)
022B 9818           JZ "GTO"
022D C300           LD 0(3)            ;GET SECONDARY ADDR
022F CAF1           ST -15(2)          ;STORE TEMP
0231 C301           LD 1(3)
0233 33             XPAL P3
0234 C2F1           LD -15(2)
0236 37             XPAH P3
0237 90E8           JMP "STST"         ;TEST AGAIN
0239 C300    READ:  LD 0(3)
```

Fig. A-2. "Subroutine Supervisor."

```
0238 CAF5            ST -11(2)
023D 90C1            JMP "RET"
023F C2F5     WRT:   LD -11(2)
0241 CB00            ST 0(3)
0243 90BB            JMP "RET"
0245 C7FF     GTO:   LD @-1(3)        ;ADJUST P3
0247 C2F5            LD -11(2)
0249 3F              XPPC P3
024A 90BF            JMP "SSUP"


 RET = 0200
SSUP = 020B
STST = 0221
READ = 0239
 WRT = 023F
 GTO = 0245
```

Instead of programming:

```
LD1     "subh"
XPAH    P3
LD1     "subl"
XPAL    P3
XPPC    P3.
```

I now have P3 set once and for all to "SSUP" and program:

```
XPPC    P3
.BYTE   03
.BYTE   "subh"
.BYTE   "subl"
```

which is a saving of 3 bytes. Or better even for a load:

Old Method:

```
LD1     newh
XPAH    P3
LD1     newl
XPAL    P3
LD      0(3)
ST      stack
LD1     oldh
XPAH    P3
LD1     oldl
XPAL    P3
LD      stack
```

New Method:

```
XPPC    P3
.BYTE   01
.BYTE   newh
.BYTE   newl
```

A saving of 14 bytes!

Fig. A-2. "Subroutine Superivsor" (continued from page 269).

MICRO EDIT

BLOCK TRANSFER FACILITY FOR SC/MP KIT

by J. F. Verrij, *Ir. Ing.*
Schillerstraat 140
5924 CS Venlo
Netherlands

When a program is being developed on a SC/MP Keyboard Kit, insertion of a few extra instructions in the program will include the tedious job of re-entering all following instructions. Block transfer, one of the features of an Editor, does this job for you. Such transfer can be implemented by a nine instruction program, which I named "micro edit". It moves a block of N bytes of RAM contents to a new location, leaving a free gap of G bytes for additional instructions.

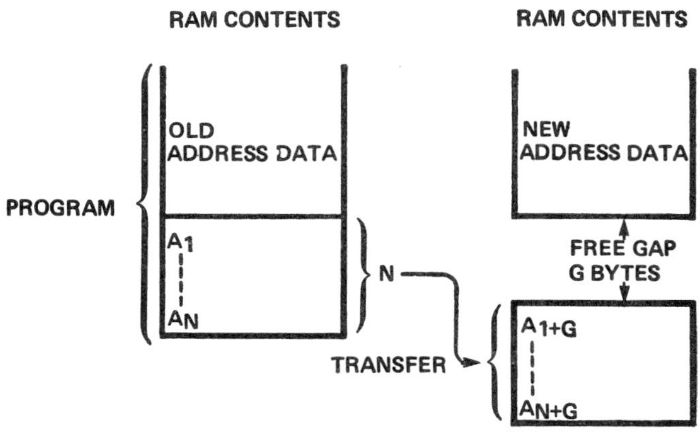

User's data: 0Fxx = (P_1): bottom byte original address plus one = A_{N+1}.
nn = number of bytes (N) to be transferred
gg = gap width (G) or transfer distance

Fig. A-3. "Block Transfer."

FLOW DIAGRAM

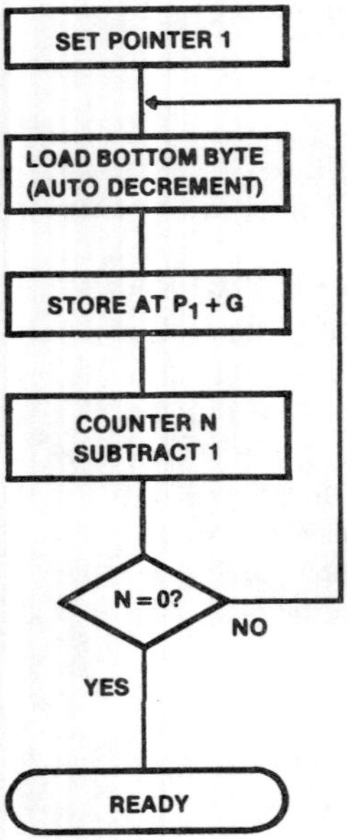

PROGRAM

ADDRESS	OPCODE	MNEMONIC	COMMENT
0F..	C4	LDI 0F	Set $(P_1) = 0Fxx$
..	0F		
..	35	XPAH 1	
..	C4	LDI xx	
..	xx		
..	31	XPAL 1	
..	C5	LD a − 1 (1)	Load bottom byte
..	FF		
..	C9	ST gg (1)	Store at new location
..	gg		
..	B8	DLD N	Counter N
..	04		subtract 1
..	9C	JNZ − 8 (0)	Next byte
..	F8		
..	3F	XPPC 3	Back to kitbug if ready
..	nn		Location of counter N

Note: Do not forget to change the displacements of jumps, etc., across the new gap. Micro Edit does not take account of labels.

With a few modifications Micro Edit can do the opposite, closing a gap after deletion of instructions. With one more pointer, transfer can be performed to other pages of RAM. With further refinements you could make the microprocessor calculate G and N from the original — and target addresses of the moved block, but as long as it has to be entered manually, Micro Edit finds its usefulness in its shortness compared to the size of the transferred block.

Fig. A-3. "Block Transfer" (continued from page 271).

HEXADECIMAL MEMORY DUMP FOR NIBL

by Thomas M. Farr, Jr.
TMF Electronics
569 Medina Dr.
Lewisville, Texas 75067

Users of NIBL (National Industrial BASIC Language) may find the following program useful. The program produces a hexadecimal listing of memory contents, formatted with an address followed by the contents of that address and the next 15 addresses on a line. The listing may be started and ended anywhere within the pages recognized by NIBL (Pages 0-7). The program requires the NIBL interpreter and the minimum NIBL hardware configuration: TTY I/O, Page 0 = NIBL ROM, Page 1 = RAM. The input parameters consist of the starting address and ending address.

The source listing of the program is shown below, followed by an example listing of the first 256 bytes of Page 0.

```
  2 REM; A=STARTING ADDRESS
  3 REM; B=ENDING ADDRESS
  4 REM; C=CURRENT ADDRESS
  5 REM; D==CHARACTER TO BE
    PRINTED
 10 PRINT "START LISTING AT ";:
    INPUT A
 20 PRINT "END LISTING AT ";:
    INPUT B
 30 FOR C=A TO B
 40 IF (C AND #000F) <>0 GO TO 110
 50 PRINT " "
 60 D=(C AND #F000)/4096: GOSUB 180
 70 D=(C AND #0F00)/256: GOSUB 180
 80 D=(C AND #00F0)/16: GOSUB 180
 90 D=(C AND #000F): GOSUB 180
100 PRINT " ";
110 D=(@C AND #F0)/16: GOSUB 180
120 D=(@C AND #0F): GOSUB 180
130 PRINT " ";
140 NEXT C
150 PRINT " "
160 PRINT "LISTING COMPLETE"
170 END
180 IF D=0 PRINT "0";: RETURN
190 IF D=1 PRINT "1";: RETURN
200 IF D=2 PRINT "2";: RETURN
210 IF D=3 PRINT "3";: RETURN
220 IF D=4 PRINT "4";: RETURN
230 IF D=5 PRINT "5";: RETURN
240 IF D=6 PRINT "6";: RETURN
250 IF D=7 PRINT "7";: RETURN
260 IF D=8 PRINT "8";: RETURN
270 IF D=9 PRINT "9";: RETURN
280 IF D=10 PRINT "A";: RETURN
290 IF D=11 PRINT "B";: RETURN
300 IF D=12 PRINT "C";: RETURN
310 IF D=13 PRINT "D";: RETURN
320 IF D=14 PRINT "E";: RETURN
330 IF D=15 PRINT "F";: RETURN
```

Fig. A-4. "Hex Memory Dump."

```
>PRINT TOP
 5490

>RUN
START LISTING AT ? #0000
END LISTING AT ? #0100

0000  08  C4  10  36  C4  1C  32  C4  11  35  C4  20  31  C4  FF  C9
0010  00  C9  01  C4  0D  C9  FF  C4  02  CA  F6  31  C4  20  35  B9
0020  02  01  C1  80  E4  0D  98  02  BA  F6  C4  20  35  C4  FF  C9
0030  00  C9  01  C4  0D  C9  FF  35  02  F4  10  E4  80  98  04  E4
0040  80  90  E9  C4  00  CA  F4  CA  F5  C4  58  CA  FB  C4  0C  CA
0050  FA  C4  00  CA  EA  01  C4  00  CA  80  AA  EA  01  C4  34  60
0060  9C  F4  C4  50  CA  FD  C4  7A  CA  FF  C4  6A  CA  FC  C4  A6
0070  CA  F9  C4  8A  CA  FE  C2  FB  33  C2  FA  37  C7  01  01  C7
0080  01  33  CA  FB  40  D4  0F  DC  00  37  CA  FA  40  D4  F0  E4
0090  20  98  2F  E4  A0  98  07  E4  C0  98  E1  3F  90  D8  C2.  F9
00A0  E4  D6  9C  04  C4  0A  90  60  E4  D6  33  CA  EA  C4  10  37
00B0  01  C2  FB  CF  01  C2  FA  CF  01  C2  EA  33  CA  F9  40  37
00C0  90  BA  CA  E7  C5  01  E4  20  98  FA  C5  FF  C2  FA  37  CA
00D0  EC  C2  FB  33  CA  ED  C7  01  01  BA  E7  40  D4  7F  E5  01
00E0  9C  07  40  94  F1  90  95  90  8D  C2  E7  01  C5  80  C2  ED
00F0  33  C2  EC  37  90  CA  C4  10  37  C2  F9  33  C7  FF  01  C7
0100  FF
LISTING COMPLETE
```

Fig. A-4. "Hex Memory Dump" (continued from page 273).

Reverse Polish

by Lawrence J. Stempnik
Senior Project Engineer
Schiller Industries, Inc.
7111 E. Eleven Mile Rd.
Warren, MI 48090

REVERSE POLISH PAPER TAPE LOADER

The Reverse Polish Notation developed by Jan Lukasiewicz results in calculator simplifications due to the parenthesis-free notation. In a similar manner, a Reverse Polish Paper Tape Loader can be written with just seven opcodes. This program will load up to 230 memory locations in the standard KITBUG configuration from a punched paper tape.

This RPPTLD program is very helpful in loading debugged programs into SC/MP from teletypes. A forward loading tape program can be used but it requires three times more typing to load it manually and it takes up more storage locations.

To use the program, a special paper tape must be prepared, either manually or with a microprocessor. Each line consists of two consecutive pairs of hex digits plus a carriage return or line feed, but not both. The tape is then punched with each four character line in reverse sequence. Punch the last four digits to be stored first, and continue until reaching the first four digits to be stored.

This RPPTLD program uses the GHEX subroutine in KITBUG. The same rules as MODIFY apply when entering more or less than four digits per line. If more than four digits are entered on one line, only the last four are used. A CR without numbers results in 00 being stored in two consecutive memory locations. If one to three digits are entered before CR, GHEX will supply leading zeros to produce four digits. For convenience in handling the paper tape and reader, enter 20 leading zeros on the first and last line of the tape.

Pointer P1 must be set to 00DF for GHEX. Pointer P2 points to one plus the last storage location (end of program to be stored). To use the program, type in the seven opcodes plus the three pointer registers using the MODIFY command. Load the paper tape in the reader and position to start of first line of data, type G, return, and turn tape reader on. After tape is loaded, hit RESET and return control to KITBUG.

```
02E7  35  XPAH    P1  H(GHEX)
02E8  37  XPAH    P3  HI=00
02E9  31  XPAL    P1  L(GHEX)-1
02EA  33  XPAL    P3=00DF=GHEX-1
02EB  3F  XPPC    P3  GET TAPE INPUT
02EC  90  JMP     REPEAT TO END OF TAPE
02ED  FD  P0-3
```

Fig. A-5. "Reverse Polish Tape Loader."

REVERSE POLISH PAPER TAPE PUNCH

This SC/MP program is used to prepare tapes to be loaded into SC/MP using the RPPTLD program. It prints (punches) four consecutive hex characters per line from storage with the lines in reverse sequence. Each line ends with CR but does not use line feed.

KITBUG subroutines used are PHEX2 for printing (punching) the hex characters and PUTC for the CR. To use the program, set P0 and P2 to 02D7 and set P1 to 1 plus the last line of the stored data or program. Turn on the paper tape punch just before running the program. The paper tape will be punched as the teletype prints the lines. About 208 bytes can be punched using memory locations 02CF through 0200.

02D0	XX	XX	USED BY SUBS
02D1	XX	XX	USED BY SUBS
02D2	XX	XX	USED BY SUBS
02D3	XX	XX	AD HI
02D4	XX	XX	AD LO
02D5	XX	XX	USED BY SUBS
02D6	XX	XX	USED BY SUBS
02D7	C4	LDI	REV POL PT PCH
02D8	01	HEX = DECIMAL + 1	
02D9	37	XPAH	P3 HI = 01
02DA	C4	LDI	L(PHEX2) − 1
02DB	43	HEX = DECIMAL + 67	
02DC	33	XPAL	P3 = 0143 = PHEX2 − 1
02DD	C5	@LD	MOVE PUNCH POINTER
02DE	FE	P1 − 2	
02DF	3F	XPPC	P3 PRT HI BYTE
02E0	C4	LDI	
02E1	43	HEX = DECIMAL + 67	
02E2	33	XPAL	P3 = 0143 = PHEX2 − 1
02E3	C1	LD	
02E4	01	P1 + 1	
02E5	3F	XPPC	P3 PRT LO BYTE
02E6	C4	LDI	
02E7	C4	HEX = DECIMAL − 60	
02E8	33	XPAL	P3 = 01C4 = PUTC − 1
02E9	C4	LDI	
02EA	0D	HEX = DECIMAL + 13	
02EB	3F	XPPC	P3 PRT CR
02EC	90	JMP	REPEAT TO START PROG
02ED	E9	P0 − 23	

Fig. A-5. "Reverse Polish Tape Loader" (continued from page 275).

Punch Complemented Binary PROM Tape on SC/MP LCDS

```
                  TITLE PUNCH, 'PUNCH COMP BIN TAPE.'
      0001 P1     =      1
      0002 P2     =      2
      0003 P3     =      3
      77D0 RAM    =      0077D0
      7B4F GHEX   =      07B4F
      7B16 MESG   =      07B16
      7AE1 PUTC   =      07AE1
      7A88 GETC   =      07A88
      0064 LEN    =      100
      7700        .=     07700
           ;      P1 MUST HAVE ADDRESS OF BUFFER.
           ;      P2 MUST HAVE RAM ADDRESS. 077D0 IS OK.
           ;      E REG = 0 PUNCH 256 WORDS.
           ;      E REG = -1 PUNCH 512 WORDS.
           START:
7700 C47A         LDI    H(GETC)
7702 37           XPAH   P3
7703 C488         LDI    L(GETC)
7705 33           XPAL   P3
7706 3F           XPPC   P3        ;WAIT FOR KEYBD HIT.
7707 C464         LDI    LEN
7709 CA01         ST     1(P2)     ;SET COUNTER FOR LEADER.
770B C47A         LDI    H(PUTC)   ;SET ADDRESS OF PUTC.
770D 37           XPAH   P3
770E C4E1         LDI    L(PUTC)
7710 33           XPAL   P3
           LDRLP:
7711 C400         LDI    0
7713 3F           XPPC   P3        ;PUNCH LEADER.
7714 BA01         DLD    1(P2)     ;UP DATE COUNTER.
7716 9CE9         JNZ    LDRLP     ;KEEP GOING.
7718 C4FF         LDI    -1        ;PUT RUBOUT.
771A 3F           XPPC   P3
           LOOP:
771B C400         LDI    0         ;SET A TO 0.
771D 02           CCL              ;CLEAR CARRY.
771E FD01         CAD    @1(P1)    ;GET COMPLEMENTED
                                   DATA. **
7720 3F           XPPC   P3        ;PUNCH DATA.
7721 AA01         ILD    1(P2)     ;UP DATE COUNT.
7723 9CF6         JNZ    LOOP
7725 40           LDE              ;FIND WHICH.
7726 9805         JZ     OUT       ;CHECK IF DONE.
7728 C400         LDI    0         ;RESET E REG.
772A 01           XAE
772B 90EE         JMP    LOOP
           OUT:
772D C464         LDI    LEN       ;SET LENGTH OF TRAILER.
772F CA01         ST     1(P2)
           TRAIL:
7731 C400         LDI    0         ;PUNCH TRAILER.
7733 3F           XPPC   P3
7734 BA01         DLD    1(P2)
7736 9CF9         JNZ    TRAIL
7738 C488         LDI    L(GETC)
773A 33           XPAL   P3
773B C47A         LDI    H(GETC)
773D 37           XPAH   P3
773E 3F           XPPC   P3        ;WAIT FOR TAPE OFF.
773F 00           HALT             ;RETURN TO DEBUG.
     0000         .END
```

Fig. A-6. "Punch Complemented Binary Tape."

QUAD SQUARE ROOT
SQRT4

by Jon Kleiser
Oslo, Norway

The program SQRT4 takes the square root of a 32-bit unsigned number in RAD1, 2, 3 and 4. The root, which is to be found in RT1 and RT2, is a 16-bit integer whose square is less than or equal to the radicand. The program also calculates the first fractional bit of the root for cases where a round-up is wanted. Then one shall add the content of the Carry/Link flag to the root after return. However, if the input value is greater than 4 294 901 760 (X'FFFF0000), the root will be X'FFFF with the CY/L set, and a round-up will give X'0000.

Register P2 is used as a stack pointer. P3 is the subroutine calling register.

Execution time is from 16140 to 18586 microcycles, i.e., 32.3 37.2 millisec. for an old SC/MP.

The algorithm used is the one for calculating the square root by hand, based on the following formula:

$$(r_1+r_2+...+r_n)^2 = r_1^2+r_2(2r_1+r_2)+...+r_n(2(r_1+r_2+...+r_{n-1})+r_n)$$

Stack usage:

Disp	Name	Function
0	RAD1	Radicand
1	RAD2	--
2	RAD3	--
3	RAD4	--
4	RT1	Root
5	RT2	--
6	RT3	Root frac.
7	DIF1	Difference
8	DIF2	--
9	DIF3	--
A	ONE1	Ones' pos.
B	ONE2	--
C	ONE3	--
D	CNT	Counter

The program:

```
0000    C400    SQRT4   LDI    0
0002    CA04            ST     RT1 (2)      ZERO ROOT
0004    CA05            ST     RT2 (2)
0006    CA06            ST     RT3 (2)
0008    CA0B            ST     ONE2 (2)
000A    CA0C            ST     ONE3 (2)
000C    01              XAE                 . ZERO RADICAND EXTENSION
000D    C440            LDI    X'40         . SET ONES' POSITION
000F    CA0A            ST     ONE1 (2)     . INDICATOR
0011    C411            LDI    17           . SET BIT COUNTER
```

Fig. A-7. "Quad Square Root."

```
0013    CA0D             ST   CNT (2)
0015    C204    NEW      LD   RT1 (2)       ADD "ONE" TO ROOT
0017    DA0A             OR   ONE1 (2)
0019    CA04             ST   RT1 (2)
001B    C205             LD   RT2 (2)
001D    DA0B             OR   ONE2 (2)
001F    CA05             ST   RT2 (2)
0021    C206             LD   RT3 (2)
0023    DA0C             OR   ONE3 (2)
0025    CA06             ST   RT3 (2)
0027    03               SCL                SET CARRY/LINK FLAG
0028    C202             LD   RAD3 (2)      . SUBTRACT ROOT FROM
002A    FA06             CAD  RT3 (2)       HIGHEST PART OF
002C    CA09             ST   DIF3 (2)      RADICAND
002E    C201             LD   RAD2 (2)
0030    FA05             CAD  RT2 (2)
0032    CA08             ST   DIF2 (2)
0034    C200             LD   RAD1 (2)
0036    FA04             CAD  RT1 (2)
0038    CA07             ST   DIF1 (2)
003A    40               LDE                DO A POSSIBLE BORROW FROM
003B    FC00             CAI  0             . RADICAND EXTENSION
003D    06               CSA                . CHECK SIGN OF DIFFERENCE
003E    9428             JP   ZERO          . JUMP IF DIFFERENCE NEG
0040    C400             LDI  0             . ZERO RADICAND EXTENSION
0042    01               XAE
0043    C207             LD   DIF1 (2)      . MOVE DIFFERENCE TO
0045    CA00             ST   RAD1 (2)      . HIGHEST PART OF
0047    C208             LD   DIF2 (2)      . RADICAND
0049    CA01             ST   RAD2 (2)
004B    C209             LD   DIF3 (2)
004D    CA02             ST   RAD3 (2)
004F    02               CCL                . CLEAR CARRY/LINK FLAG
0050    C206             LD   RT3 (2)       . ADD "ONE" TO ROOT AGAIN
0052    F20C             ADD  ONE3 (2)      . TO COMPLETE "THE
0054    CA06             ST   RT3 (2)       . DOUBLING OF THE ROOT"
0056    C205             LD   RT2 (2)
0058    F20B             ADD  ONE2 (2)
005A    CA05             ST   RT2 (2)
005C    C204             LD   RT1 (2)
005E    F20A             ADD  ONE1 (2)
0060    CA04             ST   RT1 (2)
0062    9016             JMP  SHIFT
0064    909A    SQRT4X   JMP  SQRT4         . LONG DISTANCE JUMP
0066    90AD    NEWX     JMP  NEW           . SPLIT-UPS
0068    C204    ZERO     LD   RT1 (2)       . REMOVE THE "ONE"
006A    E20A             XOR  ONE1 (2)      . ADDED PREVIOUSLY
006C    CA04             ST   RT1 (2)
006E    C205             LD   RT2 (2)
0070    E20B             XOR  ONE2 (2)
0072    CA05             ST   RT2 (2)
0074    C206             LD   RT3 (2)
0076    E20C             XOR  ONE3 (2)
0078    CA06             ST   RT3 (2)
007A    C203    SHIFT    LD   RAD4 (2)      . SHIFT RADICAND LEFT
007C    F203             ADD  RAD4 (2)      . (A CCL WAS NOT NECESSARY)
007E    CA03             ST   RAD4 (2)
0080    C202             LD   RAD3 (2)
0082    F202             ADD  RAD3 (2)
0084    CA02             ST   RAD3 (2)
0086    C201             LD   RAD2 (2)
0088    F201             ADD  RAD2 (2)
008A    CA01             ST   RAD2 (2)
008C    C200             LD   RAD1 (2)
008E    F200             ADD  RAD1 (2)
0090    CA00             ST   RAD1 (2)
0092    40               LDE
0093    70               ADE
0094    01               XAE
0095    C20A             LD   ONE1 (2)      . SHIFT ONES' POSITION
                                            . INDICATOR RIGHT
0097    1F               RRL                . (A CCL WAS NOT NECESSARY)
0098    CA0A             ST   ONE1 (2)
009A    C20B             LD   ONE2 (2)
009C    1F               RRL
009D    CA0B             ST   ONE2 (2)
009F    C20C             LD   ONE3 (2)
00A1    1D               SRL
00A2    CA0C             ST   ONE3 (2)
00A4    BA0D             DLD  CNT (2)       . DECREMENT BIT COUNTER
00A6    9CBE             JNZ  NEW           . JUMP IF NOT FINISHED
00A8    C206             LD   RT3 (2)       ; SHIFT FRACTION INTO
00AA    F206             ADD  RT3 (2)       . CARRY/LINK FLAG
00AC    3F               XPPC 3             . RETURN
00AD    90B5             JMP  SQRT4X
```

by Jon Kleiser
Oslo, Norway

The program DMPY multiplies two 16-bit unsigned numbers in MRH, MRL and MNDH, MNDL. The 32-bit product is to be found in PROD1, 2, 3 and 4.

Register P2 is used as a stack pointer. P3 is the subroutine calling register.

Stack usage:

Disp	Entry: Name	Function	Return: Name	Function
FE	SHIFT	Shift reg.	SHIFT	
FF	CNT	Counter	CNT	
00	MRH	Multiplicator	PROD1	Product
01	MRL	---	PROD2	---
02	MNDH	Multiplicand	PROD3	---
03	MNDL	---	PROD4	---

The program:

```
0000 C200 DMPY  LD   MRH (2)    GET HIGHER PART OF MULTIPLIER
0002 CAFE       ST   SHIFT (2)  SAVE IN SHIFT
0004 C201       LD   MRL (2)    GET LOWER PART OF MULTIPLIER
0006   01       XAE             SAVE IN E-REG
0007 C400       LDI  0          :ZERO HIGHEST HALF OF PRODUCT
0009 CA00       ST   PROD1 (2)
000B CA01       ST   PROD2 (2)
000D C410       LD1  16         :SAVE COUNTER
000F CAFF       ST   CNT (2)
0011 02   $LOOP CCL             :CLEAR CARRY/LINK FLAG
0012 40         LDE             :LOAD LOWER PART OF MULTIPLIER
0013 D401       ANI  1          :TEST LEAST BIT
0015 9828       JZ   NO         :JUMP IF BIT IS ZERO
0017 C201       LD   PROD2 (2)  :ADD MULTIPLICAND (16 BITS)
0019 F203       ADD  MNDL (2)   :TO 32-BIT RESULT
001B CA01       ST   PROD2 (2)
001D C200       LD   PROD1 (2)
001F F202       ADD  MNDH (2)
0021   1F NOADD RRL             :SHIFT
0022 CA00       ST   PROD1 (2)
0024 C201       LD   PROD2 (2)
0026   1F       RRL             :SHIFT
0027 CA01       ST   PROD2 (2)
0029 C2FE       LD   SHIFT (2)
002B   1F       RRL             :SHIFT (PRODUCT IS SHIFTED
002C CAFE       ST   SHIFT (2)  :INTO SHIFT/E-REG WHILE
002E   40       LDE             :MULTIPLIER IS SHIFTED OUT)
002F 1D         SRL             :SHIFT
0030   01       XAE
0031 BAFF       DLD  CNT (2)    :DECREMENT COUNTER
0033 9CDC       JNZ  $LOOP      :JUMP IF NOT FINISHED
0035 C2FE       LD   SHIFT (2)  :GET LOWEST HALF OF PRODUCT
0037 CA02       ST   PROD3 (2)
0039   40       LDE
003A CA03       ST   PROD4 (2)
003C   3F       XPPC 3          :RETURN
003D 90C1       JMP  DMPY
003F C200 NO    LD   PROD1 (2)
0041 90DE       JMP  NOADD
```

Fig. A-8. "Double Unsigned Multiply."

SC/MP Program SL0041A SCSQRT

Source paper tape $5.00 each

SCSQRT takes the square root of a 15-bit positive integer in memory. The result is an 8-bit positive integer. The largest input value is 32,768.

```
                .TITLE   SCSQRT ,    SC/MP SQUARE ROOT'
                ;SL0041A  SC/MP  COMPUTE LIBRARY  11/76
                ;
                ;        ED SCHOELL
                ;        NS ELECTRONICS AUSTRALIA
                ;        13 OCT 1975
                ;
                ;THIS ROUTINE TAKES THE SQUARE ROOT OF
                ;THE 15-BIT POSITIVE NUMBER IN LS,MS AND
                ;LEAVES THE RESULT IN LS AS AN 8-BIT
                ;POSITIVE NUMBER.
                ;
                ;THIS IS TOP OF STACK, RESULT IS ON
                ;STACK (CORRECTED) RANGE OF NUMBERS IS
                ;32768 AS INPUT GIVING 181 AS RESULT
                ;IN LS.  RESULT IS INTEGER WHOSE SQUARE
                ;IS LESS THAN ENTRY NUMBER.
                ;
                ;TIME IN 172 +N(158) +<N-128(126)FOR N>128>
                ;IE  1320 MICROSEC FOR N=1
                ;    40.8 MILLISEC FOR N=128
                ;    70.8 MILLISEC FOR N=181  (N IS RESULT)
                ;
                ;
                ;
                ;RESULT IS INTEGER,IE SQRT OF 25 IS 5
                ;AND 5 IS GIVEN AS RESULT  FOR ANY INPUT
                ;IN RANGE OF 25 TO 35. 36 GIVES 6 AS RES.
                ;
                ;P2 IS USED AS STACK POINTER TO RAM
                ;P1 AND P3 ARE NOT USED.
                ;
                ;ALGORITHM USED IS (N+1)^2-N^2=2N+1
                ;SUCCESSIVE VALUES OF 2N+1 ARE SUBTRACTED
                ;FROM THE NUMBER AS N IS ACCUMULATED UNTIL
                ;THE ORIGINAL NUMBER IS LESS THAN ZERO.
                ;
                ;EG....7 SQUARED IS 49
                ;       6 SQUARED IS 36
                ;       DIFFERENCE...13...=2*6+1
                ;
                ;
                ;EXAMPLE OF STACK USAGE.
                ;
                ;        REL       ENTRY          EXIT
                ;        ----------------------------
                ;       -1
                ;(P2)-> 0           7F             00
                ;       +1          F9       (P2)->B5
                ;
                ;    IE SQRT OF 7FF9 (32,768) IS B5 (181)
                .PAGE
        0000    MS      =     0
        0001    LS      =     1
0000    00      SQRT:   HALT
0001    02              CCL
0002    C400            LDI   0
0004    FA01            CAD   LS(2)
0006    CA01            ST    LS(2)
0008    C400            LDI   0
000A    FA00            CAD   MS(2)
```

Fig. A-9. "Fifteen-Bit Square Root."

```
000C CA00                ST     MS(2)
                ;BOTH PARTS NOW TWO'S COMPLEMENTED.
000E C400                LDI    0
0010 01                  XAE              ;EXT REG STORES N
0011 03         LOOP:    SCL              ;ADD 1
0012 40                  LDE
0013 F201                ADD    LS(2)     ;ADD N+1 TO COMP NO
0015 CA01                ST     LS(2)     ;AND SAVE IT
0017 06                  CSA              ;TEST CARRY
0018 940B                JP     NOCRY1
001A C200                LD     MS(2)     ;CARRY INTO MS
001C F400                ADI    0         ;ADD IT IN
001E CA00                ST     MS(2)
0020 06                  CSA              ;TEST FOR END
0021 D480                ANI    080       ;MASK CRY BIT
0023 9C11                JNZ    EXIT
0025 40         NOCRY1:  LDE              ;CRY IS CLEAR
0026 F201                ADD    LS(2)     ;ADD N TO COMP NO
0028 CA01                ST     LS(2)
002A 06                  CSA              ;TEST CRY
002B 9410                JP     NOCRY2
002D C200                LD     MS(2)     ;CRY TO MS
002F F400                ADI    0
0031 CA00                ST     MS(2)
0033 06                  CSA              ;TEST FOR END?
0034 9407                JP     NOCRY2
0036 C601       EXIT:    LD     @1(2)     ;ADJ STACK
0038 40                  LDE
0039 CA00                ST     MS(2)     ;PUT SQRT TO STACK
003B 9006                JMP    CONT      ;COULD BE XPPC (IE RTS)
003D 40         NOCRY2:  LDE
003E F401                ADI    1         ;INC N
0040 01                  XAE
0041 90CE                JMP    LOOP
0043 00         CONT:    HALT             ;PROG TERMINATES HERE
     0000                .END

CONT    0043     EXIT    0036
LOOP    0011     LS      0001
MS      0000     NOCRY1  0025
NOCRY2  003D     SQRT    0000 *

NO ERROR LINES
SOURCE CHECKSUM = FA02
```

Fig. A-9. "Fifthteen-Bit Square Root." (continued from page 281)

SC/MP Editor

by Jamieson Rowe

Reprinted with permission from Electronics Australia

If you've ever tried punching up a paper tape of a program in assembly or problem-orientated language using a teleprinter on "local", you'll know just how frustrating it can be. Even if you are extremely careful it seems to be impossible not to make a few errors, and Murphy's Law always

Fig. A-10. "SC/MP Editor."

seems to ensure that you rarely discover these until you have typed in at least three more lines!

Of course if you realize that you made an error immediately after having done it, you can use the back-space facility and delete the wrong character(s) with the rubout key. But if you don't discover an error until later on, you are forced to either perform cut-and-paste surgery on the tape, or try stop-and-go editing of the tape while punching a new "clean" version.

Preparing long programs in this way can be very tedious and time consuming. Small wonder, then, that the people working on minicomputers and larger machines have for years been using the computer itself to make the job easier and faster. This is done by using a software utility program known as a symbolic text editor.

Using such a program, you type your own program text into a section of the computer's memory which is set aside as the "buffer". Then with the text in the buffer, the editor lets you change lines, delete lines, insert extra lines at any desired point, inspect lines or groups of lines, and finally either type out a listing or punch out a clean tape (or both).

In short, a text editor program can be a very useful item of software, and it is well worth having one even on small microcomputer systems. The only trouble is that not too many microcomputers have been provided with editor programs as yet, particularly the systems based on the more recent microprocessor chips.

To help alleviate this situation, I have written a text editor program for the SC/MP Low Cost Development System (LCDS). I chose this system because at present the SC/MP chip and its systems appear to be growing fastest in popularity, particularly in the hobby area.

The editor program itself is written in SC/MP machine language, and occupies 864 bytes of memory. It uses the 256-byte RAM in the LCDS system base as a stack and line address buffer, but can use RAM memory at any other location in SC/MP memory space for its text buffer. The larger the available RAM, the more text the editor can handle at one time.

The editor can be written into a pair of MM5204 512-byte PROMs. With the PROMs plugged into the correct locations on a SC/MP ROM card which is programmed for the appropriate address range (hex. 3000-3FFF), you will then have the editor permanently resident in your system, and available at any time merely by calling it at its starting address (hex. 3C00).

Alternatively, you can run the editor in RAM memory, loading it in each time you need it by means of a punched paper tape or cassette. The four-digit numbers at the start of each line are addresses; as you can see the program runs from 3C00 to 3F5F, inclusive.

Note that if you do elect to run the editor in RAM, you will need at least one 2k-byte RAM card. This will give you 1k of RAM left for the text buffer — enough for small text slabs, but barely good enough for serious work. With the editor in PROMs, even a single RAM card gives you a full 2k for the text buffer, which is very much more practical.

SC/MP SYMBOLIC EDITOR PROGRAM

```
3CØØ Ø4 C4 77 36 C4 FF 32 C4 7B 37 C4 16 33 3F 3F Ø7
3C1Ø C4 4F 33 3F C6 Ø1 CA F1 C6 Ø1 CA EF 3F C6 Ø1 CA
3C2Ø EF C6 Ø1 CA ED C4 ØØ CA F9 C2 FØ CA F8 C2 EF CA
3C3Ø F7 C4 7A 37 C4 E1 33 C4 Ø7 3F C4 ØD 3F C4 ØA 3F
3C4Ø C4 ØØ CA F4 CA F3 CA F2 C4 7A 37 C4 9Ø CA F1 33
3C5Ø 3F D4 7F Ø1 C4 3F 35 C4 2B 31 C5 Ø3 98 3B 6Ø 9C
3C6Ø F9 C1 FE CA F6 C1 FF 31 C2 F6 35 C2 F9 Ø3 FA F3
3C7Ø 94 Ø2 9Ø 12 C2 F9 Ø3 FA F2 94 Ø2 9Ø Ø9 C2 F2 98
3C8Ø 1Ø Ø3 FA F3 94 ØB C4 7A 37 C4 E1 33 C4 3F 3F 9Ø
3C9Ø A9 3F D4 7F E4 ØD 9C EE 3D 4Ø D4 7Ø E4 3Ø 98 19
3CAØ 4Ø E4 2F 9C Ø9 C4 3E 35 C4 B2 31 3D 9Ø 8C 4Ø E4
3CBØ 2C 9C D3 C4 Ø1 CA F4 9Ø 8F 4Ø D4 ØF CA F5 C2 F4
3CCØ 9C Ø4 C2 F3 9Ø Ø2 C2 F2 Ø2 Ø1 4Ø 7Ø Ø1 7Ø 7Ø Ø1
3CDØ 4Ø 7Ø F2 F5 Ø1 C2 F4 9C Ø5 4Ø CA F3 9Ø D9 4Ø CA
3CEØ F2 9Ø D4 35 CA F6 31 CA F5 C2 F7 CA E9 C2 F8 CA
3CFØ EA C2 F1 E4 84 98 Ø9 C4 7A 37 C4 E1 33 C4 ØA 3F
3DØØ C4 7A 37 C2 F1 33 C2 EE Ø2 F4 Ø1 E2 F8 98 53 C2
3D1Ø F8 35 C2 F7 31 3F D4 7F Ø1 4Ø E4 ØA 98 1D 4Ø E4
3D2Ø Ø7 98 3F 4Ø E4 25 9C Ø4 C5 FF 9Ø ØF 4Ø 98 ØC E4
3D3Ø 7F 98 Ø8 4Ø E4 ØD 98 Ø1 4Ø CD Ø1 C4 77 35 CA F8
3D4Ø C2 F3 Ø2 F2 F3 31 CA F7 4Ø E4 ØD 9C B6 C2 E9 C9
3D5Ø ØØ C2 EA C9 Ø1 C2 F9 E4 74 98 Ø7 C2 F6 35 C2 F5
3D6Ø 31 3D C4 3C 35 C4 3Ø 31 C4 9Ø CA F1 3D Ø8 Ø8 Ø8
3D7Ø C4 8Ø 9Ø Ø2 C4 Ø1 CA F1 C2 F3 9C ØA C4 Ø1 CA F3
3D8Ø C2 F9 CA F2 9Ø Ø8 C2 F2 9C Ø4 C2 F3 CA F2 C4 E1
```

Fig. A-10. "SC/MP EDITOR" (continued from page 283).

```
3D9Ø 33 C4 ØA 3F C2 F1 98 15 C4 7A 37 C4 88 33 3F C4
3DAØ E1 33 C4 CØ CA F6 C4 ØØ 3F AA F6 9C F9 C4 77 35
3DBØ C2 F3 Ø2 F2 F3 31 C1 ØØ CA F6 C1 Ø1 35 C2 F6 31
3DCØ C5 Ø1 98 Ø3 3F 9Ø F9 C4 ØD 3F C4 ØA 3F C2 F3 E2
3DDØ F2 98 Ø4 AA F3 9Ø D6 C2 F1 98 1A 94 Ø3 C4 Ø7 3F
3DEØ C4 CØ CA F6 C4 ØØ 3F AA F6 9C F9 C4 7A 37 C4 88
3DFØ 33 3F C4 E1 33 C4 3C 35 C4 39 31 3D Ø8 Ø8 Ø8 Ø8
3EØØ C2 F2 9C Ø6 C2 F3 98 35 CA F2 BA F3 C2 F2 E2 F9
3E1Ø 98 2Ø AA F3 AA F2 C4 77 35 C2 F2 Ø2 F2 F2 31 C4
3E2Ø 77 37 C2 F3 Ø2 F2 F3 33 C1 ØØ CB ØØ C1 Ø1 CB Ø1
3E3Ø 9Ø DA C2 F3 CA F9 C4 3C 35 C4 3Ø 31 3D C4 3C 35
3E4Ø C4 83 31 3D C2 F3 98 F5 C2 F2 9C F1 C4 77 35 C2
3E5Ø F3 Ø2 F2 F3 31 C1 ØØ CA EB C1 Ø1 CA EC C4 3C 35
3E6Ø C4 E2 31 3D C2 F9 CA F2 AA F9 C4 77 35 C2 F2 Ø2
3E7Ø F2 F2 31 C1 ØØ C9 Ø2 C1 Ø1 C9 Ø3 C2 F2 E2 F3 98
3E8Ø Ø4 BA F2 9Ø E5 C2 EB C9 Ø2 C2 EC C9 Ø3 AA F3 9Ø
3E9Ø BB C2 F2 9C Ø6 C2 F3 98 A4 CA F2 C4 3C 35 C4 E2
3EAØ 31 3D C2 F3 E2 F2 98 Ø4 AA F3 9Ø EF C4 3C 35 C4
3EBØ 3Ø 31 3D C4 ØØ CA F6 CA F5 CA F4 C2 F9 Ø2 F4 9C
3ECØ 94 Ø5 C2 F9 Ø1 9Ø Ø3 Ø1 AA F6 4Ø Ø2 F4 F6 94 Ø2
3EDØ 9Ø Ø5 Ø1 AA F5 9Ø F3 4Ø Ø2 F4 FF 94 Ø2 9Ø Ø5 Ø1
3EEØ AA F4 9Ø F3 C4 7A 37 C4 E1 33 C4 3D 3F C2 F6 98
3EFØ Ø3 C4 31 3F C2 F5 98 Ø3 DC 3Ø 3F C2 F4 DC 3Ø 3F
3FØØ C4 3C 35 C4 39 31 3D ØD ØA 45 44 49 54 4F 52 2Ø
3F1Ø 52 45 41 44 59 2E ØD ØA 47 49 56 45 2Ø 42 55 46
3F2Ø 46 45 52 2Ø 52 41 4E 47 45 3A ØØ 41 3F 4A 52 3F
3F3Ø 46 4C 3D 75 43 3E 9Ø 49 3E 43 44 3D FF 4B 3C 24
3F4Ø 5Ø 3D 73 42 3D 6F ØØ C4 84 9Ø Ø2 C4 9Ø CA F1 C2
3F5Ø F9 CA F3 AA F3 C4 3C 35 C4 E2 31 3D AA F9 9Ø EF
```

Use this complete hexadecimal listing of the program if you wish to prepare a paper tape or cassette to run the editor in RAM, or if you are able to burn your own PROMs.

BASIC COMMANDS AND THEIR FUNCTIONS:
(THE CLOSING BRACKET ")" SYMBOLIZES A CARRIAGE RETURN)

COMMAND	FUNCTION
A)	APPEND LINES TO BUFFER
R)	READ TAPE INTO BUFFER
L) , ML) , M , NL)	LIST ALL LINES, OR LINE M, OR LINES M-N
MC) , M , NC)	CHANGE LINE M, OR LINES M-N
MI)	INSERT LINE OR LINES BEFORE LINE M
MD) , M , ND)	DELETE LINE M, OR LINES M-N
K)	KILL TEXT IN BUFFER
P) , MP) , M , NP)	PUNCH ALL LINES, LINE M, OR LINES M-N
B) , MB) , M , NB)	AS FOR PUNCH, BUT PUNCHES A BELL CHAR AT END OF TEXT
/	EDITOR PRINTS NUMBER OF LINES CURRENTLY HELD IN TEXT BUFFER, IN DECIMAL
BELL	WHEN IN TEXT MODE, RETURNS EDITOR TO COMMAND MODE

(M AND N REPRESENT DECIMAL LINE NUMBERS)

Here is the basic command set of the editor, showing the various command letters, the possible arguments for each, and their functions. In text mode, a percent sign acts as a backspace.

When the editor is called, it announces itself and then asks you to give the available text buffer range in memory. This must be supplied as two hexadecimal numbers, separated by any non-hex character other than space. Leading zeroes are not required, but the second number must end with a carriage return.

The editor then enters its command mode, signalling this by ringing the bell. The user may then type in a command letter, followed by a carriage return. If text is to be fed in via the keyboard, the command letter "A" is appropriate, while "R" tells the editor to read in a previously punched tape via the reader.

Once the text is in the buffer, you can edit it using the commands shown in the table. Note that the L, C, I, D, P and B commands may all have arguments, to specify individual lines or a group of lines. In fact the I command must have one argument, to indicate where the insertion is to take place.

The argument number or numbers must precede the command letter. Thus to list lines 12 to 15, for example, you simply type 12, 15L followed by a carriage return.

To change a line, say line 34, you simply type 34C, a carriage return, and then type in the new line text. Similarly to insert a new line or lines before an existing line, say line 17, type 17I followed by a carriage return and then type in the extra lines. To delete lines, say lines 20, 21 and 22, type 20, 22D and then a carriage return.

A single argument implies that the command should affect only the one line. Two arguments imply that the command should affect all lines between the two corresponding lines, inclusively. Thus 15, 20C implies that six new lines are to be fed in, replacing the existing lines 15, 16, 17, 18, 19 and 20.

Fig. A-10. "SC/MP EDITOR" (continued from page 285).

Note that following the input of an A, R, or I command, the editor enters its text-input mode of operation. It will normally remain in this mode until it detects a "bell" character, whereupon it returns to the command mode, ringing the bell to indicate its response. If a tape being read in following the R command is terminated with a bell character (having been punched previously using the B command), this will take place automatically at the end of the tape.

All commands other than A, R or I cause the editor to return to command code automatically when the command function has been performed.

Note that the editor will also return to the command mode automatically during the text input (including during editing which involves additional text), if either of the internal line or text buffers become full.

The number of text lines held in the editor's buffer may be found at any time by typing an oblique (/) while in command mode. The editor will immediately type the current line total, in decimal.

When accepting text from the keyboard, the editor will detect a "percent" character (%) and interpret this as a destructive backspace. This may be used to correct character errors if they are noticed very shortly after entry (i.e., before the end of the line). Each percent character is interpreted as a one-character backspace, and the character may be used repeatedly to backspace right back to the start of the current line, if necessary.

Before punching a tape following the P or B commands, the editor pauses to allow the user to turn on the tape punch. Typing any character on the keyboard then causes it to punch a leader, the desired output text (followed by a bell character in the case of the B command), and finally a trailer. It then pauses once more, to allow the user to turn off the tape punch. After doing so, the user must then signal the editor to continue by typing any character via the keyboard.

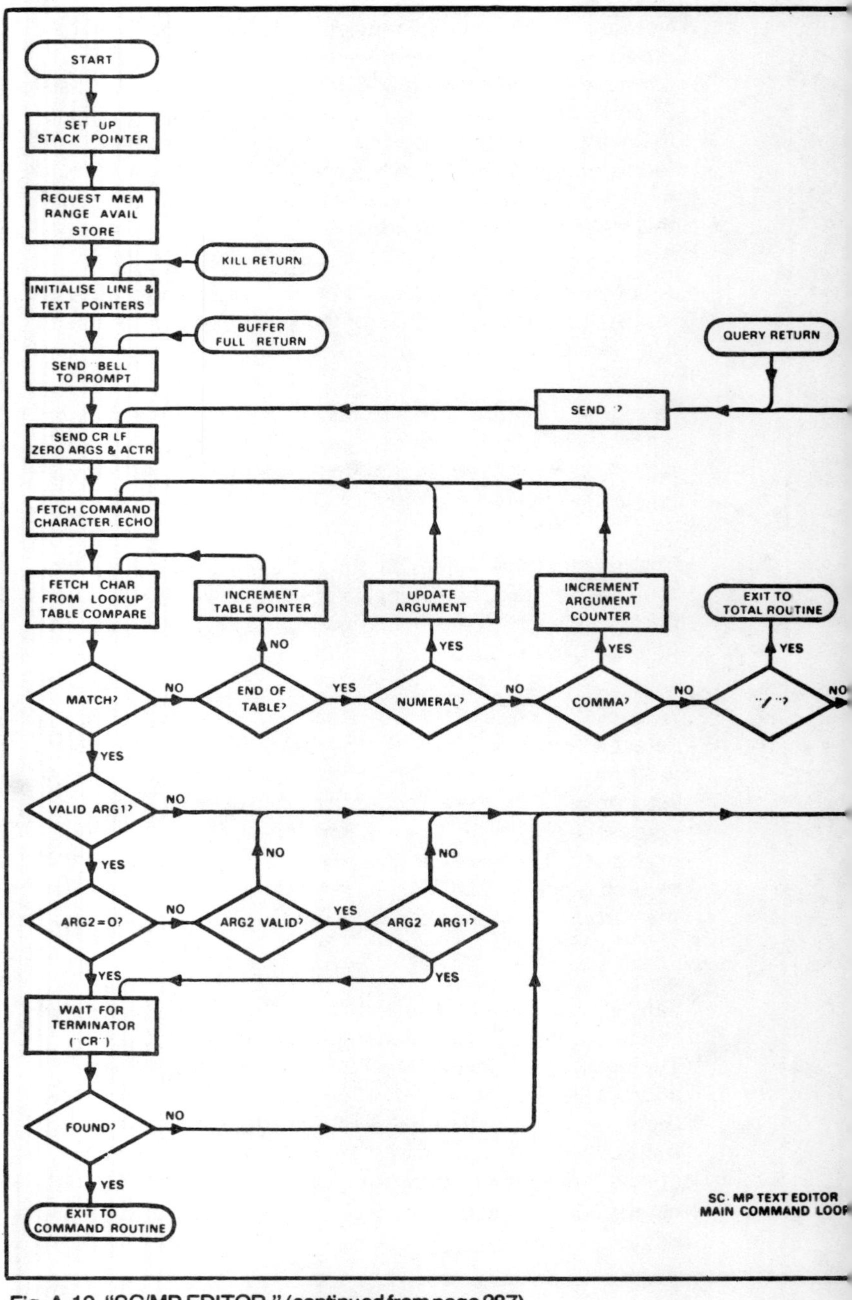

Fig. A-10. "SC/MP EDITOR" (continued from page 287).

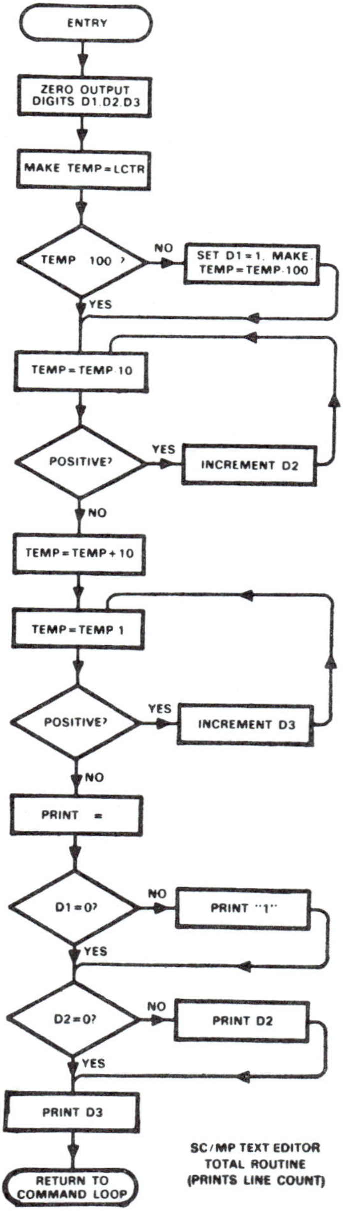

SC/MP TEXT EDITOR
TOTAL ROUTINE
(PRINTS LINE COUNT)

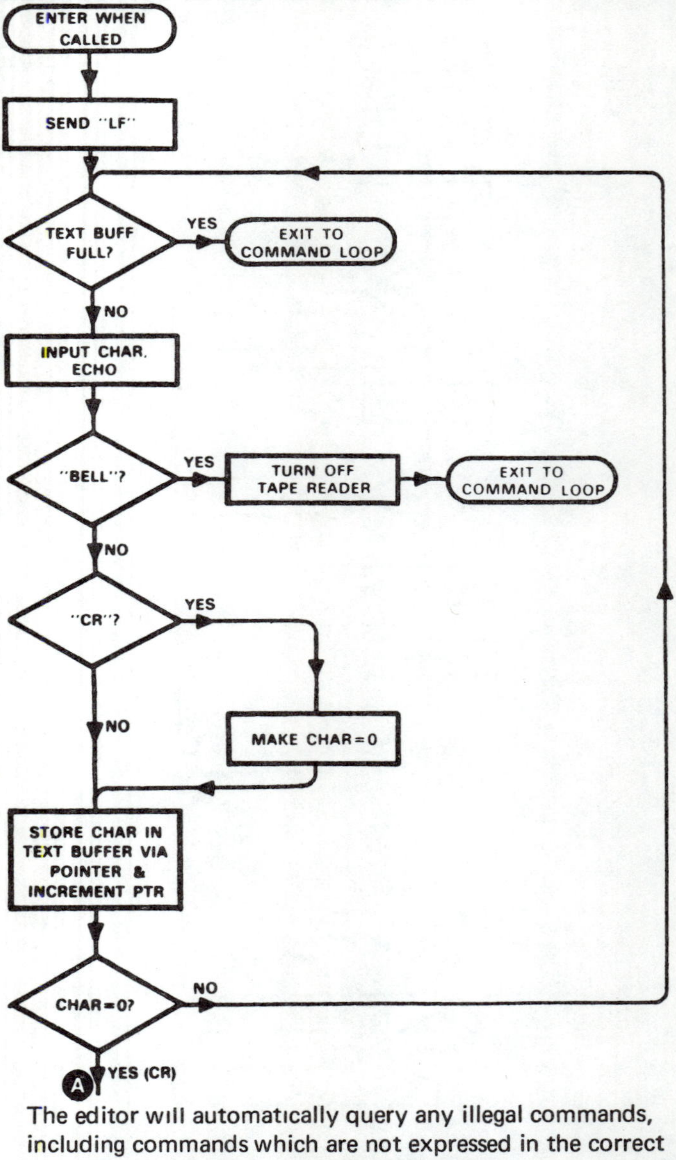

The editor will automatically query any illegal commands, including commands which are not expressed in the correct form. It will also query you if the arguments in a command are either inappropriate to the command, or not valid for the text currently held in the buffer. For example if you

Fig. A-10. "SC/MP EDITOR" (continued from page 289).

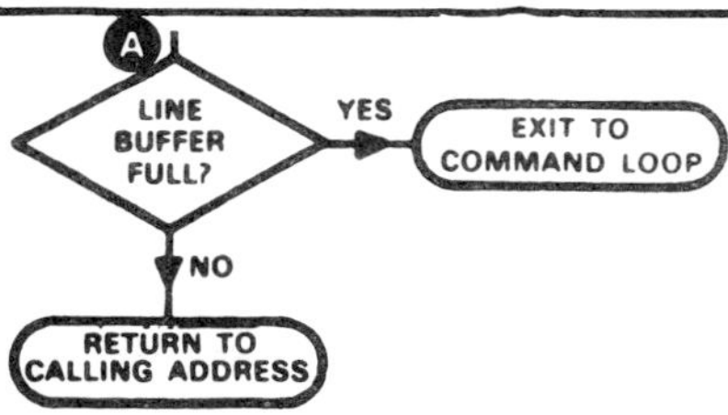

SC MP TEXT EDITOR
TEXTIN SUBROUTINE

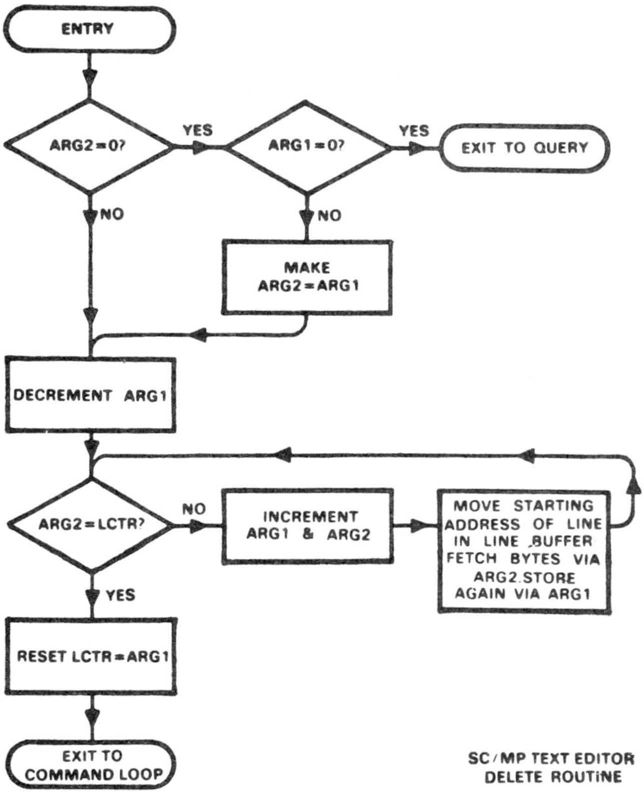

SC/MP TEXT EDITOR
DELETE ROUTINE

have only 30 lines of text and you give it a command like 45L, it will promptly reply with a query.

The editor has been fairly extensively tested, and to the best of my knowledge it is now free of bugs. However, if you should find a bug, I would appreciate being advised so that all users can be informed.

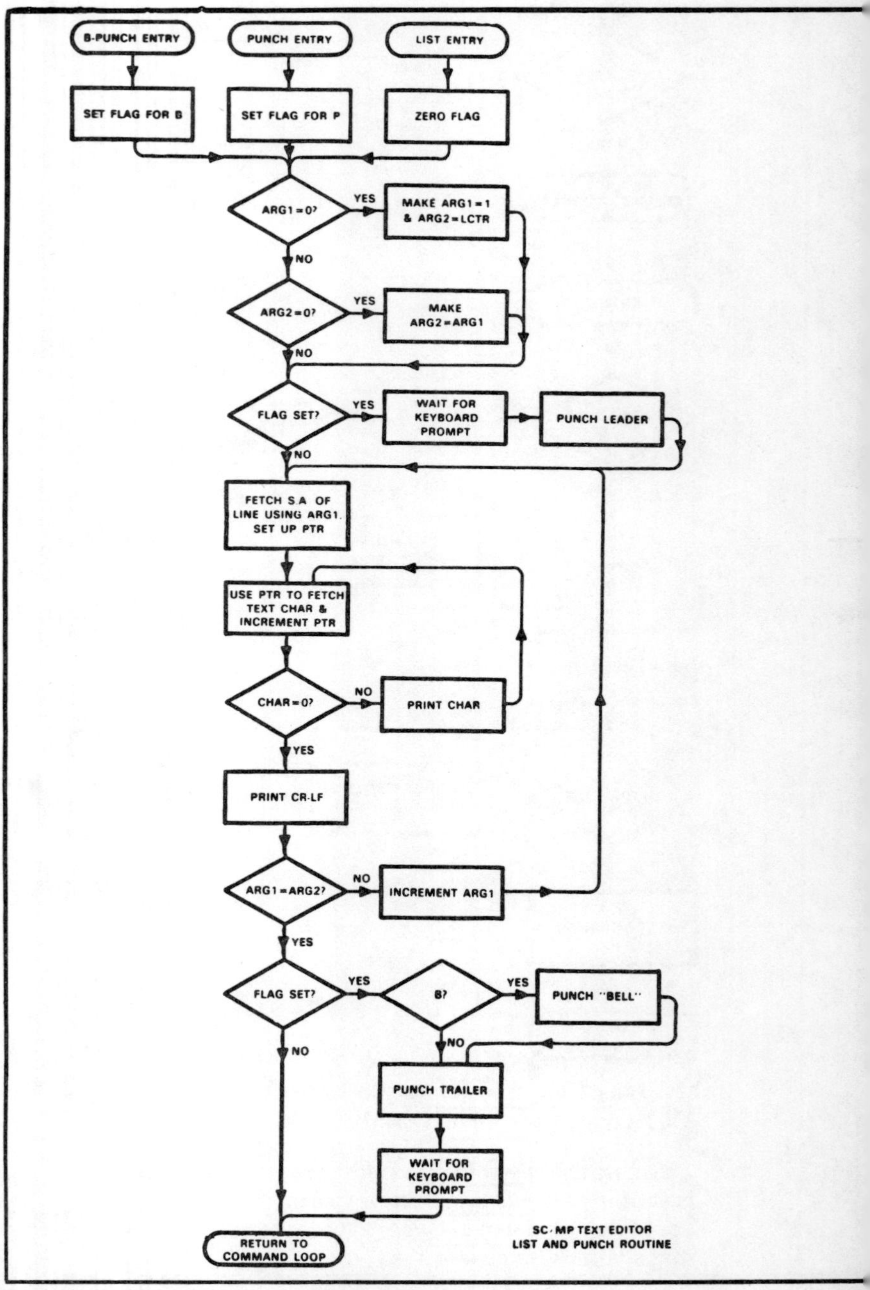

Fig. A-10. "SC/MP EDITOR" (continued from page 291).

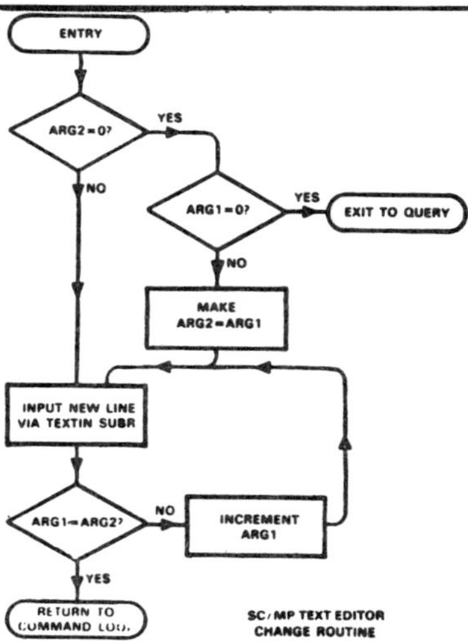

SC/MP TEXT EDITOR
CHANGE ROUTINE

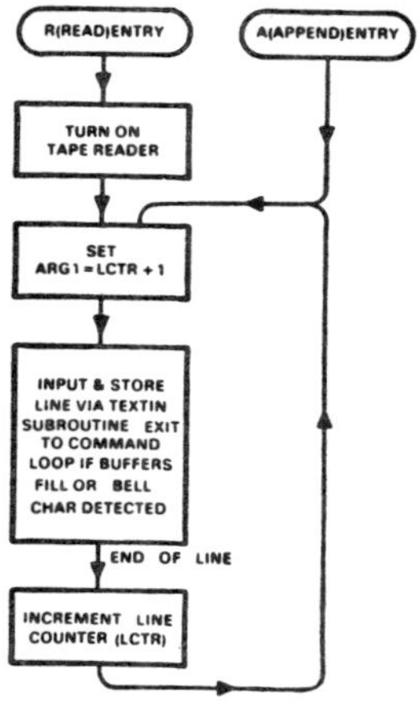

SC MP TEXT EDITOR
READ & APPEND ROUTINE

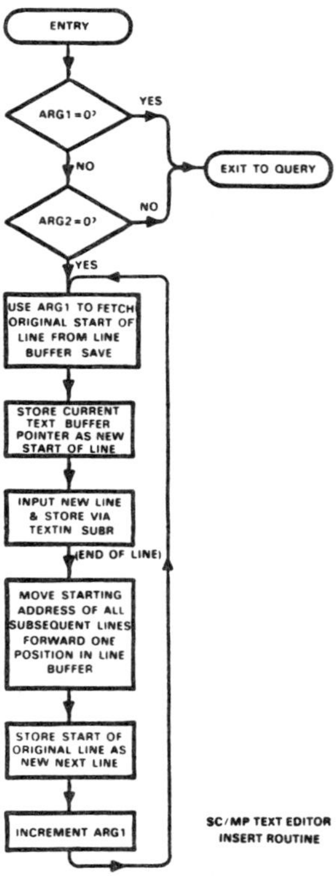

SC/MP TEXT EDITOR
INSERT ROUTINE

NEW LIBRARY PROGRAM

SC/MP Program SL0047A-PLOT
(from Jermyn MicroComputer Center, Kent, England)
Program Listing — see page 3
Source paper tape — $5.00 each

A sample plot of a SINE wave drawn on a Teletype by the PLOT routine is shown below.

GRAPH PLOTTER FROM RAM A JERMYN UTILITY PROGRAM

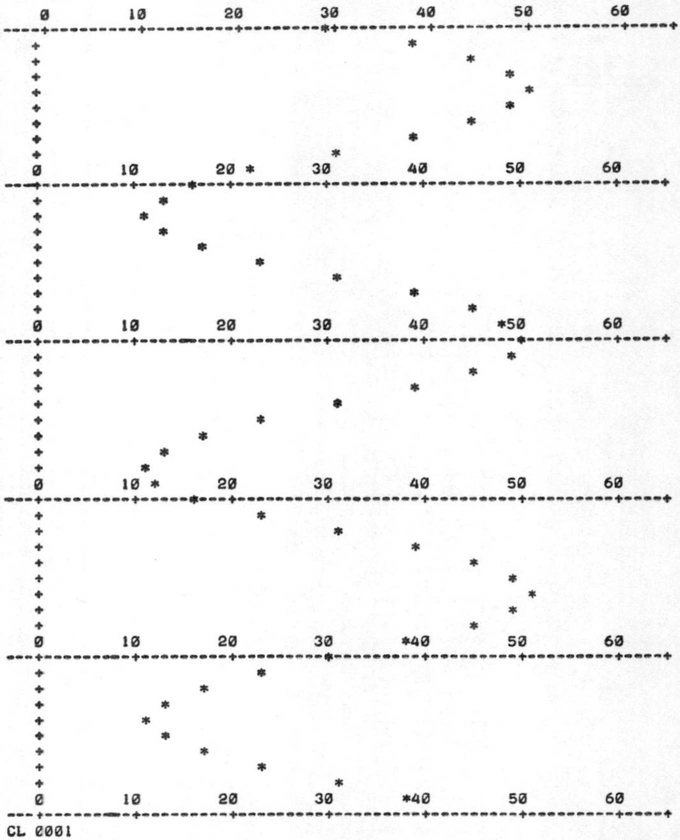

Fig. A-11. "PLOT."

```
                .TITLE PLOT,'SC/MP PLOTTER FROM RAM'
                ;LIBRARY PROGRAM SL0047A
                ;
                ; THIS PROGRAM FETCHES DATA STORED IN RAM
                ; AND DISPLAYS THE CONTENT IN GRAPHICAL
                ; FORM ON THE TELETYPE CONSOLE.
                ; AFTER LOADING SET THE TARGET PROGRAM
                ; COUNTER TO HEX 0001 (INITIALISE) AND
                ; THEN RUN.
                ;
                ; THIS PROGRAM WILL RUN WITTHIN AN L.C.D.S
                ; CONTAINING 2K BYTES OF RAM (ISP-8C/002).
                ; DATA FOR PLOTTING IS FETCHED FROM LOCATION
                ; MDATA UPWARDS AND LOOPS AROUND AT LOCATION
                ; DATEND.
                ;
          0001  P1      =       1
          0002  P2      =       2
          0003  P3      =       3
          07FD  VECTOR  =       X'07FD
          07FF  COUNT   =       X'07FF
          0000  ZERO    =       X'00
          0004  EOT     =       X'04
          07FC  STACK   =       X'07FC
          7AE2  PUTC    =       X'7AE2
          000D  CR      =       X'0D
          000A  LF      =       X'0A
          0008  BS      =       X'08
          FFFF  MINONE  =       -1
          FFF6  MINTEN  =       -10
          FFBF  MIN65   =       -65
          0020  SP      =       X'20
          0007  BEL     =       X'07
          003F  RSTCYL  =       X'3F

    0000                        .=X'0001
    0001 C4FC   BEGIN:  LDI     L(STACK)
    0003 32             XPAL    P2
    0004 C407           LDI     H(STACK)
    0006 36             XPAH    P2
    0007 C4FD           LDI     L(VECTOR)
    0009 31             XPAL    P1
    000A C407           LDI     H(VECTOR)
    000C 35             XPAH    P1
    000D C402           LDI     H(MDATA)
    000F CD01           ST      @1(P1)
    0011 C42F           LDI     L(MDATA)
    0013 C900           ST      (P1)
    0015 C454           LDI     L(MIAXIS)
    0017 31             XPAL    P1
    0018 C401           LDI     H(MIAXIS)
    001A 35             XPAH    P1
    001B C400           JS      P3,PRINT
    001D 37C4
    001F D333
    0021 3F
    0022 C400           LDI     L(MCRLF)
    0024 31             XPAL    P1
    0025 C401           LDI     H(MCRLF)
    0027 35             XPAH    P1
    0028 C400           JS      P3,PRINT
    002A 37C4
    002C D333
    002E 3F
    002F C4FF   CLEAR:  LDI     L(COUNT)
    0031 31             XPAL    P1
    0032 C407           LDI     H(COUNT)
    0034 35             XPAH    P1
```

```
0035 C400            LDI    ZERO
0037 C900            ST     (P1)
0039 C4D8            LDI    L(MCURSR)
003B 31              XPAL   P1
003C C401            LDI    H(MCURSR)
003E 35              XPAH   P1
003F 9033            JMP    PRICYC
0041 9CEC   NOSAX:   JNZ    CLEAR
0043 C401            LDI    H(MAXIS)
0045 35              XPAH   P1
0046 C491            LDI    L(MAXIS)
0048 31              XPAL   P1
0049 9029            JMP    PRICYC
004B C407   OVER:    LDI    L(MOVER)
004D 31              XPAL   P1
004E C401            LDI    H(MOVER)
0050 35              XPAH   P1
0051 9006            JMP    PRIRET
0053 C403   POINT:   LDI    L(MPOINT)
0055 31              XPAL   P1
0056 C401            LDI    H(MPOINT)
0058 35              XPAH   P1
0059 C400   PRIRET:  JS     P3,PRINT
005B 37C4
005D D333
005F 3F
0060 C407   RETURN:  LDI    H(COUNT)
0062 35              XPAH   P1
0063 C4FF            LDI    L(COUNT)
0065 31              XPAL   P1
0066 A900            ILD    (P1)
0068 C900            ST     (P1)
006A F4F6            ADI    MINTEN
006C 94D3            JP     NOSAX
006E C402            LDI    H(MSAXIS)
0070 35              XPAH   P1
0071 C424            LDI    L(MSAXIS)
0073 31              XPAL   P1
0074 C400   PRICYC:  JS     P3,PRINT
0076 37C4
0078 D333
007A 3F
007B C4FD   CYCLE:   LDI    L(VECTOR)
007D 31              XPAL   P1
007E C407            LDI    H(VECTOR)
0080 35              XPAH   P1
0081 C501            LD     @1(P1)
0083 01              XAE
0084 C100            LD     (P1)
0086 31              XPAL   P1
0087 40              LDE
0088 35              XPAH   P1
0089 40              LDE
008A E402            XRI    H(DATEND)
008C 9C0F            JNZ    FETCH
008E 31              XPAL   P1
008F 01              XAE
0090 40              LDE
0091 31              XPAL   P1
0092 40              LDE
0093 E43F            XRI    L(DATEND)
0095 9C06            JNZ    FETCH
0097 C402            LDI    H(MDATA)
0099 35              XPAH   P1
009A C42F            LDI    L(MDATA)
009C 31              XPAL   P1
009D C501   FETCH:   LD     @1(P1)     ;DATA FETCH
009F CEFF            ST     @-1(P2)    ;AND TEST
```

Fig. A-11. "PLOT" (continued from page 295).

```
00A1 C4FD            LDI     L(VECTOR) ;ROUTINE
00A3 31              XPAL    P1
00A4 01              XAE
00A5 C407            LDI     H(VECTOR)
00A7 35              XPAH    P1
00A8 CD01            ST      @1(P1)
00AA 01              XAE
00AB C900            ST      (P1)
00AD C601            LD      @1(P2)
00AF 98A2            JZ      POINT
00B1 01              XAE
00B2 06              CSA
00B3 D43F            ANI     RSTCYL
00B5 07              CAS
00B6 40              LDE
00B7 F4BF            ADI     MIN65
00B9 9490            JP      OVER
00BB C4FE   SPACE:   LDI     L(MSPACE)
00BD 31              XPAL    P1
00BE C400            LDI     H(MSPACE)
00C0 35              XPAH    P1
00C1 C400            JS      P3,PRINT
00C3 37C4
00C5 D333
00C7 3F
00C8 06              CSA
00C9 D43F            ANI     RSTCYL
00CB 07              CAS
00CC 40              LDE
00CD F4FF            ADI     MINONE
00CF 9882            JZ      POINT
00D1 01              XAE               ;DATA TO EXT
00D2 90E7            JMP     SPACE

00D4 01     PRINT:   XAE               ;MESSAGE PRINT
00D5 CEFF            ST      @-1(P2)   ;ROUTINE INDEXED
00D7 06              CSA               ;BY POINTER 1
00D8 CEFF            ST      @-1(P2)
00DA 37              XPAH    P3
00DB CEFF            ST      @-1(P2)
00DD 33              XPAL    P3
00DE CEFF            ST      @-1(P2)
00E0 C501   OBTAIN:  LD      @1(P1)
00E2 01              XAE
00E3 C404            LDI     EOT
00E5 60              XRE
00E6 9809            JZ      EXIT
00E8 C4E1            LDI     L(PUTC)-1 ;CALL PUTC IN
                                       ;L. C. D. S.  MONITOR
00EA C47A            LDI     H(PUTC)   ;TO OUTPUT
00EC 37              XPAH    P3        ;CHARACTER
00ED 40              LDE
00EE 3F              XPPC    P3
00EF 90EF            JMP     OBTAIN
00F1 C601   EXIT:    LD      @1(P2)
00F3 33              XPAL    P3
00F4 C601            LD      @1(P2)
00F6 37              XPAH    P3
00F7 C601            LD      @1(P2)
00F9 07              CAS
00FA C601            LD      @1(P2)
00FC 01              XAE
00FD 3F              XPPC    P3

00FE 20     MSPACE:  .BYTE   SP,EOT
00FF 04
0100 0A     MCRLF:   .BYTE   LF,CR,EOT
0101 0D
```

```
     0102 04
     0103 2A      MPOINT: .ASCII '*'          ;OR OPTIONS
     0104 0A              .BYTE  LF,CR,EOT
     0105 0D
     0106 04
     0107 2D2D    MOVER:  .ASCII '----------------'
     0109 2D2D
     010B 2D2D
     010D 2D2D
     010F 2D2D
     0111 2D2D
     0113 2D2D
     0115 2D2D
     0117 2D2D             ASCII '----------------'
     0119 2D2D
     011B 2D2D
     011D 2D2D
     011F 2D2D
     0121 2D2D
     0123 2D2D
     0125 2D2D
     0127 2D2D            .ASCII '----------------'
     0129 2D2D
     012B 2D2D
     012D 2D2D
     012F 2D2D
     0131 2D2D
     0133 2D2D
     0135 2D2D
     0137 2D56            .ASCII '-VALUE OVER RANGE---->'
     0139 414C
     013B 5545
     013D 204F
     013F 5645
     0141 5220
     0143 5241
     0145 4E47
     0147 452D
     0149 2D2D
     014B 2D3E
     014D 07              .BYTE  BEL,BEL,CR,LF,BEL,BEL,EOT
     014E 07
     014F 0D
     0150 0A
     0151 07
     0152 07
     0153 04
     0154 0D      MIAXIS:  BYTE  CR,LF,LF,BEL,BEL,LF,LF
     0155 0A
     0156 0A
     0157 07
     0158 07
     0159 0A
     015A 0A
     015B 2020            .ASCII '    GRAPH PLOTTER FROM RAM
     015D 2020
     015F 4752
     0161 4150
     0163 4820
     0165 504C
     0167 4F54
     0169 5445
     016B 5220
     016D 4652
     016F 4F4D
     0171 2052
```

Fig. A-11. "PLOT" (continued from page 297).

```
    0173 414D
    0175 20
197 0176 4120           .ASCII 'A JERMYN UTILITY PROGRAM'
    0178 4A45
    017A 524D
    017C 594E
    017E 2055
    0180 5449
    0182 4C49
    0184 5459
    0186 2050
    0188 524F
    018A 4752
    018C 414D
198 018E 0A             .BYTE  LF,CR,LF
    018F 0D
    0190 0A
199 0191 2020   MAXIS:  .ASCII '    0         10'
    0193 2020
    0195 3020
    0197 2020
    0199 2020
    019B 2020
    019D 2031
    019F 30
200 01A0 2020           .ASCII '        20'
    01A2 2020
    01A4 2020
    01A6 2020
    01A8 3230
201 01AA 2020           .ASCII '        30'
    01AC 2020
    01AE 2020
    01B0 2020
    01B2 3330
202 01B4 2020           .ASCII '        40'
    01B6 2020
    01B8 2020
    01BA 2020
    01BC 3430
203 01BE 2020           .ASCII '        50'
    01C0 2020
    01C2 2020
    01C4 2020
    01C6 3530
204 01C8 2020           .ASCII '        60'
    01CA 2020
    01CC 2020
    01CE 2020
    01D0 3630
205 01D2 0D             .BYTE  CR,SP,SP,SP,SP,EOT
    01D3 20
    01D4 20
    01D5 20
    01D6 20
    01D7 04
206 01D8 2D2D   MCURSR: .ASCII '----+---------'
    01DA 2D2D
    01DC 2B2D
    01DE 2D2D
    01E0 2D2D
    01E2 2D2D
    01E4 2D2D
207 01E6 2B2D           .ASCII '+---------'
    01E8 2D2D
    01EA 2D2D
    01EC 2D2D
    01EE 2D2D
```

```
01F0 2B2D             .ASCII '+---------'
01F2 2D2D
01F4 2D2D
01F6 2D2D
01F8 2D2D
01FA 2B2D             .ASCII '+---------'
01FC 2D2D
01FE 2D2D
0200 2D2D
0202 2D2D
0204 2B2D             .ASCII '+---------'
0206 2D2D
0208 2D2D
020A 2D2D
020C 2D2D
020E 2B2D             .ASCII '+---------'
0210 2D2D
0212 2D2D
0214 2D2D
0216 2D2D
0218 2B2D             .ASCII '+----+'
021A 2D2D
021C 2D2B
021E 0D               .BYTE  CR,SP,SP,SP,SP,EOT
021F 20
0220 20
0221 20
0222 20
0223 04
0224 2020   MSAXIS:   .ASCII '    +'
0226 2020
0228 2B
0229 0D               .BYTE  CR,SP,SP,SP,SP,EOT
022A 20
022B 20
022C 20
022D 20
022E 04
            ;
            ;SINE WAVE /\/\/\/
            ;TEST PATTERN FOR PLOTTER PROGRAM
022F 1E     MDATA:    .BYTE  30,38,44,48,50,48,44,38
0230 26
0231 2C
0232 30
0233 32
0234 30
0235 2C
0236 26
0237 1E               .BYTE  30,22,16,12,10,12,16,22
0238 16
0239 10
023A 0C
023B 0A
023C 0C
023D 10
023E 16
     023F   DATEND  =                  ;EQU TO PGM CNTR
            ;
     0000             .END

BEGIN   0001 *   BEL     0007     BS      0008 *
CLEAR   002F     COUNT   07FF     CR      000D
CYCLE   007B *   DATEND  023F     EOT     0004
EXIT    00F1     FETCH   009D     LF      000A
MAXIS   0191     MCRLF   0100     MCURSR  01D8
MDATA   022F     MIAXIS  0154     MIN65   FFBF
```

Fig. A-11. "PLOT" (continued from page 299).

```
MINONE FFFF     MINTEN FFF6     MOVER  0107
MPOINT 0103     MSAXIS 0224     MSPACE 00FE
NOSAX  0041     OBTAIN 00E0     OVER   004B
P1     0001     P2     0002     P3     0003
POINT  0053     PRICYC 0074     PRINT  00D4
PRIRET 0059     PUTC   7AE2     RETURN 0060 *
RSTCYL 003F     SP     0020     SPACE  00BB
STACK  07FC     VECTOR 07FD     ZERO   0000

NO ERROR LINES
SOURCE CHECKSUM = CA1E
```

SC/MP DISASSEMBLER

The SC/MP disassembler is written in NIBL, National's industrial basic language, and requires RAM at X 1ØØØ- 1FFF. The code to be disassembled can be loaded into another RAM card by the SC/MP monitor or can be read from a PROM stuck in an empty ROM socket on the memory board. The operator runs the program and types the start address for the disassembler to begin.

The disassembler uses the standard instruction codes. One difference, however, is that all jumps also have the decimal equivalent of the jump printed in <>.

Jim Derosa
NSC, FAE Boston

```
1  PRINT"SC/MP DISASSEMBLER"
2  PRINT""
3  GOTO130
5  IFX=0THENPRINT"HALT";
6  IFX=1THENPRINT"XAE";
7  IFX=2THENPRINT"CCL";
8  IFX=3THENPRINT"SCL";
9  IFX=4THENPRINT"DINT";
10 IFX=5THENPRINT"IEN";
11 IFX=6THENPRINT"CSA";
12 IFX=7THENPRINT"CAS";
13 IFX=8THENPRINT"NOP";
14 IFX=25THENPRINT"SIO";
15 IFX=28THENPRINT"SR";
16 IFX=29THENPRINT"SRL";
17 IFX=30THENPRINT"RR";
18 IFX=31THENPRINT"RRL";
19 IFX=48THENPRINT"XPAL PC";
20 IFX=49THENPRINT"XPAL P1";
21 IFX=50THENPRINT"XPAL P2";
```

Fig. A-12. "SC/MP Dissembler (in NIBL)."

```
22  IFX=51THENPRINT"XPAL P3";
23  IFX=52THENPRINT"XPAH PC";
24  IFX=53THENPRINT"XPAH P1";
25  IFX=54THENPRINT"XPAH P2";
26  IFX=55THENPRINT"XPAH P3";
27  IFX=60THENPRINT"XPPC PC";
28  IFX=61THENPRINT"XPPC P1";
29  IFX=62THENPRINT"XPPC P2";
30  IFX=63THENPRINT"XPPC P3";
31  IFX=64THENPRINT"LDE";
32  IFX=80THENPRINT"ANE";
33  IFX=88THENPRINT"ORE";
34  IFX=96THENPRINT"DAE";
35  IFX=104THENPRINT"ADE";
36  GOTO120
37  IFX=143THENPRINT"DLY";
38  IFX=143THENGOTO227
39  IFX>THENGOTO41
40  GOTO47
41  IFX<148THENGOTO43
42  GOTO47
43  PRINT"JMP ";
44  X=X-144
45  GOSUB194
46  GOTO140
47  IFX>147THENGOTO49
48  GOTO55
49  IFX<152THENGOTO51
50  GOTO55
51  PRINT"JP ";
52  X=X-148
53  GOSUB194
54  GOTO140
55  IFX>151THENGOTO57
56  GOTO63
57  IFX<156THENGOTO59
58  GOTO63
59  PRINT"JZ ";
60  X=X-148
61  GOSUB194
62  GOTO140
63  IFX>155THENGOTO65
64  GOTO71
65  IFX<159THENGOTO67
66  GOTO71
67  PRINT"JNZ ";
68  X=X-156
69  GOSUB194
70  GOTO140
71  IF X>167THENGOTO73
72  GOTO79
73  IFX<172THENGOTO75
74  GOTO79
75  PRINT"ILD ";
76  X=X-168
77  GOSUB194
78  GOTO227
```

Fig. A-12. "SC/MP DISSEMBLER (in NIBL)" (continued from page 301).

```
79   IFX>183THENGOTO81
80   GOTO87
81   IFX<188THENGOTO83
82   GOTO87
83   PRINT"DLD ";
84   X=X-184
85   GOSUB194
86   GOTO227
87   IFX=196THENPRINT"LDI";
88   IFX=196THENGOTO227
89   IFX=212THENPRINT"ANI";
90   IFX=212THENGOTO227
91   IFX=220THENPRINT"ORI";
92   IFX=220THENGOTO227
93   IFX=228THENPRINT"XRI";
94   IFX=228THENGOTO227
95   IFX=236THENPRINT"DAI";
96   IFX=236THENGOTO227
97   IFX=244THENPRINT"ADI";
98   IFX=244THENGOTO227
99   IFX=252THENPRINT"CAI";
100  IFX=252THENGOTO227
101  N=192
102  Z=1
103  C=1
104  I=0
105  P=0
106  IFX=NTHENGOTO160
107  N=N+1
108  IFN=256THENGOTO239
109  Z=Z+1
110  IFZ=9THENGOTO112
111  GOTO116
112  Z=1
113  C=C+1
114  I=0
116  IFZ=4THENI=1
117  P=P+1
118  IFP=4THENP=0
119  GOTO106
120  IFX=120THENPRINT"CAE";
121  IFX<143THENGOTO230
122  GOTO37
130  PRINT"DEC. START ADDRESS";
131  INPUTA
132  PRINT""
133  PRINT"DEC.   HEX.   CODE INSTRUCTION"
134  GOTO240
140  A=A+1
141  X=@A
142  GOSUB199
143  PRINT"    <";
144  X=@A
145  IFX>127THENX=X-256
146  PRINTX,">";
147  GOTO230
150  IFX>127THENX=X-256
```

```
151   PRINTX;
152   RETURN
154   X=@(A+1)
155   GOSUB200
156   PRINT"   ".
157   X=@A
158   GOTO37
160   IFC=1THENPRINT"LD ";
161   IFC=2THENPRINT"ST ";
163   IFC=3THENPRINT"AND ";
164   IFC=4THENPRINT"OR ";
165   IFC=5THENPRINT"XOR ";
166   IFC=6THENPRINT"DAD ";
167   IFC=7THENPRINT"ADD ";
168   IFC=8THENPRINT"CAD ";
169   IFI=1THENPRINT"@";
170   A=A+1
171   X=@A
172   GOSUB150
173   PRINT" (";
174   X=P
175   GOSUB194
176   PRINT")";
177   GOTO230
180   IFA<10THENPRINT" ";
181   IFA<100THENPRINT" ";
182   IFA<1000THENPRINT" ";
183   IFA<10000THENPRINT" ";
184   RETURN
194   IFX=0THENPRINT"PC";
195   IFX=1THENPRINT"P1";
196   IFX=2THENPRINT"P2";
197   IFX=3THENPRINT"P3";
198   RETURN
199   PRINT"    ";
200   Y=15
201   IFX>=Y*16THENGOTO204
202   Y=Y-1
203   GOTO201
204   X=X-Y*16
205   GOSUB210
206   Y=15
207   IFX>=YTHENGOTO210
208   Y=Y-1
209   GOTO207
210   IFY=15THENPRINT"F";
211   IFY=14THENPRINT"E";
212   IFY=13THENPRINT"D";
213   IFY=12THENPRINT"C";
214   IFY=11THENPRINT"B";
215   IFY=10THENPRINT"A";
216   IFY=9THENPRINT"9";
217   IFY=8THENPRINT"8";
218   IFY=7THENPRINT"7";
219   IFY=6THENPRINT"6";
220   IFY=5THENPRINT"5";
221   IFY=4THENPRINT"4";
```

Fig. A-12. "SC/MP DISSEMBLER (in NIBL)" (continued from page 303).

```
222  IFY=3THENPRINT"3";
223  IFY=2THENPRINT"2";
224  IFY=1THENPRINT"1";
225  IFY=0THENPRINT"0";
226  RETURN
227  A=A+1
228  X=@A
229  GOSUB199
230  PRINT""
239  A=A+1
240  PRINTA;
241  B=A/255
242  GOSUB180
243  X=B
244  GOSUB200
245  X=A-256*B
246  GOSUB200
248  X=@A
249  GOSUB199
251  X=@A
252  IFX>142THENGOTO154
253  PRINT"   ";
254  GOTO5
255  END
```

```
                        .TITLE WORDG,'SC/MP WORD GAME'
                 ;
                 ;THIS PROGRAM TAKES ABOUT 1K BYTES OF PROGRAM STORAG
                 ;WITH LOCATION ZERO.  IT ALSO USES ABOUT 34 RAM LOCAT
                 ;TEMPORARY STORAGE PURPOSES BETWEEN X'416 AND X'437
                 ;HAS BEEN LOADED INTO A PAIR OF MM5204 ROMS FOR PERM
                 ;AGE.  THE PROGRAM WAS WRITTEN AROUND A TV TYPEWRITTE
                 ;HAS A 32 CHARACTER BY 16 LINE/SCREEN FORMAT.
                 ;
      0001              P1      =1         ;POINTER OR INDEX REGISTER
      0002              P2      =2         ;  "     "    "      "
      0003              P3      =3         ;  "     "    "      "
      0821              WORD   =0821
      0070              PSPTG   =070
      0000              .=0
0000  08                NOP
                 ;
                 ;PRINT BEGINNING OF GAME MESSAGE.
                 ;
0001  C408       BEG:   LDI     8
0003  36                XPAH    P2         ;SET P2 TO 820H, TEMPORARY
0004  C420              LDI     020        ;STORAGE LOCATION.
0006  32                XPAL    P2
0007  C402              LDI     H(PUTC)    ;SET P3 TO TELETYPE CHARAC
0009  37                XPAH    P3         ;OUTPUT ROUTINE, PUTC.
000A  C49B              LDI     L(PUTC)-1
000C  33                XPAL    P3
000D  C4C8              LDI     L(MBEG)    ;SET P1 TO BEGINNING MESSA
000F  31                XPAL    P1
0010  C402              LDI     H(MBEG)
0012  35                XPAH    P1
0013  C501       N1:    LD      @1(P1)     ;TAKE AN ASCII CHARACTER
0015  3F                XPPC    P3         ;PRINT A CHARACTER
0016  31                XPAL    P1
0017  CAFA              ST      -6(P2)     ;TEST FOR END OF MESSAGE
0019  E432              XRI     L(MEXC)
001B  9805              JZ      TAG        ;JUMP IF DONE.
001D  C2FA              LD      -6(P2)
001F  31                XPAL    P1
0020  90F1              JMP     N1
                 ;
                 ;BEGINNING OF GAME.  ALSO THE POINT OF RETURN AT END
                 ;
```

Fig. A-13. "Word Game."

```
0022 C40D   TAG:    LDI    0D          ;PRINT A CR AND LF
0024 3F             XPPC   P3
0025 C40A           LDI    0A
0027 3F             XPPC   P3
0028 C420           LDI    020
002A 3F             XPPC   P3
002B C420           LDI    020         ,PRINT A SPACE
002D 3F             XPPC   P3
002E C400           LDI    0           ,CLEAR A COUNTER
0030 CAF6           ST     -10(P2)
                    ;
                    ;GET CHARACTER, DO NOT ECHO  REFER TO SC/MP LCDS LIS
                    ;COMMENTS. THIS IS A SOFTWARE IMPLIMENTED RECEIVING
                    ;OF A UART  IT RECEIVES 8 BITS IN SERIAL, BEGINNING
                    ;START BIT FOLLOWED BY 7 BITS AND A PARITY BIT  THE
                    ,BIT IS STRIPED OFF
                    ,
0032 C408   GETC.   LDI    8           ,SET BIT COUNT TO 8
0034 CAFF           ST     -1(P2)
0036 06     GRET·   CSA                ;GET BIT
0037 D420           ANI    020         ;MASK OUT UNDESIRABLE BITS
0039 9CFB           JNZ    GRET        ;TEST FOR A ZERO, START BI
003B C457           LDI    87
003D 8F04           DLY    4           ,WAIT FOR 1/2 BIT DURATION
003F 06             CSA                ,SAMPLE BIT AGAIN AND TEST
0040 D420           ANI    020
0042 9CF2           JNZ    GRET        ;MUST HAVE BEEN A NOISE  G
0044 C47E   GLOOP:  LDI    126         ;START BIT DETECTED  WAIT
0046 8F08           DLY    8           ;BIT TIME
0048 06             CSA                ,AND SAMPLE INPUT
0049 D420           ANI    020
004B 9802           JZ     BZ          ;BIT=0
004D C401           LDI    1           ;BIT=1
004F CAFE   BZ:     ST     -2(P2)      ;STORE IN TEMP. REGISTER
0051 1F             RRL                ;MOVE BIT TO LINK BIT POSI
0052 01             XAE                ,LOAD EXTENSION REGISTER
0053 1D             SRL                ,MOVE ONE RIGHT. NOW LINK
0054 01             XAE                ;IS MOVED TO MSB
0055 BAFF           DLD    -1(P2)      ;STORE BACK IN E REGISTER.
0057 9CEB           JNZ    GLOOP       ;DECREMENT BIT COUNT
0059 40             LDE                ,IF ALL BITS ARE IN, THEN
005A D47F           ANI    07F         ;STRIP MSB OFF, PARITY BIT
005C 01             XAE
005D C2F6           LD     -10(P2)     ;TEST REGISTER -10  IF 0 T
005F 9C14           JNZ    NENTRY      ;INPUT CHARACTER IS THE FI
                                       ,ACTER
                    ;
                    ;FIRST CHARACTER OF SECRET WORD IS ENTERED. ERASE SC
                    ;HOMEUP CURSOR. THE SCREEN IS ERASED AND THE CURSOR
                    ;AT THE BEGINNING OF THE GAME
                    ;
0061 C455   FENTRY· LDI    055
0063 01             XAE
0064 19             SIO                ;SEND OUT 1,0,1,0 THROUGH
0065 19             SIO                ;OUTPUT PIN WHICH IS CONNE
0066 19             SIO                ;TO HOMEUP AND ERASE SCREE
0067 19             SIO
0068 01             XAE
0069 CAF6           ST     -10(P2)     ;STORE NONZERO VALUE IN RE
006B C421           LDI    L(WORD)     ;-10 TO SIGNIFY THAT THE F
006D 31             XPAL   P1
006E C408           LDI    H(WORD)
0070 35             XPAH   P1
0071 C400           LDI    0
0073 CAF9           ST     -7(P2)
                    ,
                    ,REMAINING CHARACTERS OF SECRET WORD ARE ENTERED
                    ;TEST FOR SPACE, $, AND ALPHABET
                    ;
0075 C420   NENTRY  LDI    020
0077 60             XRE
0078 985A           JZ     TSPACE      ,TEST FOR SPACE, JUMP IF S
007A C424           LDI    '$'         ,TEST FOR $, THIS WILL RES
007C 60             XRE
007D 98A3           JZ     TAG
007F C440           LDI    '@'         ,TEST FOR NON ALPHABETIC C
0081 03             SCL                ,BELOW A
0082 78             CAE
0083 943A           JP     NOALP
0085 03             SCL
0086 C45B           LDI    'Z'+1       ,ABOVE Z
0088 78             CAE
```

Fig. A-13. "WORD GAME" (continued from page 305).

```
0089 9402              JP     ALPHA
008B 9032              JMP    NOALP
                ;
                ;STORE CHARACTERS IN MEMORY.  TEST FOR LIMIT IN WORD
                ;PRINT 1ST CHARACTER AND '?' FOR ALL OTHERS.  E.G.  G?
                ;GAME.
                ;
008D 40         ALPHA: LDE
008E CD01              ST     @1(P1)      ;STORE CHARACTER IN MEMORY
0090 AAF9              ILD    -7(P2)
0092 E40B              XRI    11          ;TEST IF INPUT WORD >11 CH
0094 980C              JZ     EXCED
0096 E40A              XRI    10
0098 9C04              JNZ    NOF         ;TEST IF IT WAS A FIRST CH
009A 40                LDE                ; IF IT IS THEN ECHO, OR S
009B 3F         PGEC:  XPPC   P3          ;JUMP TO PUT CHARACTER ROU
009C 9094              JMP    GETC        ;THEN JUMP TO GET CHARACTE
009E C43F       NOF:   LDI    '?'         ;LOOKING FOR MORE CHARACTE
00A0 90F9              JMP    PGEC
                ;
                ;WORDE SIZE EXCEDED, PRINT ERROR MESSAGE AND RETURN
                ;BEGINNING.
                ;
00A2 C432       EXCED: LDI    L(MEXC)     ;SET P1 TO MESSAGE ADDRESS
00A4 31                XPAL   P1          ;AND PRINT MESSAGE
00A5 C403              LDI    H(MEXC)     ;THE MESSAGE IS "LIMIT EXC
00A7 35                XPAH   P1          ;WISE GUY".
00A8 C501       N2:    LD     @1(P1)
00AA 3F                XPPC   P3
00AB 31                XPAL   P1
00AC CAFA              ST     -6(P2)
00AE E44C              XRI    L(MNAL)
00B0 9805              JZ     TAG1        ;THEN JUMP TO THE BEGINNIN
00B2 C2FA              LD     -6(P2)      ;OF THE GAME.
00B4 31                XPAL   P1
00B5 90F1              JMP    N2
00B7 C400       TAG1:  LDI    0           ;SET POINTER 1 AND USE THI
00B9 35                XPAH   P1          ;JUMP TO TAG, THE BEGINNIN
00BA C470              LDI    070
00BC 31                XPAL   P1
00BD 91B1              JMP    TAG-PSPTG(P1)
                ;
                ;CHARACTERS OTHER THAN ALPHABET ENTERED, PRINT ERROR
                ;
00BF C44C       NOALP: LDI    L(MNAL)     ;LOAD MESSAGE ADDRESS
00C1 31                XPAL   P1          ;INTO P1.
00C2 C403              LDI    H(MNAL)
00C4 35                XPAH   P1
00C5 C501       N3:    LD     @1(P1)      ;LOAD CHARACTER.
00C7 3F                XPPC   P3          ;PRINT CHARACTER.
00C8 31                XPAL   P1
00C9 CAFA              ST     -6(P2)      ;LOCATION -6 HOLDS THE NUM
00CB E46B              XRI    L(MIMW)     ;OF CHARACTERS.
00CD 98E8              JZ     TAG1        ;IF MESSAGE COMPLETED JUMP
00CF C2FA              LD     -6(P2)      ;TO TAG 1
00D1 31                XPAL   P1
00D2 90F1              JMP    N3
                ,
                ,FIRST SECRET LETTER ENTERED IS SPACE, PRINT ERROR M
                ,IF NOT PRINT NUMBER OF CHARACTERS
00D4 C2F9       TSPACE: LD    -7(P2)      ;LOCATION -7 HOLDS THE NUM
00D6 E400              XRI    0           ;INPUT CHARACTERS.
00D8 98E5              JZ     NOALP       ;IF THE SPACE WAS ON THE F
00DA 02                CCL                ;CHARICTER POSITION, THEN
00DB DC30              ORI    030         ;TO PRINT ERROR MESSAGE,
00DD 3F                XPPC   P3          ;OTHERWISE CONVERT TO ASCI
00DE C400              LDI    0           ;ADDING 30H AND PRINT.
00E0 CAF6              ST     -10(P2)
                ,
                ,ENTER SPACE AT END OF SECRET WORD.  UPDATE LINE COUN
                ,BER OF GUESS.) AND PRINT.
                ,
00E2 C40D       SPACE: LDI    0D          ;CARRAGE RETURN
00E4 3F                XPPC   P3
00E5 C40A              LDI    0A          ;LINE FEED
00E7 3F                XPPC   P3
00E8 AAF6              ILD    -10(P2)     ;LINE NUMBER UPDATED.
00EA 02                CCL
00EB F430              ADI    030
00ED E43A              XRI    03A
```

```
00EF 9805            JZ       TEN
00F1 E43A            XRI      03A
00F3 3F              XPPC     P3
00F4 9003            JMP      COUNT
00F6 C430   TEN      LDI      '0'           ;PRINT 0 IN PLACE OF LINE
00F8 3F              XPPC     P3
            ;
            ;TEST GUESS WORD FOR $, SPACE, AND SIZE. PUT IN TEMP
            ;
00F9 C420   COUNT    LDI      020
00FB 3F              XPPC     P3
00FC C400            LDI      0             ;INITALIZE
00FE CAF8            ST       -8(P2)        ;CHARACTER COUNT AND
0100 CAF7            ST       -9(P2)        ;
0102 C408            LDI      H(WORD)       ;SET POINTER 1 TO THE ADDR
0104 35              XPAH     P1            ;GUESS WORD STORAGE
0105 C42B            LDI      L(WORD)+10
0107 31              XPAL     P1
0108 C402   N4       JS       P3,GECO       ;GO OUT AND LOOK FOR CHARA
010A 37C4
010C 5C33
010E 3F
010F C47F            LDI      07F           ;INPUT. STORE INPUT CHARAC
0111 60              XRE                    ;IN E REGISTER
0112 98F4            JZ       N4            ;TEST FOR RUBOUT, IF TRUE,
0114 C424            LDI      '$'           ;TEST FOR $, IF TRUE JUMP
0116 60              XRE                    ;THE BEGINNING OF THE GAME
0117 989E            JZ       TAG1
0119 C420            LDI      020           ;TEST FOR SPACE
011B 60              XRE
011C 980B            JZ       NSPCE
011E 40              LDE
011F CD01            ST       @1(P1)        ;STORE CHARACTER IN MEMORY
0121 AAF8            ILD      -8(P2)        ;INCREMENT CHARACTER COUNT
0123 E40F            XRI      15            ;IF MORE THAN 15 CHARACTER
0125 984D            JZ       IMW           ;THEN JUMP TO PRINT MESSAG
0127 90DF            JMP      N4
            ;
            ;COMPARE SIZE OF SECRET AND GUESS WORD THEN COMPARE
            ;BY CHARACTER. PRINT CHARACTER IF MATCHING AND '?' I
            ;MATCHING

0129 C2F9   NSPCE    LD       -7(P2)        ;STORE SECRET WORD CHARACT
012B 01              XAE                    ;IN E REGISTER
012C C2F8            LD       -8(P2)        ;AND COMPARE WITH INPUT WO
012E 60              XRE                    ;TER COUNT
012F 9C43            JNZ      IMW           ;IF NOT MATCHING, PRINT ME
0131 C400            LDI      0             ;INITIALIZE INPUT WORD CHA
0133 CAF8            ST       -8(P2)        ;COUNTER AGAIN
0135 C408            LDI      H(WORD)       ;SET P1 TO THE ADDRESS OF
0137 35              XPAH     P1            ;SECRET WORD
0138 C421            LDI      L(WORD)
013A 31              XPAL     P1
013B C402            LDI      H(PUTC)       ;SET P3 TO ADDRESS OF CHAR
013D 37              XPAH     P3            ;PRINT SUBROUTINE
013E C49B            LDI      L(PUTC)-1
0140 33              XPAL     P3
0141 C42F            LDI      02F           ;PRINT '1'
0143 3F              XPPC     P3
0144 C501   N6:      LD       @1(P1)        ;LOAD SECRET WORD CHARACTE
0146 01              XAE                    ;AND STORE IN E REGISTER
0147 C109            LD       9(P1)         ;LOAD INPUT WORD CHARACTER
0149 60              XRE                    ;COMPARE
014A 9824            JZ       CMATCH        ;MATCH
014C C43F            LDI      '?'           ;PRINT '?' IN PLACE OF NON
014E 3F              XPPC     P3            ;CHARACTER
014F C2F7            LD       -9(P2)        ;TEST FOR COMPARE STATUS
0151 9C13            JNZ      TLASC
0153 C1FF            LD       -1(P1)        ;LOAD SECRET CHARACTER AGA
0155 01              XAE                    ;AND STORE IN E REGISTER
0156 C109            LD       9(P1)         ;LOAD GUESS CHARACTER
0158 03              SCL                    ;COMPARE IF INPUT CHARACTE
0159 78              CAE                    ;HIGHER OR LOWER THAN SECR
015A 9406            JP       HIGH          ;WORD CHARACTER
015C C401            LDI      1             ;SET THE STATUS AFTER THE
015E CAF7            ST       -9(P2)        ;COMPARISON HAS BEEN MADE
0160 9004            JMP      TLASC         ;STORE 1 IF INPUT< SECRET
0162 C402   HIGH:    LDI      2             ;STORE 2 IF INPUT > SECRET
0164 CAF7            ST       -9(P2)
            ;
            ;TEST FOR COMPLETION OF COMPARISON.
            ;
```

Fig. A-13. "WORD GAME" (continued from page 307).

```
0166 AAF8   TLASC:   ILD      -8(P2)      ;UPDATE INPUT CHARACTER CO
0168 01              XAE                  ;AND COMPARE WITH SECRET W
0169 C2F9            LD       -7(P2)      ;CHARACTER COUNT. IF MATCH
016B 60              XRE                  ;COMPARISON IS FINISHED
016C 9821            JZ       PMESS
016E 90D4            JMP      N6
0170 40     CMATCH:  LDE                  ;PRINT THE MATCHING CHARAC
0171 3F              XPPC     P3
0172 90F2            JMP      TLASC
            ;
            ;ERROR MESSAGE FOR INCORRECT WORD SIZE
            ;
0174 C46B   IMW:     LDI      L(MIMW)     ;LOAD P1 WITH THE ADDRESS
0176 31              XPAL     P1          ;OF THE MESSAGE
0177 C403            LDI      H(MIMW)
0179 35              XPAH     P1
017A C402            LDI      H(PUTC)     ;LOAD P3 WITH CHARACTER
017C 37              XPAH     P3          ;OUTPUT SUBROUTINE ADDRESS
017D C49B            LDI      L(PUTC)-1
017F 33              XPAL     P3
0180 C501   N5:      LD       @1(P1)      ;PRINT MESSAGE
0182 3F              XPPC     P3
0183 31              XPAL     P1
0184 CAFA            ST       -6(P2)
0186 E487            XRI      L(MGL)
0188 981E            JZ       NGC
018A C2FA            LD       -6(P2)
018C 31              XPAL     P1
018D 90F1            JMP      N5
            ;
            ;PRINT 'LO , HI' MESSAGE, GOOD LUCK MESSAGE AT THE E
            ;TRY AND SORRY YOU LOOSE MESSAGE AT THE END OF 10TH
            ;
018F C42D   PMESS:   LDI      02D         ;PRINT '-'
0191 3F              XPPC     P3
0192 C2F7            LD       -9(P2)      ;IF -9=0 THEN THE GUESS IS
0194 987A            JZ       CORRA       ;CORRECT
0196 E401            XRI      1           ;IF -9=1 PRINT LO
0198 9C08            JNZ      MHI
019A C44C            LDI      'L'
019C 3F              XPPC     P3
019D C44F            LDI      'O'
019F 3F              XPPC     P3
01A0 9006            JMP      NGC
01A2 C448   MHI      LDI      'H'         ;IF -9 DOES NOT = 1 PRINT
01A4 3F              XPPC     P3
01A5 C449            LDI      'I'
01A7 3F              XPPC     P3
01A8 C2F6  NGC     LD       -10(P2)     ;PRINT GOOD LUCK MESSAGE
01AA E407            XRI      7           ;AT END OF LINE 7
01AC 980E            JZ       GL
01AE C2F6            LD       -10(P2)     ;PRINT SORRY U LOSE MESSAG
01B0 E40A          XRI      10
01B2 981D          JZ       SUL
           ;
           ;JUMP BACK AND WAIT FOR NEXT GUESS WORD ENTRY.
           ;
01B4 C400  SPACE1: LDI      0
01B6 35            XPAH     P1
01B7 C470          LDI      070
01B9 31            XPAL     P1
01BA 9171          JMP      SPACE-PSPTG(P1)
           ;
           ;PRINT '-------GOOD LUCK' MESSAGE.
           ;
01BC C487  GL:     LDI      L(MGL)
01BE 31            XPAL     P1
01BF C403          LDI      H(MGL)
01C1 35            XPAH     P1
01C2 C501  N7:     LD       @1(P1)
01C4 3F            XPPC     P3
01C5 31            XPAL     P1
01C6 CAFA          ST       -6(P2)
01C8 E4A8          XRI      L(MSUL)
01CA 98E8          JZ       SPACE1
01CC C2FA          LD       -6(P2)
01CE 31            XPAL     P1
01CF 90F1          JMP      N7
           ;
           ;PRINT 'SORRY YOU LOSE-------' MESSAGE
           ;
```

```
01D1 C4A8  SUL     LDI    L(MSUL)
01D3 31            XPAL   P1
01D4 C403          LDI    H(MSUL)
01D6 35            XPAH   P1
01D7 C501  N8:     LD     @1(P1)
01D9 3F            XPPC   P3
01DA 31            XPAL   P1
01DB CAFA          ST     -6(P2)
01DD E4CB          XRI    L(MCOA1)+2
01DF 9805          JZ     SECW
01E1 C2FA          LD     -6(P2)
01E3 31            XPAL   P1
01E4 90F1          JMP    N8
                   ;
                   ;PRINT 'THE SECRET WORD IS *SECRET*' MESSAGE.
                   ;
01E6 C408  SECW:   LDI    H(WORD)
01E8 35            XPAH   P1
01E9 C421          LDI    L(WORD)
01EB 31            XPAL   P1
01EC C42A          LDI    '*'
01EE 3F            XPPC   P3
01EF C501  N9.     LD     @1(P1)
01F1 3F            XPPC   P3
01F2 31            XPAL   P1
01F3 CAFA          ST     -6(P2)
01F5 C2F9          LD     -7(P2)
01F7 02            CCL
01F8 F421          ADI    L(WORD)
01FA 01            XAE
01FB C2FA          LD     -6(P2)
01FD 60            XRE
01FE 9805          JZ     ESEC
0200 C2FA          LD     -6(P2)
0202 31            XPAL   P1
0203 90EA          JMP    N9
0205 C42A  ESEC.   LDI    '*'
0207 3F            XPPC   P3
                   ;
                   ;SET POINTER TO JUMP TO BEGINNING OF GAME.
                   ;
0208 C400  TAG2·   LDI    0
020A 35            XPAH   P1
020B C470          LDI    070
020D 31            XPAL   P1
020E 91B1          JMP    TAG-PSPTG(P1)
                   ,
                   ;PRINT 'YOU ANSWERED CORRECTLY' MESSAGE.
                   ;
0210 C4C9  CORRA   LDI    L(MCOA1)
0212 31            XPAL   P1
0213 C403          LDI    H(MCOA1)
0215 35            XPAH   P1
0216 C501  N10.    LD     @1(P1)
0218 3F            XPPC   P3
0219 31     XPAL    P1
021A CAFA          ST     -6(P2)
021C E4E4          XRI    L(MCOA2)
021E 9805          JZ     SCORE
0220 C2FA          LD     -6(P2)
0222 31            XPAL   P1
0223 90F1          JMP    N10
                   ,
                   ,PRINT ''YOU ARE A GENIUS' MESSAGE IF GUESSED
                   ,IN LESS THAN 3 TRIES
                   ,
0225 C2F6  SCORE   LD     -10(P2)
0227 03            SCL
0228 FC03          CAI    3
022A 9415          JP     NSCORE
022C C4E4          LDI    L(MCOA2)
022E 31            XPAL   P1
022F C403          LDI    H(MCOA2)
0231 35            XPAH   P1
0232 C501  N11     LD     @1(P1)
0234 3F            XPPC   P3
0235 31            XPAL   P1
0236 CAFA          ST     -6(P2)
0238 E4F5          XRI    L(MCOA3)
023A 98CC          JZ     TAG2
```

Fig. A-13. "WORD GAME" (continued from page 309).

```
0230 C2FA            LD      -6(P2)
023E 31              XPAL    P1
023F 90F1            JMP     N11

                PRINT TERRIFIC  MESSAGE IF GUESSED IN LESS THAN 7 T

0241 C2F6    NSCORE  LD      -10(P2)
0243 03              SCL
0244 FC06            CAI     6
0246 94C0            JP      TAG2
0248 C4F5            LDI     L(MCOA3)
024A 31              XPAL    P1
024B C403            LDI     H(MCOA3)
024D 35              XPAH    P1
024E C501    N12     LD      @1(P1)
0250 3F              XPPC    P3
0251 31              XPAL    P1
0252 CAFA            ST      -6(P2)
0254 E4FE            XRI     L(MEND)
0256 98B0            JZ      TAG2
0258 C2FA            LD      -6(P2)
025A 31              XPAL    P1
025B 90F1            JMP     N12
                      PAGE   GECO
                      LOCAL
025D C408    GECO    LDI     8              ,SET COUNT = 8
025F CAFF            ST      -1(P2)
0261 06      $2      CSA                    ;WAIT FOR START BIT
0262 D420            ANI     020
0264 9CFB            JNZ     $2             ,NOT FOUND
0266 C457            LDI     87             ,DELAY 1/2 BIT TIME
0268 8F04            DLY     4
026A 06              CSA                    ,IS START BIT STILL THERE?
026B D420            ANI     020
026D 9CF2            JNZ     $2             ,NO
026F 07              CAS                    ,SEND START BIT (NOTE THAT
0270 DC01            ORI     1              ;OUT PUT IS INVERTED)
0272 07              CAS
0273 C47E    $LOOP   LDI     126            ,DELAY 1 BIT TIME
0275 8F08            DLY     8
0277 06              CSA                    ,GET BIT (SENSEB)
0278 D420            ANI     020
027A 9802            JZ      $3
027C C401            LDI     1
027E CAFE    $3      ST      -2(P2)         ,SAVE BIT VALUE (0 OR 1)
0280 1F              RRL                    ,ROTATE INTPO LINK
0281 01              XAE

0282 1D              SRL                    ,SHIFT INTO CHARACTER
0283 01              XAE                    ,RETURN CHAR TO E
0284 06              CSA                    ,ECHO BIT TO OUTPUT
0285 DC01            ORI     1
0287 E2FE            XOR     -2(P2)
0289 07              CAS
028A BAFF            DLD     -1(P2)         ;DECREMENT BIT COUNT
028C 9CE5            JNZ     $LOOP          ,LOOP UNTIL 0
028E 06              CSA                    ,SET STOP BIT
028F D4FE            ANI     0FE
0291 07              CAS
0292 8F08            DLY     8
0294 40              LDE                    ,AC HAS INPUT CHARACTER
0295 D47F            ANI     07F
0297 01              XAE
0298 40              LDE
0299 3F              XPPC    P3             ,RETURN
029A 90C1            JMP     GECO
                      PAGE   'PUTC'
                     .LOCAL
029C 01      PUTC    XAE
029D C4FF            LDI     255
029F 8F17            DLY     23
02A1 06              CSA                    ,SET OUTPUT BIT TO LOGIC 0
02A2 DC01            ORI     1              ,FOR START BIT  (NOTE INVE
02A4 07              CAS
02A5 C409            LDI     9              ,INITIALIZE BIT COUNT
02A7 CAFF            ST      -1(P2)
02A9 C48A    $1      LDI     138            ,DELAY 1 BIT TIME
02AB 8F08            DLY     8
02AD BAFF            DLD     -1(P2)         ,DECREMENT BIT COUNT
02AF 9810            JZ      $EXIT
02B1 40              LDE                    ,PREPARE NEXT BIT
02B2 D401            ANI     1
02B4 CAFE            ST      -2(P2)
```

```
02B6 01                  XAE                      ,SHIFT DATA RIGHT 1 BIT
02B7 1C                  SR
02B8 01                  XAE
02B9 06                  CSA                      ;SET UP OUTPUT BIT
02BA DC01                ORI      1
02BC E2FE                XOR      -2(P2)
02BE 07                  CAS                      ;PUT BIT TO TTY
02BF 90E8                JMP      $1
02C1 06        $EXIT     CSA                      ,SET STOP BIT
02C2 D4FE                ANI      0FE
02C4 07                  CAS
02C5 3F                  XPPC     P3
02C6 90D4                JMP      PUTC
02C8 0D0A      MBEG       DBYTE X'0D0A
02CA 574F                .ASCII  'WORD GAME'
02CC 5244
02CE 2047
02D0 414D
02D2 45
02D3 0D0A                 DBYTE X'0D0A
02D5 504C                .ASCII  'PLAYER-A ENTER A WORD OF UP TO'
02D7 4159
02D9 4552
02DB 2D41
02DD 3A45
02DF 4E54
02E1 4552
02E3 2041
02E5 2057
02E7 4F52
02E9 4420
02EB 4F46
02ED 2055
02EF 5020
02F1 544F
02F3 0D0A                .DBYTE X'0D0A
02F5 3130                .ASCII  '10 LETTERS, THEN HIT SPACE BAR'
02F7 204C
02F9 4554
02FB 5445
02FD 5253
02FF 2C54
0301 4845
0303 4E20
0305 4849
0307 5420
0309 5350
030B 4143
030D 4520
030F 4241
0311 52
0312 0D0A                .DBYTE X'0D0A
0314 504C                .ASCII  'PLAYER-B ENTER YOUR BEST GUESS'
0316 4159
0318 4552
031A 2D42
031C 3A45
031E 4E54
0320 4552
0322 2059
0324 4F55
0326 5220
0328 4245
032A 5354
032C 2047
032E 5545
0330 5353
0332 0D0A      MEXC      .DBYTE X'0D0A
0334 4C49                .ASCII  'LIMIT EXCEEDED, WISE GUY.
0336 4D49
0338 5420
033A 4558
033C 4345
033E 4544
0340 4544
0342 2C57
0344 4953
0346 4520
0348 4755
034A 592E
034C 0D0A      MNAL:      DBYTE X'0D0A
034E 414C                .ASCII  'ALPHABET ONLY TRY AGAIN CUTIE'
0350 5048
```

Fig. A-13. "WORD GAME" (continued from page 311).

```
0352 4142
0354 4554
0356 204F
0358 4E4C
035A 592E
035C 5452
035E 5920
0360 4147
0362 4149
0364 4E20
0366 4355
0368 5449
036A 45
036B 0D0A  MIMW     DBYTE X'0D0A
036D 494E           ASCII  INCORRECT WORD SIZE,DUMMY
036F 434F
0371 5252
0373 4543
0375 5420
0377 574F
0379 5244
037B 2053
037D 495A
037F 452C
0381 4455
0383 4D4D
0385 592E
0387 0D0A  MGL      DBYTE X'0D0A
0389 5520           ASCII 'U HAVE 3 CHANCES LEFT GOOD LUCK'
038B 4841
038D 5645
038F 2033
0391 2043
0393 4841
0395 4E43
0397 4553
0399 204C
039B 4546
039D 542E
039F 474F
03A1 4F44
03A3 204C
03A5 5543
03A7 4B
03A8 0D0A  MSUL     DBYTE X'0D0A
03AA 534F           ASCII 'SORRY U LOSE THE SECRET WORD IS'
03AC 5252
03AE 5920
03B0 5520
03B2 4C4F
03B4 5345
03B6 2E54
03B8 4845
03BA 2053
03BC 4543
03BE 5245
03C0 5420
03C2 574F
03C4 5244
03C6 2049
03C8 53
03C9 0D0A  MCOA1    DBYTE X'0D0A
03CB 594F           ASCII 'YOU ANSWERED CORRECTLY '
03CD 5520
03CF 414E
03D1 5357
03D3 4552
03D5 4544
03D7 2043
03D9 4F52
03DB 5245
03DD 4354
03DF 4C59
03E1 2E
03E2 0D0A           DBYTE X'0D0A
03E4 594F  MCOA2:   ASCII 'YOU ARE A GENIUS.
03E6 5520
03E8 4152
03EA 4520
03EC 4120
03EE 4745
03F0 4E49
```

```
    03F2 5553
    03F4 2E
563 03F5 5445  MCOA3:   .ASCII 'TERRIFIC!'
    03F7 5252
    03F9 4946
    03FB 4943
    03FD 21
564      0000  MEND:     END



ALPHA    008D          BEG      0001 *        BZ       004F
CMATCH   0170          CORRA    0210          COUNT    00F9
ESEC     0205          EXCED    00A2          FENTRY   0061 *
GECO     025D          GETC     0032          GL       01BC
GLOOP    0044          GRET     0036          HIGH     0162
IMW      0174          MBEG     02C8          MCOA1    03C9
MCOA2    03E4          MCOA3    03F5          MEND     03FE
MEXC     0332          MGL      0387          MHI      01A2
MIMW     036B          MNAL     034C          MSUL     03A8
N1       0013          N10      0216          N11      0232
N12      024E          N2       00A8          N3       00C5
N4       0108          N5       0180          N6       0144
N7       01C2          N8       01D7          N9       01EF
NENTRY   0075          NGC      01A8          NOALP    00BF
NOF      009E          NSCORE   0241          NSPCE    0129
P1       0001          P2       0002          P3       0003
PGEC     009B          PMESS    018F          PSPTG    0070
PUTC     029C          SCORE    0225          SECW     01E6
SPACE    00E2          SPACE1   01B4          SUL      01D1
TAG      0022          TAG1     00B7          TAG2     0208
TEN      00F6          TLASC    0166          TSPACE   00D4
WORD     0821          $1       02A9          $2       0261
$3       027E          $EXIT    02C1          $LOOP    0273



NO ERROR LINES
SOURCE CHECKSUM=A314
FIRST INPUT SECTOR HEX -  0315
FINAL INPUT SECTOR HEX -  032F
```

Fig. A-13. "WORD GAME" (continued from page 313).

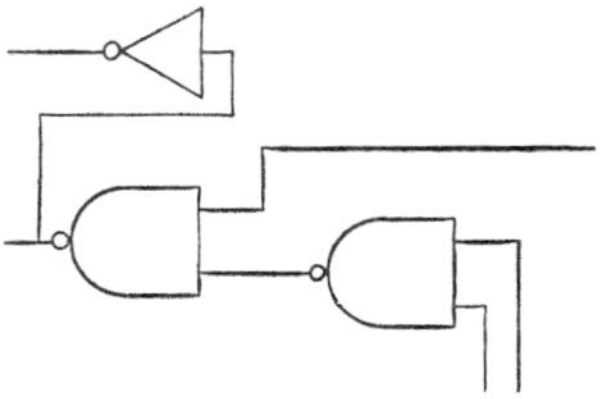

References

Advanced Micro Devices. "AM 9511 Arithmetic Processing Unit Specification Sheet," and "Algorithm Details for the AM 9511 Arithmetic Processing Unit." Sunnyvale, CA.

Allen, David M. *Byte*, January 1977

ARRL 1975. *Specialized Communications Techniques* ARRL, Newington, Ct.

Barden, Wm. Jr. 1978. *The Z80 Microcomputer Handbook*, Sams, Indianapolis.

Boehm, C. and Jacopini, G. Flow Diagrams, Turing Machines and Languages with only Two Formation Rules, *Communications of the ACM*, 1979. (May, 1966) p. 366-71.

Ciarcia, Steve. "Joystick Interfaces", *Byte*, September, 1979.

Dahmke, Mark. "Introduction to Multi-Programming," *Byte*, September, 1979.

Declerq, Michael J., "Plug-in Module Automates On-site EPROM Loading," *Electronic Design*, August 2, 1980: v. 28, n. 16.

Digi-Key Corp. *"The Nibbler" Handbook*, Thief River Falls.

Dynabyte, Inc. 1978. *BASIC Controller Data Manual*, Palo Alto, CA, 1978.

Editors of 73 Magazine 1979. *The Giant Handbook of Computer Projects*, TAB, Blue Ridge Summit.

Feldman, Len. "Understanding Tape Specs," *Radio Electronics*. March 1976: v. 47. n. 3.

Green, Wayne. 1972. *RTTY Handbook*, TAB, Blue Ridge Summit, PA.

Harron, Brian A. "Inexpensive Optical Paper-Tape Reader," *Byte*, September 1979.

Haviland, R.P. 1977. *The "Computator" Book*, TAB, Blue Ridge Summit.

Haviland, R.P., 1979. *How to Design, Build and Program Your Own Working Computer System*, TAB, Blue Ridge Summit, PA.

Hordeski, Michael 1980. *Microprocessor Cookbook*, TAB, Blue Ridge Summit, PA.

Kapps, Charles A. 1979. "Five Useful Programs for the SC/MP," *Byte*, November, 1979.

Lancaster, Don. 1978. *The Cheap Video Cook Book*, Sams, Indianapolis.

Lancaster, Don. 1977. *TV Typewriter Cook Book*, Indianapolis.

Lancaster, Don. "Cheap Video For Your Heathkit H8," *Kilobaud Microcomputing*, March, 1979.

McFarland, Ward J. Jr. "The 'El Cheapo' EPROM Programmer," *Kilobaud Microcomputing*, March, 1977.

McNatt, Michael S. "A Guide to Baudot Machines," *Byte*, Part 1, April, 1977., Part 2, May, 1977.

National Semiconductor, Santa Clara, CA.

PROM Power-Down Circuits, Application Note AN-171.

SC/MP NIBL Reference Guide, Publication 42030505398-001X, January, 1977.

SC/MP Instruction Guide, Publication 4200110A, 1976.

INS-8073 NSC Tiny BASIC Microinterpreter, July, 1980.

INS-8298A 8080A LLL BASIC Interpreter, October, 1977.

NSC Tiny BASIC Users Manual, June, 1980.

INS-8060 Single-Chip Eight-Bit N-Channel Microprocessor (SC/MP) January, 1978.

INS-8070-Series Microprocessor Family, April, 1979.

70-Series Microprocessor Users Manual, June, 1980.

SC/MP Technical Description, January, 1976.

SC/MP Microprocessor Applications Handbook, October, 1976.

SC/MP Microprocessor Assembly Language Programming Manual, January, 1977.

INS-8080A-SC/MP Multiprocessor System Concept, June, 1976.

Number Cruncher (NM 57109) Interface to Microprocessor, July, 1977.

SC/MP as an Analog to Digital Converter, July, 1977.

ISP-8A/600 Single-Chip N-Channel Microprocessor (SC/MP II)

Peoples Computer Company. 1977. *PCCs Reference Book*, Menlo Park.

Poe, Elmer C. and Goodwin, James C. 1979. *The S-100 and Other Microbuses*, Sams, Indianapolis.

Pilipovich, George. "Multi-Tasking the 8080: FAMOS, an On-Line System," *Interface Age*, February 1979.

Pond, Robert M. "Let Microprocessors Communicate," *Electronics Design 23*, November 8, 1977.

Radio Shack. *TRS-80 Microcomputer Technical Reference Handbook*.

Stark, Craig. "Tape Recorder Hygiene," *Popular Electronics* July, 1977.

Sudding, Robert. 1977. "Why Wait? Build a FAST Cassette Interface," *Best of Byte*, v. 1, p. 190-196, Ahl and Helmers, ed., *Creative Computing, Morristown, New Jersey.*

Gigus, Christopher A. 1979. *TEA, an 8080/8085 Coresident Editor/Assembler*, Sams, Indianapolis.

Warme, Paul, 1979. "BASEX, A Simple Language and Compiler for 8080 Systems," a Paperbyte Book, *Byte*, Peterborough, *New Hampshire.*

Warren, Carl L., 1979 ed., *Best of Interface Age*, v. 1, "Software in BASIC (including NIBL and LLL Basic), Dilithium," Portland.

ZILOG, Cupertino, CA.

Z80-CPU Product Specification, March, 1978.

Z80 Family Program Interrupt Structure, May, 1978.

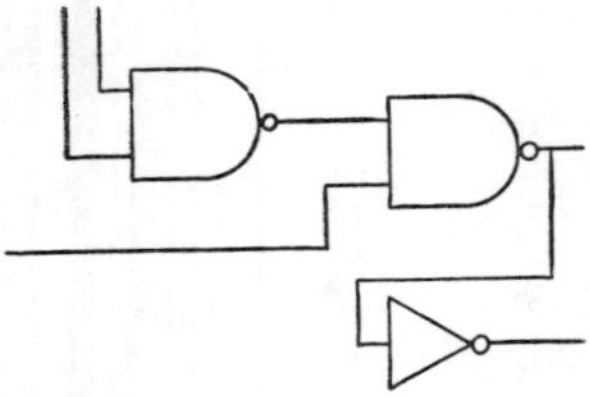

Index